Introduction to MATLAB®
with Numerical Preliminaries

Alexander Stanoyevitch

WILEY-
INTERSCIENCE

A JOHN WILEY & SONS, INC., PUBLICATION

Published by John Wiley & Sons, Inc., Hoboken, New Jersey.
Published simultaneously in Canada.

For general information on our other products and services please contact our Customer Care
Department within the U.S. at 877-762-2974, outside the U.S. at 317-572-3993 or fax 317-572-4002.

Wiley also publishes its books in a variety of electronic formats. Some content that appears in print,
however, may not be available in electronic format.

Library of Congress Cataloging-in-Publication Data:

Stanoyevitch, Alexander.
    Introduction to MATLAB with numerical preliminaries / Alexander Stanoyevitch.
      p. cm.
    Includes bibliographical references and index.
    ISBN 0-471-69737-0 (pbk.)
     1. Engineering mathematics—Data processing. 2. MATLAB. 3. Numerical
analysis—Data processing. I. Title.

TA345.S75 2004
620'.001'51—dc22   2004053006

Printed in the United States of America

10 9 8 7 6 5 4 3 2 1

Contents

---

[1] An asterisk that precedes a section indicates that the section may be skipped without a significant loss of continuity to the main development of the text.

PREFACE

MATLAB® is an abbreviation for MATrix LABoratory and it is ideally suited for computations involving matrices. Since all of the sciences routinely collect data in the form of (spreadsheet) matrices, MATLAB turns out to be particularly suitable for the analysis of mathematical problems in an assortment of fields. MATLAB is very easy to learn how to use and has tremendous graphical capabilities. Many schools have site licenses and student editions of the software are available at special affordable rates. MATLAB is perhaps the most commonly used mathematical software in the general scientific fields (from biology, physics, and engineering to fields like business and finance) and is used by numerous universities in mathematics departments.

 This book is an undergraduate-level textbook giving a thorough introduction to the use of the MATLAB software with an emphasis on scientific computing. It is largely self-contained, with the prerequisite of a basic course in single-variable calculus and it covers many topics in numerical analysis. Numerous applications are given to an assortment of fields including linear algebra, probability, finance, ecology, and discrete mathematics.

MATERIAL

The book includes seven chapters, whose contents we now briefly summarize. Chapter 1 introduces the reader to some of MATLAB's basic functionality in a tutorial fashion. The important concept of a "while loop" is introduced rather early through a problem in finance (paying off a loan). Chapter 2 covers the basic notions of numerical analysis including approximations, errors, and relative errors. The problem of approximating a function using Taylor polynomials is used as a prototypical problem in numerical analysis. Taylor's theorem readily provides quality control (error bounds) on such approximations, and MATLAB will make it easy to see the effectiveness of such bounds. In Chapter 3, we introduce the reader to M-files, which are programs in MATLAB. Examples of M-files are created that highlight the distinctions between the two types of M-files, scripts and functions. MATLAB has many built-in M-files, and the final section for this chapter presents several that are particularly useful for (pre-)calculus-related tasks. Subsequently, in Chapter 4, we cover all of the logical features that are available for creating M-files. Through a series of interesting problems, this chapter introduces the reader to various strategies for writing effective programs. When MATLAB does computations, it needs to translate the numbers that the user inputs into computer numbers, and uses a corresponding floating point arithmetic to carry out the needed computations. Details of this process are covered in Chapter 5. One important topic in this chapter is the propagation of the roundoff errors, and strategies on how to control them.   The problem of finding a root of an equation goes back to antiquity.   Chapter 6 begins with a historical introduction to this interesting problem, and then goes on to introduce several effective numerical schemes for rootfinding problems.   M-files are constructed and practical experiments are conducted to test their relative effectiveness. The final section of Chapter 6 provides some important theory on these methods.

Chapter 7 covers matrices and linear systems, and is, by far, the longest chapter of the book.   Since the basic structure in MATLAB is a matrix and it was originally conceived as a software for solving associated linear systems, there are many features that are presented.   The chapter begins with an introduction to matrices and their arithmetic. Next, in Section 7.2, some beautiful applications of matrices to the generation of computer graphics are given. This section will show, among other things, how to create movies in MATLAB and how to create high-resolution graphics of complicated objects such as fractals. In Sections 7.3 and 7.4, the important problem of solving linear systems is introduced, and it is shown how to take advantage of MATLAB's built-in solvers.   Included among the numerous applications of linear systems are polynomial interpolation problems, traffic logistics, input-output analysis, and combinatorics. In Section 7.5, the algorithm of Gaussian elimination is discussed in detail from a numerical and algorithmic perspective.   Related topics such as complexity and LU decompositions are also included. Section 7.6 covers topics from numerical linear algebra concerning norms, error analysis and eigenvalues.   In Section 7.7 we develop iterative methods. Codes are written for several of these methods and then they are compared along with their rates of convergence. Some interesting theory is also presented, as well as a detailed discussion on sparse matrices.

The book also includes two appendices. Appendix A gives a brief introduction to the use of MATLAB's Symbolic Toolbox, and Appendix B provides solutions to all of the Exercises for the Reader that are scattered throughout the text. The text is punctuated with numerous historical profiles (and photographs) of some of the scientists who have made significant contributions to the areas under investigation.

INTENDED AUDIENCE AND STYLE OF THIS BOOK

This book is written for students or professionals wishing to learn about either MATLAB or scientific computation.   It is part of a larger book that the author has written (also published by John Wiley & Sons), entitled *Introduction to Numerical Ordinary and Partial Differential Equations Using MATLAB*. Explanations are written in a style that is rigorous yet user-friendly. We do not gloss over any needed details. Many illustrative examples are included throughout the text, and in addition, there are many "Exercises for the Reader" interspersed throughout the text that test the reader's understanding of the important concepts being introduced. Solutions to all of these Exercises for the Reader are included in an appendix. In addition, the individual sections conclude with extensive exercise sets, making this book suitable for use as a textbook. As a textbook, it could serve either as a supplementary text to any math/science course wishing to make use of MATLAB (such as a numerical analysis or linear algebra course), or as a primary text for an introductory course in scientific computation.   To facilitate readability of the text, we employ the following font conventions: Regular text is printed in the (current) Times New Roman font, MATLAB inputs and commands appear in `Courier New` font, whereas MATLAB output is printed in Ariel font. Key terms are set in **bold type**, while less essential vocabulary is set in *italics*.

Some sections are marked with an asterisk to indicate that they should be considered as optional; their deletion would cause no major disruption to the main themes of the text. Some of these optional sections are more theoretical than the others (e.g., Section 6.5: Error Analysis and Comparison of Rootfinding Methods), while others present applications in a particular related area (e.g., Section 7.2: Introduction to Computer Graphics and Animation). The programs and codes in the book have all been developed to work with the latest versions of MATLAB (Student Versions or Professional Versions).[1]   All of the M-files developed in the text and the Exercises for the Reader can be downloaded from book's ftp site:

`ftp://ftp.wiley.com/public/sci_tech_med/numerical_preliminaries/`

Although it is essentially optional throughout the book, when convenient we occasionally use MATLAB's Symbolic Toolbox that comes with the Student Version (but is optional with the Professional Version).

ACKNOWLEDGMENTS

Many individuals and groups have assisted in various ways that have led to the development of this book and I would like to express my appreciation. I would like to thank my students who have taken my courses (very often as electives) and have read through preliminary versions of the book. Their feedback has resulted in an improved pedagogy of this text. The people at MathWorks (the company that develops MATLAB), in particular, Courtney Esposito, have been very supportive in providing me with software and high-quality technical support.

I have had many wonderful teachers throughout my life and I would like to express my appreciation to all of them, and make special mention of some of them. Firstly, back in middle school, I spent a year in a parochial school with a teacher, Sister Jarlath, who really had a tremendous impact in kindling my interest in mathematics; my experience with her led me to develop a newfound respect for education. Although sister Jarlath has passed, her kindness and caring for students

---

[1] Every code and M-file in this book has been tested on MATLAB versions 5, 6, and 7. The (very) rare instances where a version-specific issue arises are carefully explained. One added feature of later versions is the extended menu options that make many tasks easier than they used to be. A good example of this is the improvements in the MATLAB graphics window. Many features of a graph can be easily modified directly using (user-friendly) menu options. In older versions, such editing had to be done by entering the correct "handle graphics" commands into the MATLAB command window.

and the learning process will live on with me forever. It was her example that made me decide to become a mathematics professor as well as a teacher who cares. Several years later when I arrived in Ann Arbor, Michigan, for the mathematics PhD program, I had intended to complete my PhD in the area of abstract algebra, in which I was very well prepared and interested. During my first year, however, I was so enormously impressed and enlightened by the analysis courses that I needed to take, that I soon decided to change my area of focus to analysis. I would particularly like to thank my analysis professors Peter Duren, Fred Gehring, M. S. ("Ram") Ramanujan, and the late Allen Shields. Their cordial, rigorous, and elegant lectures replete with many historical asides were a most delightful experience.

Portions of this book were completed while the author was spending semesters at the National University of Ireland and at the University of Missouri–Columbia. I would like to thank my hosts and the mathematics departments at these institutions for their hospitality and for providing such stimulating atmospheres in which to work.

Feedback from reviewers of this book has been very helpful. These reviewers include: Chris Gardiner (Eastern Michigan University) Mark Gockenbach (Michigan Tech), Murli Gupta (George Washington University), Jenny Switkes (Cal Poly Pomona), Robin Young (University of Massachusetts), and Richard Zalik (Auburn University). Among these, I owe special thanks to Drs. Gockenbach and Zalik; each read carefully through major portions of the text (Gockenbach read through the entire manuscript) and have provided extensive suggestions, scholarly remarks, and corrections. I would like to thank Robert Krasny (University of Michigan) for several useful discussions on numerical linear algebra. Also, the historical accounts throughout the text have benefited from the extensive MacTutor website. I thank Professor Benoit Mandelbrot for permitting the inclusion of his photograph.

Last, but certainly not least, I have two more individuals to thank. My mother, Christa Stanoyevitch, has encouraged me throughout the project and has done a superb job proofreading the entire book. Her extreme conscientiousness and ample corrections and suggestions have significantly improved the readability of this book. I would like to also thank my friend Sandra Su-Chin Wu for assistance whenever I needed it with the many technical aspects of getting this book into a professional form. Inevitably, there will remain some typos and perhaps more serious mistakes. I take full responsibility for these and would be grateful to readers who could direct my attention to any such oversights.

Chapter 1: MATLAB Basics

1.1: WHAT IS MATLAB?

As a student who has already taken courses at least up through calculus, you most likely have seen the power of graphing calculators and perhaps those with symbolic capabilities. MATLAB adds a whole new exciting set of capabilities as a powerful computing tool. Here are a few of the advantages you will enjoy when using MATLAB, as compared to a graphing calculator:

1. It is easy to learn and use. You will be entering commands on your big, familiar computer keyboard rather than on a tiny little keypad where sometimes each key has four different symbols attached.
2. The graphics that MATLAB produces are of very high resolution. They can be easily copied to other documents (with simple clicks of your mouse) and printed out in black/white or color format. The same is true of any numerical and algebraic MATLAB inputs and outputs.
3. MATLAB is an abbreviation for MATrix LABoratory. It is ideally suited for calculations and manipulations involving matrices. This is particularly useful for computer users since the spreadsheet (the basic element for recording data on a computer) is just a matrix.
4. MATLAB has many built-in programs and you can interactively use them to create new programs to perform your desired tasks. It enables you to take advantage of the full computing power of your computer, which has much more memory and speed than a graphing calculator.
5. MATLAB's language is based on the C-family of computer languages. People experienced with such languages will find the transition to MATLAB natural and people who learn MATLAB without much computer background will, as a fringe benefit, be learning skills that will be useful in the future if they need to learn more computer languages.
6. MATLAB is heavily used by mathematicians, scientists and engineers and there is a tremendous amount of interesting programs and information available on the Internet (much of it is free). It is a powerful computing environment that continues to evolve.

We wish here and now to present a disclaimer. MATLAB is a spectacularly vast computing environment and our plan is not to discuss all of its capabilities, but rather to give a decent survey of enough of them so as to give the reader a powerful new arsenal of uses of MATLAB for solving a variety of problems in mathematics and other sciences. Several good books have been written just on using MATLAB; see, for example, references [HiHi-00], [HuLiRo-01], [PSMI-98], and [HaLi-00].[1]

1.2: STARTING AND ENDING A MATLAB SESSION

We assume that MATLAB has been installed on the system that you are using.[2] Instructions for starting MATLAB are similar to those for starting any installed software on your system.   For example, on most windows-based systems, you should be able to simply double click on MATLAB's icon.   Once MATLAB is started, a command window should pop up with a **prompt**: >> (or EDU>> if you are using the student version).   In what follows, if we tell you to enter something like >> 2+2 (on the command window), you enter 2+2 only at the prompt—which is already there waiting for you to type something.   Before we begin our first brief tutorial, we point out that there is a way to create a file containing all interactions with a particular MATLAB session.   The command `diary` will do this. Here is how it works: Say you want to save the session we are about to start to your floppy disk, which you have inserted in the a:/-drive. After the prompt type:

```
>> diary a:/tutorl.txt
```

NOTE: If you are running MATLAB in a computer laboratory or on someone else's machine, you should always save things to your portable storage device or personal account. This will be considerate to the limitations of hard drive space on the machines you are using and will give you better assurance that the files still will be available when you need them.

This causes MATLAB to create a text file called `tutorl.txt` in your a:/- drive called `tutorl.txt`, which, until you end the current MATLAB session, will be a carbon copy of your entire session on the command window.   You can later open it up to edit, print, copy, etc. It is perhaps a good idea to try this out once to see how it works and how you like it (and we will do this in the next section), but in practice, most MATLAB users will often just copy the important parts of their MATLAB session and paste them appropriately in an open word processing window of their choice.

On most platforms, you can end a MATLAB session by clicking down your left mouse button after you have moved the cursor to the "File" menu (located on the upper-left corner of the MATLAB command window). This will cause a menu of commands to appear that you can choose from. With the mouse button still held down, slide the cursor down to the "Exit MATLAB" option and release it.   This

---

[1] Citations in square brackets refer to items in the References section in the back of this book.
[2] MATLAB is available on numerous computing platforms including PC Windows, Linux, MAC, Solaris, Unix, HP-UX. The functionality and use is essentially platform independent although some external interface tasks may vary.

will end the session. Another way to accomplish the same would be to simply click (and release) the left mouse button after you have slid it on top of the "X" button at the upper-right corner of the command window. Yet another way is to simply enter the command:

```
>> quit
```

Any diary file you created in the session will now be accessible.

1.3: A FIRST MATLAB TUTORIAL

As with all tutorials we present, this is intended to be worked by the reader on a computer with MATLAB installed. Begin by starting a MATLAB session as described earlier. If you like, you may begin a diary as shown in the previous section on which to document this session. MATLAB will not respond to or execute any command until you press the "enter key," and you can edit a command (say, if you made a typo) and press enter no matter where the cursor is located in a given command line. Let us start with some basic calculations: First enter the command:

```
>> 5+3
→ ans = 8
```

The arrow ($\rightarrow$) notation indicates that MATLAB has responded by giving us ans = 8. As a general rule we will print MATLAB input in a different font (Courier New) than the main font of the text (Times New Roman). It does not matter to MATLAB whether you leave spaces around the + sign.[3] (This is usually just done to make the printout more legible.) Instead of adding, if we wanted to divide 5 by 3, we would enter (the operation ÷ is represented by the keyboard symbol / in MATLAB)

```
>> 5/3
→ ans =1.6667
```

The output "1.6667" is a four-decimal approximation to the unending decimal approximation. The exact decimal answer here is 1.666666666666... (where the 6's go on forever). The four-decimal display mode is the default format in which MATLAB displays decimal answers. The previous example demonstrates that if the inputs and outputs are integers (no decimals), MATLAB will display them as such. MATLAB does its calculations using about 16 digits—we shall discuss this in greater detail in Chapters 2 and 5. There are several ways of changing how your outputs are displayed. For example, if we enter:

```
>> format long
```

---

[3] The format of actual output that MATLAB gives can vary slightly depending on the platform and version being used. In general it will take up more lines and have more blank spaces than as we have printed it. We adopt this convention throughout the book in order to save space.

```
>> 5/3
→ ans =1.66666666666667
```

we will see the previous answer displayed with 15 digits. All subsequent calculations will be displayed in this format until you change it again. To change back to the default format, enter >> format short. Other popular formats are >> format bank (displays two decimal places, useful for applications to finance) and >> format rat (approximates all answers as fractions of small integers and displays them as such). It is not such a good idea to work in format rat unless you know for sure the numbers you are working with are fractions as opposed to irrational numbers, like π = 3.14159265..., whose decimals go on forever without repetition and are impossible to express via fractions.

In MATLAB, a single equals sign (=) stands for "is assigned the value." For example, after switching back to the default format, let us store the following constants into MATLAB's workspace memory:

```
>> format short
>> a = 2.5
→ a = 2.5000
>> b = 64
→ b=64
```

Notice that after each of these commands, MATLAB will produce an output of simply what you have inputted and assigned. You can always suppress the output on any given MATLAB command by tacking on a semicolon (;) at the end of the command (before you press enter). Also, you can put multiple MATLAB commands on a single line by separating them with commas, but these are not necessary after a semicolon. For example, we can introduce two new constants aa and bb without having any output using the single line:

```
>> aa = 11; bb = 4;
```

Once variables have been assigned in a MATLAB session, computations involving them can be done using any of MATLAB's built-in functions. For example, to evaluate $aa + a\sqrt{b}$, we could enter

```
>> aa + a*sqrt(b)
→ ans=31
```

Note that aa stands for the single variable that we introduced above rather than a^2, so the output should be 31. MATLAB has many built-in functions, many of which are listed in MATLAB Command Index at the end of this book.

MATLAB treats all numerical objects as matrices, which are simply rectangular arrays of numbers. Later we will see how easy and flexible MATLAB is in

manipulating such arrays. Suppose we would like to store in MATLAB the
following two matrices:

$$A = \begin{bmatrix} 2 & 4 \\ -1 & 6 \end{bmatrix}, \quad B = \begin{bmatrix} 2 & 5 & -3 \\ 1 & 0 & -1 \\ 8 & 4 & 0 \end{bmatrix}.$$

We do so using the following syntax:

```
>> A = [2 4 ; -1 6]
→ A= 2    4
     -1   6


>> B = [2 5 -3; 1 0 -1; 8 4 0]
→ B= 2    5   -3
     1    0   -1
     8    4    0
```

(note that the rows of a matrix are entered in order and separated by semicolons;
also, adjacent entries within a row are given at least one space between). You can
see from the outputs that MATLAB displays these matrices pretty much in their
mathematical form (but without the brackets).

In MATLAB it is extremely simple to edit a previous command into a new one.
Let's say in the matrix B above, we wish to change the bottom-left entry from
eight to three. Since the creation of matrix B was the last command we entered,
we simply need to press the up-arrow key (↑) once and magically the whole last
command appears at the cursor (do this!). If you continue to press this up-arrow
key, the preceding commands will continue to appear in order. Try this now!
Next press the down arrow key (↓) several times to bring you back down again to
the most recent command you entered (i.e., where we defined the matrix B). Now
simply use the mouse and/or left- and right-arrow keys to move the cursor to the 8
and change it to 3, then press enter. You have now overwritten your original
matrix for B with this modified version. Very nice indeed! But there is more. If
on the command line you type a sequence of characters and then press the up-
arrow key, MATLAB will then retrieve only those input lines (in order of most
recent occurrence) that begin with the sequence of characters typed. Thus for
example, if you type a and then up-arrow twice, you would get the line of input
where we set aa = 11.

A few more words about "variables" are in order. Variable names can use up to
19 characters, and must begin with a letter, but after this you can use digits and
underscores as well. For example, two valid variable names are
diffusion22time and Shock_wave_index; however, Final$Amount
would not be an acceptable variable name because of the symbol $. Any time that
you would like to check on the current status of your variables, just enter the
command who:

```
>> who
```

→Your variables are:
A B a aa ans b bb

For more detailed information about all of the variables in the workspace (including the size of all of the matrices) use the command `whos`:

```
>> whos
```

→	Name	Size	Bytes	Class
	A	2x2	32 double	array
	B	3x3	72 double	array
	a	1x1	8 double	array
	aa	1x1	8 double	array
	ans	1x1	8 double	array
	b	1x1	8 double	array
	bb	1x1	8 double	array

You will notice that MATLAB retains both the number a and the matrix A. MATLAB is case-sensitive. You will also notice that there is the variable ans in the workspace. Whenever you perform an evaluation/calculation in MATLAB, an automatic assignment of the variable ans is made to the most recent result (as the output shows). To clear any variable, say aa, use the command

```
>>clear aa
```

Do this and check with `who` that `aa` is no longer in the workspace. If you just enter `clear`, all variables are erased from the workspace. More importantly, suppose that you have worked hard on a MATLAB session and would like to retain all of your workspace variables for a future session. To save (just) the workspace variables, say to your floppy a:\ drive, make sure you have your disk inserted and enter:

```
>> save a:/tutvars
```

This will create a file on your floppy called `tutvars.mat` (you could have called it by any other name) with all of your variables. To see how well this system works, go ahead and quit this MATLAB session and start a new one. If you type `who` you will get no output since we have not yet created any variables in this new session. Now (making sure that the floppy with `tutvars` is still inserted) enter the command:

```
>> load a:/tutvars
```

If you enter `who` once again you will notice that all of those old variables are now in our new workspace. You have just made it through your first MATLAB tutorial. End the session now and examine the diary file if you have created one.

If you want more detailed information about any particular MATLAB command, say `who`, you would simply enter:

```
>> help who
```

and MATLAB would respond with some usage information and related commands.

1.4: VECTORS AND AN INTRODUCTION TO MATLAB GRAPHICS

On any line of input of a MATLAB session, if you enter the percent symbol (%), anything you type after this is ignored by MATLAB's processor and is treated as a comment.[4] This is useful, in particular, when you are writing a complicated program and would like to enhance it with some comments to make it more understandable (both to yourself at a later reading and to others who will read it). Let us now begin a new MATLAB session.

A **vector** is a special kind of matrix with only one row or one column. Here are examples of vectors of each type:

$$x = [1 \quad 2 \quad 3] \qquad y = \begin{bmatrix} 2 \\ -3 \\ 5 \end{bmatrix}.$$

```
>> % We create the above two vectors and one more as variables in our
>> % MATLAB session.
>>   x = [1 2 3], y = [2 ; -3 ; 5], z = [4   -5   6]
→ x = 1    2    3      y = 2            z = 4  -5   6
                           -3
                           5
```

```
>> %  Next we perform some simple array operations.
>> a = x + z
→a = 5  -3   9
>> b = x + y  %MATLAB needs arrays to be the same size to add/subtract
→??? Error using ==> +
Matrix dimensions must agree.
>> c=x.*z    %term by term multiplication, notice the dot before the *
→ c = 4 -10    18
```

The **transpose** of any matrix A, denoted as A^T or A', consists of the matrix whose rows are (in order) the columns of A and vice versa. For example the transpose of the 2 by 3 matrix

$$A = \begin{bmatrix} 2 & 4 & 9 \\ 1 & -2 & 5 \end{bmatrix}$$

is the 3 by 2 matrix

$$A' = \begin{bmatrix} 2 & 1 \\ 4 & -2 \\ 9 & 5 \end{bmatrix}.$$

---

[4] MATLAB's windows usually conform to certain color standards to make codes easier to look through. For example, when a comment is initiated with %, the symbol and everything appearing after it will be shown in green. Also, warning/error messages (as we will soon experience on the next page) appear in red. The default color for input and output is black.

In particular, the transpose of a row vector is a column vector and vice versa.

```
>> y'    %MATLAB uses the  prime ' for the transpose operation
→ ans = 2  -3   5
>> b=x+y'   %cf. with the result for x +  y
→ b = 3  -1   8
>> % We next give some other useful ways to create vectors.
>> % To create a (row) vector having 5 elements linearly spaced
>> % between 0 and 10 you could enter
>> linspace(0,10,5)     %Do this!
→ ans = 0   2.5000  5.0000  7.5000  10.0000
```

We indicate the general syntax of `linspace` as well as another useful way to create vectors (especially big ones!):

v=linspace(F,L,N) →	If F and L are real numbers and N is a positive integer, this command creates a row vector v with: first entry = F, last entry = L and having N equally spaced entries.
v = F:G:L →	If F and L are real numbers and G is a nonzero real number, this command creates a vector v with: first entry = F, last (possible) entry = L and gap between entries = G. G is optional with default value 1.

To see an example, enter

```
>> x = 1:.25:2.5  %will overwrite previously stored value of x
→ x = 1.0000   1.2500   1.5000   1.7500  2.0000  2.2500   2.5000
>> y = -2:.5:3
→ y = -2.0000  -1.5000  -1.0000  -0.5000      0    0.5000   1.0000   1.5000   2.0000
2.5000  3.0000                        .
```

EXERCISE FOR THE READER 1.1: Use the `linspace` command above to recreate the vector y that we just built.

The basic way to plot a graph in MATLAB is to give it the x-coordinates (as a vector a) and the corresponding y-coordinates (as a vector b of the same length) and then use the `plot` command.

plot(a,b) →	If a and b are vectors of the same length, this command will create a plot of the line segments connecting (in order) the points in the xy-plane having x-coordinates listed in the vector a and corresponding y-coordinates in the vector b.

To demonstrate how this works, we begin with some simple vector plots and work our way up to some more involved function plots. The following commands will produce the plot shown in Figure 1.1.

```
>> x = [1 2 3 4]; y = [1 -3 3 0];
>> plot(x,y)
```

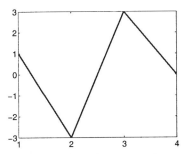

FIGURE 1.1: A simple plot resulting from the command plot(x,y) using the vector $x = [1\ 2\ 3\ 4]$ for x-coordinates and the vector $y = [1\ -3\ 3\ 0]$ for corresponding y-coordinates. [5]

Next, we use the same vector approach to graph the function $y = \cos(x^2)$ on $[0,5]$. The finer the grid determined by the vectors you use, the greater the resolution. To see this first hand, enter:

```
>> x = linspace(0,5,5);    % I will be supressing a lot of output, you
>>                         % can drop the ';'  to see it
>> y = cos(x.^2);
```

Note the dot (.) before the power operator (^). The dot before an operator changes the default matrix operation to a **component-wise operation**. Thus x.^2 will create a new vector of the same size as x where each of the entries is just the square of the corresponding entry of x. This is what we want. The command x^2 would ask MATLAB to multiply the matrix (or row vector) x by itself, which (as we will explain later) is not possible and so would produce an error message.

```
>> plot(x,y)   % produces our first very rough plot of the function
>>             % with only 5 plotting points
```

See Figure 1.2(a) for the resulting plot. Next we do the same plot but using 25 points and then 300 points. The editing techniques of Section 1.2 will be of use as you enter the following commands.

```
>> x = linspace(0,5,25);
>> y = cos(x.^2);
>> plot(x,y)    % a better plot with 25 points.
>> x = linspace(0,5,300);
>> y = cos(x.^2);
>> plot(x,y)    % the plot is starting to look good with 300 points.
```

---

[5] Numerous attributes of a MATLAB plot or other graphic can be modified using the various (very user-friendly) menu options available on the MATLAB graphics window. These include font sizes, line styles, colors, and thicknesses, axis label and tick locations, and numerous other items. To improve readability of this book we will use such features without explicit mention (mostly to make the fonts more readable to accommodate reduced figure sizes).

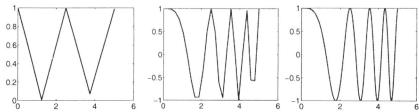

FIGURE 1.2: Plots of the function $y = \cos(x^2)$ on $[0,5]$ with increasing resolution: (a) (left) 5 plotting points, (b) (middle) 25 plotting points and (c) (right) 300 plotting points.

If you want to add more graphs to an existing plot, enter the command:

```
>> hold on     %do this!
```

All future graphs will be added to the existing one until you enter `hold off`. To see how this works, let's go ahead and add the graphs of $y = \cos(2x)$ and $y = \cos^2 x$ to our existing plot of $y = \cos(x^2)$ on $[0,5]$. To distinguish these plots, we might want to draw the curves in different styles and perhaps even different colors. Table 1.1 is a summary of the codes you can use in a MATLAB plot command to get different plot styles and colors:

TABLE 1.1: MATLAB codes for plot colors and styles.

Color/Code		Plot Style/Code	
black / k	red / r	solid / –	stars / *
blue / b	white / w	dashed / – –	x-marks / x
cyan / c	yellow / y	dotted / :	circles / o
green / g		dash-dot / – .	plus-marks / +
magenta / m		points / .	pentacles/ p

Suppose that we want to produce a dashed cyan graph of $y = \cos(2x)$ and a dotted red graph of $y = \cos^2 x$ (to be put in with the original graph). We would enter the following:

```
>> y1 = cos (2*x);
>> plot(x,y1,'c--')      %will plot with cyan dashed curve
>> y2 = cos(x).^2;       % cos(x)^2 would produce an error
>> plot(x,y2,'r:')       %will plot in dotted red style
>> hold off              %puts an end to the current graph
```

You should experiment now with a few other options. Note that the last four of the plot styles will put the given object (stars, x-marks, etc.) around each point that is actually plotted. Since we have so many points (300) such plots would look like very thick curves. Thus these last four styles are more appropriate when the density of plot points is small. You can see the colors on your screen, but unless you have a color printer you should make use of the plot styles to distinguish between multiple graphs on printed plots.

Many features can be added to a plot. For example, the steps below show how to label the axes and give your plot a title.

```
>> xlabel('x')
>> ylabel('cos(x.^2), cos(2*x), cos(x).^2')
>> title('Plot created by yourname')
```

Notice at each command how your plot changes; see Figure 1.3 for the final result.

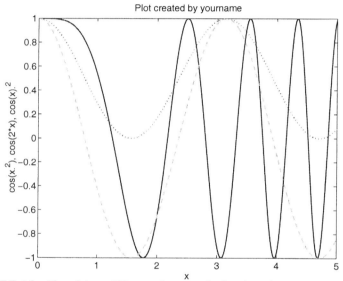

FIGURE 1.3: Plot of three different trigonometric functions done using different colors and styles.

In a MATLAB plot, the points and connecting line segments need not define the graph of a function. For example, to get MATLAB to draw the unit square with vertices (0,0), (1,0), (1,1), (0,1), we could key in the x- and y-coordinates (in an appropriate order so the connecting segments form a square) of these vertices as row vectors. We need to repeat the first vertex at the end so the square gets closed off. Enter:

```
>> x=[0 1 1 0 0]; y=[0 0 1 1 0];
>> plot(x,y)
```

Often in mathematics, the variables x and y are given in terms of an auxiliary variable, say t (thought of as time), rather than y simply being given in terms of (i.e., a function of) x. Such equations are called **parametric equations**, and are easily graphed with MATLAB. Thus parametric equations (in the plane) will look

like: $\begin{cases} x = x(t) \\ y = y(t) \end{cases}$.

These can be used to represent any kind of curve and are thus much more versatile than functions $y = f(x)$ whose graphs must satisfy the vertical line test. MATLAB's plotting format makes plotting parametric equations a simple task. For example, the following parametric equations

$$\begin{cases} x = 2\cos(t) \\ y = 2\sin(t) \end{cases}$$

represent a circle of radius 2 and center (0,0). (Check that they satisfy the equation $x^2 + y^2 = 4$.) To plot the circle, we need only let t run from 0 to 2π (since the whole circle gets traced out exactly once as t runs through these values). Enter:

```
>> t = 0:.01:2*pi;   % a lot of points for decent resolution, as you
>>                   % guessed, 'pi' is how MATLAB denotes π
>> x = 2*cos(t);
>> y = 2*sin(t);
>> plot(x,y)
```

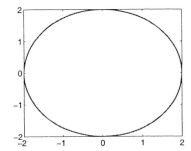 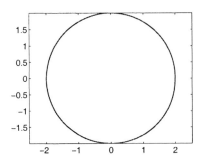

FIGURE 1.4: Parametric plots of the circle $x^2 + y^2 = 4$, (a) (left) first using MATLAB's default rectangular axis setting, and then (b) (right) after the command axis('equal') to put the axes into proper perspective.

You will see an ellipse in the figure window (Figure 1.4(a)). This is because MATLAB uses different scales on the x- and y-axes, unless told otherwise. If you enter: >>axis('equal'), MATLAB will use the same scale on both axes so the circle appears as it should (Figure 1.4(b)). Do this!

EXERCISE FOR THE READER 1.2: In the same fashion use MATLAB to create a plot of the more complicated parametric equations:

$$\begin{cases} x(t) = 5\cos(t/5) + \cos(2t) \\ y(t) = 5\sin(t/5) + \sin(3t) \end{cases} \quad \text{for } 0 \le t \le 10\pi$$

Caution: Do not attempt to plot this one by hand!

If you use the `axis('equal')` command in Exercise for the Reader 1.2, you should be getting the plot pictured in Figure 1.5:

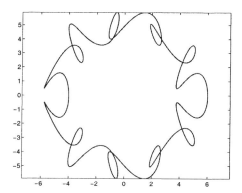

FIGURE 1.5: A complicated MATLAB parametric plot.

EXERCISES 1.4:

1. Use MATLAB to plot the graph of $y = \sin(x^4)$ for $0 \le x \le 2\pi$, first (a) using 200 plotting points, and (b) using 5000 plotting points.

2. Use MATLAB to plot the graph of $y = e^{-1/x^2}$ for $-3 \le x \le 3$, first using (a) 50 plotting points, and then (b) using 10,000 plotting points.

NOTE: When MATLAB does any plot, it automatically tries to choose the axes to exactly accommodate all of the plot points. For functions with vertical asymptotes (like the ones in the next two exercises), you will see that this results in rather peculiar-looking plots.   To improve the appearance of the plots, you can rescale the axes. This is done by using the following command:

	Resets the axis range for plots to be:
`axis([xmin xmax ymin ymax])` →	$$xmin \le x \le xmax$$ $$ymin \le y \le ymax ,$$
	Here, the four vector entries can be any real numbers with xmin < xmax, and ymin < ymax.

3. Use MATLAB to produce a nice plot of the graph of $y = \dfrac{2 - x^2}{x^2 + x - 6}$ on the interval $[-5, 5]$.
    Experiment a bit with the `axis` command as explained in the above note.

4. Use MATLAB to plot the graph of $y = \dfrac{x^4 - 16}{x^3 + 2x^2 - 6}$ on the interval $[-1, 5]$. Adjust the axes, as explained in the note preceding Exercise 3, so as to get an attractive plot.

5. Use MATLAB to plot the circle of radius 3 and center $(-2,1)$.

6. Use MATLAB to obtain a plot of the *epicycloids* that are given by the following parametric

equations:

$$\begin{cases} x(t) = (R+r)\cos t - r\cos\left(\dfrac{R+r}{r}t\right) \\ y(t) = (R+r)\sin t - r\sin\left(\dfrac{R+r}{r}t\right) \end{cases}, \quad 0 \le t \le 2\pi$$

using first the parameters $R = 4$, $r = 1$, and then $R = 12$, $r = 5$. Use no less than 1000 plotting points.

Note: An epicycloid describes the path that a point on the circumference of a smaller circle (of radius r) makes as it rolls around (without slipping) a larger circle (of radius R).

7.   Use MATLAB to plot the parametric equations:

$$\begin{cases} x(t) = e^{-\sqrt{t}}\cos(t) \\ y(t) = e^{-\sqrt{2t}}\sin(t) \end{cases}, \quad 0 \le t \le 100.$$

8.   Use MATLAB to produce a plot of the linear system (two lines):

$$\begin{cases} 2x + 3y = 13 \\ 2x - y = 1 \end{cases}.$$

Include a label for each line as well as a label of the solution (that you can easily find by hand), all produced by MATLAB.

Hints: You will need the `hold on` command to include so many things in the same graph. To insert the labels, you can use either of the commands below to produce the string of text label at the coordinates (x, y).

`text(x,y,'label')` →	Inserts the text string `label` in the current graphic window at the location of the specified point (x, y).
`gtext('label')` →	Inserts the text string `label` in the current graphic window at the location of exactly where you click your mouse.

9.   Use MATLAB to draw a regular octagon (stop-sign shape). This means that all sides have the same length and all interior angles are equal.   Scale the axes accordingly.

10.   By using the `plot` command (repeatedly and appropriately), get MATLAB to produce a circle inscribed in a triangle that is in turn inscribed in another circle, as shown in Figure 1.6.

FIGURE 1.6: Illustration for Exercise 10.

11.   By using the `plot` command (repeatedly and appropriately), get MATLAB to produce something as close as possible to the familiar figure on the right. Do not worry about the line/curve thickness for now, but try to get it so that the eyes (dots) are reasonably visible.

1.5: A TUTORIAL INTRODUCTION TO RECURSION ON MATLAB

Getting a calculator or computer to perform a single task is interesting, but what really makes computers such powerful tools is their ability to perform a long series of related tasks. Such multiple tasks often require a program to tell the computer what to do. We will get more into this later, but it is helpful to have a basic idea at this point on how this works. We will now work on a rather elementary problem from finance that will actually bring to light many important concepts. There are several programming commands in MATLAB, but this tutorial will focus on just one of them (while) that is actually quite versatile.

PROBLEM: To pay off a $100,000.00 loan, Beverly pays $1,000.00 at the end of each month after having taken out the loan. The loan charges 8% annual interest (= 8/12 % monthly interest) compounded monthly on the unpaid balance. Thus, at the end of the first month, the balance on Beverly's account will be (rounded to two decimals): $100,000 (prev. balance) + $666.27 (interest rounded to two decimals) – $1,000 (payment) = $99,666.67. This continues until Beverly pays off the balance; her last payment might be less than $1,000 (since it will need to cover only the final remaining balance and the last month's interest).
(a) Use MATLAB to draw a plot of Beverly's account balances (on the y-axis) as a function of the number of months (on the x-axis) until the balance is paid off.
(b) Use MATLAB to draw a plot of the accrued interest (on the y-axis) that Beverly has paid as a function of the number of months (on the x-axis)
(c) How many years + months will it take for Beverly to completely pay off her loan? What will her final payment be? How much interest will she have paid off throughout the course of the loan?
(d) Use MATLAB to produce a table of values, with one column being Beverly's outstanding balance given in yearly (12 month) increments, and the second column being her total interest paid, also given in yearly increments. Paste the data you get into your word processor to produce a cleaner table of this data.
(e) Redo part (c) if Beverly were to increase her monthly payments to $1,500.

Our strategy will be as follows: We will get MATLAB to create two vectors B and TI that will stand for Beverly's account balances (after each month) and the total interest accrued. We will set it up so that the last entry in B is zero, corresponding to Beverly's account finally being paid off.

There is another way to construct vectors in MATLAB that will suit us well here. We can simply assign the entries of the vector one by one. Let's first try it with the simple example of the vector $x = [1 \quad 5 \quad -2]$. Start a new MATLAB session and enter:

```
>>x(1)  = 1 %specifies the first entry of the vector x, at this point
>>          %x will only have one entry
>>x(2) = 5  %you will see from the output x now has the first two of
>>          %its three components
>>x(3) = -2
```

The trick will be to use **recursion formulas** to automate such a construction of B and TI. This is possible since a single formula shows how to get the next entry of

B or TI if we know the present entry. Such formulas are called recursion formulas and here is what they look like in this case:

$$B(i+1) = B(i) + (.08/12)B(i) - 1000$$
$$TI(i+1) = TI(i) + (.08/12)B(i)$$

In words: The next month's account balance ($B(i+1)$) is the current month's balance ($B(i)$) plus the month's interest on the unpaid balance (($.08/12)B(i)$) less Beverly's monthly payment. Similarly, the total interest accrued for the next month equals that of the current month plus the current month's interest.

Since these formulas allow us to use the information from any month to get that for the next month, all we really need are the initial values $B(1)$ and $TI(1)$ which are the initial account balance (after zero months) and total interest accrued after zero months. These are of course $100,000.00 and $0.00, respectively.

Caution: It is tempting to call these initial values $B(0)$ and $TI(0)$, respectively. However this cannot be done since they are, in MATLAB, vectors (remember, as far as numerical data is concerned: everything in MATLAB is a matrix [or a vector]!) rather than functions of time, and indices of matrices and vectors must be positive integers ($i = 1, 2, ...$). This takes some getting used to since i, the index of a vector, often gets mixed up with t, an independent variable, especially by novice MATLAB users.

We begin by initializing the two vectors B and TI as well as the index i.

```
>> B(1)=100000;  TI(1)=0; i=1;
```

Next, making use of the recursion formulas, we wish to get MATLAB to figure out all of the other entries of these vectors. This will require a very useful device called a "while loop". We want the while loop to keep using the recursion formulas until the account balance reaches zero. Of course, if we did not stop using the recursion formulas, the balance would keep getting more and more negative and we would get stuck in what is called an **infinite loop**. The format for a while loop is as follows:

```
>>while   <condition>
...MATLAB commands...
>>end
```

The way it works is that if the <condition> is met, as soon as you enter end, the "...MATLAB commands..." within the loop are executed, one by one, just as if you were typing them in on the command window. After this the <condition> is reevaluated. If it is still met, the "...MATLAB commands..." are again executed in order. If the <condition> is not met, nothing more is done (this is called *exiting the loop*). The process continues. Either it eventually terminates (exits the loop) or it goes on forever (an infinite loop—a bad program). Let's do a simple example before returning to our problem. Before you enter the following commands try to guess, based on how we just explained while loops, what exactly MATLAB's output will be. Then check your answer with MATLAB's actual output on your screen. If you get it right you are starting to understand the concept of while loops.

```
>> a=1;
>> while a^2 < 5*a
     a=a+2, a^2
   end
```

EXERCISE FOR THE READER 1.3: Analyze and explain each iteration of the above while loop. Note the equation a=a+2 in mathematics makes no sense at all. But remember, in MATLAB the single equal sign means "assignment." So for example, initially $a = 1$. The first run through the while loop the condition is met ($1 = a^2 < 5a = 5$) so a gets reassigned to be $1 + 2 = 3$, and in the same line a^2 is also called to be computed (and listed as output).

Now back to the solution of the problem. We want to continue using the above recursion formulas as long as the balance $B(i)$ remains positive. Since we have already initialized $B(1)$ and $TI(1)$, one possible MATLAB code for creating the rest of these vectors would look like:

```
>> while B(i) > 0
       B(i+1)=B(i)+ 8/12/100*B(i)-1000; % This and the next are just
       %our recursion formulas.
       TI(i+1)=TI(i)+ 8/12/100*B(i);
         i=i+1;   % this bumps the vector index up by one at each
                  % iteration.
   end
```

Notice that MATLAB does nothing, and the prompt does not reappear again, until the while loop is closed off with an end (and you press enter). Although we have suppressed all output, MATLAB has done quite a lot; it has created the vectors B and TI. Observe also that the final balance of zero should be added on as a final component. There is one subtle point that you should be aware of: The value of i after the termination of the while loop is precisely one more than the number of entries of the vectors B and TI thus far created. Try to convince yourself why this is true! Thus we can add on the final entry of B to be zero as follows:

```
>> n=i; B(n)=0;   %We could have just typed 'B(i)=0' but we wanted to
>>                % call 'n' the length of the vector B.
```

Another subtle point is that B(n) was already assigned by the while loop (in the final iteration) but was not a positive balance. This is what caused the while loop to end. So what actually will happen at this stage is that Beverly's last monthly payment will be reduced to cover exactly the outstanding balance plus the last month's interest. Also in the final iteration, the total interest was correctly given by the while loop. To do the required plots, we must first create the time vector. Since time is in months, this is almost just the vector formed by the indices of B (and TI), i.e., it is almost the vector $[1\ \ 2\ \ 3\ \ \cdots\ \ n]$. But remember there is one slight subtle twist. Time starts off at zero, but the vector index must start off at 1. Thus the time vector will be $[0\ \ 1\ \ 2\ \ \cdots\ \ n-1]$. We can easily construct it in MATLAB by

```
>> t=0:n-1;    %this is shorthand for 't=0:1:n-1', by default the
>>                  %gap size is one.
```

Since we now have constructed all of the vectors, plotting the needed graphs is a simple matter.

```
>> plot(t,B)
>> xlabel('time in months'), ylabel('unpaid balance in dollars')
>> %we add on some descriptive labels on the horizontal and vertical
>> %axis. Before we go on we copy this figure or save it (it is
>> %displayed below)
>> plot(t, TI)
>> xlabel('time in months'), ylabel('total interest paid in dollars')
```

See Figure 1.7 for the MATLAB graphics outputs.

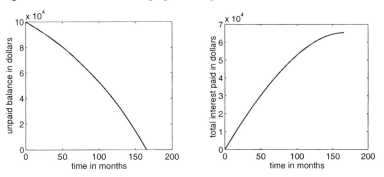

FIGURE 1.7: (a) (top) Graph of the unpaid balance in dollars, as a function of elapsed months in the loan of $100,000 that is being analyzed. (b) (bottom) Graph of the total interest paid in dollars, as a function of elapsed months in the loan of $100,000 that is being analyzed.

We have now taken care of parts (a) and (b). The answer to part (c) is now well within reach. We just have to report the correct components of the appropriate vectors. The time it takes Beverly to pay off her loan is given by the last value of the time vector, i.e.,

```
>> n-1
```
→166.00 =13 years + 10 months (<u>time of loan period</u>).

Her final payment is just the second-to-last component of B, with the final month's interest added to it (that's what Beverly will need to pay to totally clear her account balance to zero):

```
>> format bank  % this puts our dollar answers to the nearest cent.
>> B(n-1)*(1+8/12/100)
→$341.29 (last payment),
```

The total interest paid is just:

```
>> TI(n)
→$65,341.29 (total interest paid)
```

Part (d): Here we simply need to display parts of the two vectors, corresponding to the ends of the first thirteen years of the loan and finally the last month (the 10th month after the thirteenth year). To get MATLAB to generate these two vectors, we could use a while loop as follows:[6]

```
>> k=1; i=1;  %we will use two indices, k will be for the original
>>%              vectors, i will be for the new ones.
>> while k<167
    YB(i)=B(k);  YTI(i)=TI(k);  %we    create    the    two    new    "yearly"
    vectors.
    k=k+12; i=i+1;  %at each iteration, the index of the original
                    %vectors gets bumped up by 12, but that for
                    %the new vectors gets bumped up only by one.
end
```

We next have to add the final component onto each vector (it does not correspond to a year's end). To do this we need to know how many components the yearly vectors already have. If you think about it you will see it is 14, but you could have just as easily asked MATLAB to tell you:

```
>> size(YB) %this command gives the size of any matrix or vector
>>          % (# of rows, # of columns).
→ans = 1.00      14.00
```

```
>>   YB(15)=B(167);  YTI(15)=TI(167);
>>   YB=YB';  YTI=YTI';  %this command reassigns both yearly vectors to
>>                       %be column vectors
```

  Before we print them, we would like to print along the left side the column vector of the corresponding years' end. This vector in column form can be created as follows:

```
>>years   = 0:14; years = years' %first we create it as a row vector
>>                               %and then transpose it.
```

We now print out the three columns:

---

[6] A slicker way to enter these vectors would be to use MATLAB's special vector-creating construct that we mentioned earlier as follows: YB = B(1:12:167), and similarly for YTI.

```
>> years, YB, YTI   % or better [years, YB, YTI]
years =            YB =                YTI =
```

years	YB	YTI
0	100000.00	0
1.00	95850.02	7850.02
2.00	91355.60	15355.60
3.00	86488.15	22488.15
4.00	81216.69	29216.69
5.00	75507.71	35507.71
6.00	69324.89	41324.89
7.00	62628.90	46628.90
8.00	55377.14	51377.14
9.00	47523.49	55523.49
10.00	39017.99	59017.99
11.00	29806.54	61806.54
12.00	19830.54	63830.54
13.00	9026.54	65026.54
14.00	0.00	65341.29

Finally, by making use of any decent word processing software, we can embellish this rather raw data display into a more elegant form such as Table 1.2:

TABLE 1.2: Summary of annual data for the $100,000 loan that was analyzed in this section.

Years Elapsed:	Account Balance:	Total Interest Paid:
0	$100000.00	$ 0
1	95850.02	7850.02
2	91355.60	15355.60
3	86488.15	22488.15
4	81216.69	29216.69
5	75507.71	35507.71
6	69324.89	41324.89
7	62628.90	46628.90
8	55377.14	51377.14
9	47523.49	55523.49
10	39017.99	59017.99
11	29806.54	61806.54
12	19830.54	63830.54
13	9026.54	65026.54
13 + 10 months	0.00	65341.29

Part (e): We can run tAhe same program but we need only modify the line with the recursion formula for the vector B: It now becomes: B(i+1)= B(i)+I(i+1)-1500; With this done, we arrive at the following data:

```
>> i-1,  B(i-1)*(1+8/12/100), TI(i)
```
→ 89 (7 years + 5 months), $693.59(last pmt) , $32693.59 (total interest paid).

EXERCISES 1.5:

1.   Use a while loop to add all of the odd numbers up: $1+3+5+7+\cdots$ until the sum exceeds 5 million. What is the actual sum? How many odd numbers were added?

2.   Redo Exercise 1 by adding up the even numbers rather than the odd ones.

3.   (*Insects from Hell*) An insect population starts off with one pair at year zero. The insects are immortal (i.e., they never die!) and after having lived for two years each pair reproduces another pair (assumed male and female). This continues on indefinitely. So at the end of one year the insect population is still 1 pair (= 2 insects); after two years it is $1 + 1 = 2$ pairs (= 4 insects), since the original pair of insects has reproduced.   At the end of the third year it is $1 + 2 = 3$ pairs (the new generation has been alive for only one year, so has not yet reproduced), and after 4 years the population becomes $2 + 3 = 5$ pairs.   (a) Find out the insect population (in pairs) at the end of each year from year 1 through year 10. (b) What will the insect population be at the end of 64 years?

HISTORICAL ASIDE: The sequence of populations in this problem: 1, 1, $1 + 1 = 2$, $1 + 2 = 3$, $2 + 3$ $= 5$, $3 + 5 = 8$, ... was first introduced in the middle ages by the Italian mathematician Leonardo of Pisa (ca. 1180–1250), who is better known by his nickname: Fibonacci (Italian meaning: son of Bonaccio). This sequence has numerous applications and has made Fibonacci quite famous. It comes up, for example, in hereditary effects in incest, growth of pineapple cells, and electrical engineering. There is even a mathematical journal named in Fibonacci's honor (the *Fibonacci Quarterly*).

4.   Continuing Exercise 3, (a) produce a chart of the insect populations at the end of each 10th year until the end of year 100. (b) Use a while loop to find out how many years it takes for the insect population (in pairs) to exceed 1,000,000,000 pairs.

5.   (*Another Insect Population Problem*)   In this problem, we also start off with a pair of insects, this time mosquitoes. We still assume that after having lived for two years, each pair reproduces another pair. But now, at the end of three years of life, each pair of mosquitoes reproduces one more pair and then immediately dies. (a) Find out the insect population (in pairs) for each year up through year 10. (b) What will the insect population be at the end of 64 years?

6.   Continuing Exercise 5, (a) plot the mosquito (pair) population from the beginning through the end of year 500, as a function of time. (b) How many years does it take for the mosquito population (in pairs) to exceed 1,000,000,000 pairs?

7.   When their daughter was born, Mr. and Mrs. de la Hoya began saving for her college education by investing $5,000 in an annuity account paying 10% interest per year. Each year on their daughter's birthday they invest $2,000 more in the account. (a) Let A_n denote the amount in the account on their daughter's *n*th birthday. Show that A_n satisfies the following recursion formulas:

$$A_0 = 5000$$
$$A_n = (1.1)A_{n-1} + 2000.$$

(b) Find the amount that will be in the account when the daughter turns 18.

(c) Print (and nicely label) a table containing the values of n and A_n as n runs from 0 to 18.

8.   Louise starts an annuity plan at her work that pays 9% annual interest compounded monthly. She deposits $200 each month starting on her 25th birthday. Thus at the end of the first month her account balance is exactly $200. At the end of the second month, she puts in another $200, but her first deposit has earned her one month's worth of interest. The 9% interest per year means she gets 9%/12 = 0.75% interest per month. Thus the interest she earns in going from the first to second month is .75% of $200 or $1.50 and so her balance at the end of the second month is 401.50. This continues, so at the end of the 3rd month, her balance is $401.50 (old

balance) + .75 percent of this (interest) + $200 (new deposit) = $604.51. Louise continues to do this throughout her working career until she retires at age 65.

(a) Figure out the balances in Louise's account at her birthdays: 26th, 27th, …, up through her 65th birthday. Tabulate them neatly in a table (either cut and paste by hand or use your word processor—do not just give the raw MATLAB output, but rather put it in a form so that your average citizen could make sense of it).

(b) At exactly what age (to the nearest month) is Louise when the balance exceeds $100,000? Note that throughout these 40 years Louise will have deposited a total of $200/month × 12 months/yr. × 40 years = $96,000.

9. In saving for retirement, Joe, a government worker, starts an annuity that pays 12% annual interest compounded monthly. He deposits $200.00 into it at the end of each month. He starts this when he is 25 years old. (a) How long will it take for Joe's annuity to reach a value of $1 million? (b) Plot Joe's account balance as a function of time.

10. The **dot product** of two vectors of the same length is defined as follows:
If $x = [x(1) \ x(2) \ \cdots \ x(n)]$, $y = [y(1) \ y(2) \ \cdots \ y(n)]$ then

$$x \cdot y = \sum_{i=1}^{n} x(i) y(i).$$

The dot product appears and is useful in many areas of math and physics. As an example, check that the dot product of the vectors [2 0 6] and [1 −1 4] is 26. In MATLAB, if x and y are stored as row vectors, then you can get the dot product by typing x*y'(the prime stands for transpose, as in the previous section; Chapter 7 will explain matrix operations in greater detail). Let x and y be the vectors each with 100 components having the forms:

$$x = [1, -1, \ 1, -1, \ \ 1, \ -1, \ \cdots],$$
$$y = [1, \ \ 4, 9, 16, \ \ 25, \ 36, \ \cdots]$$

Use a while loop in MATLAB to create and store these vectors and then compute their dot product.

Chapter 2: Basic Concepts of Numerical Analysis with Taylor's Theorem

2.1: WHAT IS NUMERICAL ANALYSIS?

Outside the realm of pure mathematics, most practicing scientists and engineers are not concerned with finding exact answers to problems. Indeed, living in a finite universe, we have no way of exactly measuring physical quantities and even if we did, the exact answer would not be of much use. Just a single irrational number, such as

$$\pi = 3.14159265358979323846264338327950288419716939937510582097749...,$$

where the digits keep going on forever without repetition or any known pattern, has more information in its digits than all the computers in the world could possibly ever store. To help motivate some terminology, we bring forth a couple of examples.

Suppose that Los Angeles County is interested in finding out the amount of water contained in one of its reserve drinking water reservoirs. It hires a contracting firm to measure this amount. The firm begins with a large-scale pumping device to take care of most of the water, leaving only a few gallons. After this, they bring out a more precise device to measure the remainder and come out with a volume of 12,564,832.42 gallons. To get a second opinion, the county hires a more sophisticated engineering firm (that charges 10 times as much) and that uses more advanced measuring devices. Suppose this latter firm came up with the figure 12,564,832.3182. Was the first estimate incorrect? Maybe not, perhaps some evaporation or spilling took place—so there cannot really be an exact answer to this problem. Was it really worth the extra cost to get this more accurate estimate? Most likely not—even an estimate to the nearest gallon would have served equally well for just about any practical purposes.

Suppose next that the Boeing Corporation, in the design and construction of a new 767 model jumbo jet, needs some wing rivets. The engineers have determined the rivets should have a diameter of 2.75 mm with a tolerance (for error) of .000025mm. Boeing owns a precise machine that will cut such rivets to be of diameter $2.75 \pm .000006$ mm. But they can purchase a much more expensive machine that will produce rivets of diameters $2.75 \pm .0000001$ mm (60 times as accurate). Is it worth it for Boeing to purchase and use this more expensive machine? The aeronautical engineers have determined that such an improvement in rivets would not result in any significant difference in the safety and reliability of the wing and plane; however, if the error exceeds the given tolerance, the wings may become unstable and a safety hazard.

In mathematics, there are many problems and equations (algebraic, differential, and partial differential) whose exact solutions are known to exist but are difficult, very time consuming, or impossible to solve exactly. But for many practical purposes, as evidenced by the previous examples, an estimate to the exact answer will do just fine, provided that we have a guarantee that the error is not too large. So, here is the basic problem in numerical analysis: We are interested in a solution x (= **exact answer**) to a problem or equation. We would like to find an estimate x^* (= **approximation**) so that $|x - x^*|$ (= the **actual error**) is no more the maximum **tolerance** $(= \varepsilon)$, i.e., $|x - x^*| \le \varepsilon$. The maximum tolerated error will be specified ahead of time and will depend on the particular problem at hand. What makes this approximation problem very often extremely difficult is that we usually do not know x and thus, even after we get x^*, we will have no way of knowing the actual error. But regardless of this, we still need to be able to guarantee that it is less than ε. Often more useful than the actual error is the **relative error**, which measures the error as a ratio in terms of the magnitude of the actual quantity, i.e., it is defined by

$$\text{relative error} = \frac{|x - x^*|}{|x|},$$

provided, of course, that $x \ne 0$.

EXAMPLE 2.1: In the Los Angeles reservoir measurement problem given earlier, suppose we took the exact answer to be the engineering firm's estimate: x = 12,564,832.3182 gallons, and the contractor's estimate as the approximation x^* = 12,564,832.42. Then the error of this approximation is $|x - x^*|$ = 0.1018 gallons, but the relative error (divide this answer by x) is only 8.102×10^{-9}.

EXAMPLE 2.2: In the Boeing Corporation's rivet problem above, the maximum tolerated error is .000025 mm, which translates to a maximum relative error of (divide by x = 2.75) 0.000009. The machine they currently have would yield a maximum relative error of 0.000006/2.75 = 0.000002 and the more expensive machine they were considering would guarantee a maximum relative error of no more than 0.0000001/2.75 = 3.6364×10^{-8}.

For the following reasons, we have chosen Taylor's theorem as a means to launch the reader into the realm of numerical analysis. First, Taylor's theorem is at the foundation of many numerical methods. Second, it covers one of those rare situations in numerical analysis where quality error estimates are readily available and thus errors can be controlled and estimated quite effectively. Finally, most readers should have some familiarity with Taylor's theorem from their calculus courses.

Most mathematical functions are very difficult to compute by just using the basic mathematical operations: $+, -, \times, \div$. How, for example, would we compute

$\cos(27°)$ just using these operations? One type of function that is possible to compute in this way is a polynomial. A **polynomial** in the variable x is a function of the form:

$$p(x) = a_n x^n + \cdots + a_2 x^2 + a_1 x + a_0,$$

where $a_n, \cdots a_2, a_1, a_0$ are any real numbers. If $a_n \neq 0$, then we say that the **degree** of $p(x)$ equals n. Taylor's theorem from calculus shows how to use polynomials to approximate a great many mathematical functions to any degree of accuracy. In Section 2.2, we will introduce the special kind of polynomial (called Taylor polynomials) that gets used in this theorem and in Section 2.3 we discuss the theorem and its uses.

EXERCISES 2.1:

1. If $x = 2$ is approximated by $x^* = 1.96$, find the actual error and the relative error.

2. If $\pi \, (= x \,)$ is approximated by $x^* = 3\frac{1}{8}$ (as was done by the ancient Babylonians, c. 2000 BC), find the actual error and the relative error.

3. If $x = 10000$ is approximated by $x^* = 9999.96$, find the actual error and the relative error.

4. If $x = 5280$ feet (one mile) is approximated by $x^* = 5281$ feet, find the actual and relative errors.

5. If $x = 0.76$ inches and the relative error of an approximation is known to be 0.05, find the possible values for x^*.

6. If $x = 186.4$ and the relative error of an approximation is known to be 0.001, find the possible values for x^*.

7. A civil engineering firm wishes to order thick steel cables for the construction of a span bridge. The cables need to measure 2640 feet in length with a maximum tolerated relative error of 0.005. Translate this relative error into an actual tolerated maximum discrepancy from the ideal 2640-foot length.

2.2: TAYLOR POLYNOMIALS

Suppose that we have a mathematical function $f(x)$ that we wish to approximate near $x = a$. The **Taylor polynomial of order** n, $p_n(x)$, for this function **centered at** (or **about**) $x = a$ is that polynomial of degree of at most n that has the same values as $f(x)$ and its first n derivatives at $x = a$. The definition requires that $f(x)$ possess n derivatives at $x = a$. Since derivatives measure rates of change of functions, the Taylor polynomials are designed to mimic the behavior of the function near $x = a$. The following example will demonstrate this property.

EXAMPLE 2.3: Find formulas for, and interpret, the order-zero and order-one Taylor polynomials $p_0(x)$ and $p_1(x)$ of a function $f(x)$ (differentiable) at $x = a$.

SOLUTION: The zero-order polynomial $p_0(x)$ has degree at most zero, and so must be a constant function. But by its definition, we must have $p_0(a) = f(a)$. Since $p_0(x)$ is constant this means that $p_0(x) = f(a)$ (a horizontal line function). The first-order polynomial $p_1(x)$ must satisfy two conditions:

$$p_1(a) = f(a) \text{ and } p_1'(a) = f'(a). \tag{1}$$

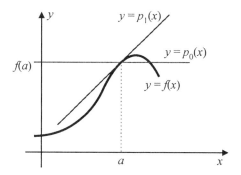

FIGURE 2.1: Illustration of a graph of a function $y = f(x)$ (heavy black curve) together with its zero-order Taylor polynomial $p_0(x)$ (horizontal line) and first-order Taylor polynomial $p_1(x)$ (slanted tangent line).

Since $p_1(x)$ has degree at most one, we can write $p_1(x) = mx + b$, i.e., $p_1(x)$ is just a line with slope m and y-intercept b. If we differentiate this equation and use the second equation in (1), we get that $m = f'(a)$. We now substitute this in for m, put $x = a$ and use the first equation in (1) to find that $f(a) = p_1(a) = f'(a)a + b$. Solving for b gives $b = f(a) - f'(a)a$. So putting this all together yields that $p_1(x) = mx + b = f'(a)x + f(a) - f'(a)a = f(a) + f'(a)(x - a)$. This is just the tangent line to the graph of $y = f(x)$ at $x = a$. These two polynomials are illustrated in Figure 2.1.

In general, it can be shown that the Taylor polynomial of order n is given by the following formula:

$$p_n(x) = f(a) + f'(a)(x - a) + \frac{1}{2}f''(a)(x - a)^2 + \frac{1}{3!}f'''(a)(x - a)^3 + \cdots + \frac{1}{n!}f^{(n)}(a)(x - a)^n, \tag{2}$$

where we recall that the factorial of a positive integer k is given by:

$$k! = \begin{cases} 1, & \text{if } k=0 \\ 1 \cdot 2 \cdot 3 \cdots (k-1) \cdot k, & \text{if } k = 1,2,3,\cdots. \end{cases}$$

Since $0! = 1! = 1$, we can use Sigma-notation to rewrite this more simply as:

$$p_n(x) = \sum_{k=0}^{n} \frac{1}{k!} f^{(k)}(a)(x-a)^k . \tag{3}$$

We turn now to some specific examples:

EXAMPLE 2.4: (a) For the function $f(x) = \cos(x)$, compute the following Taylor polynomials at $x = 0$: $p_1(x)$, $p_2(x)$, $p_3(x)$, and $p_8(x)$.
(b) Use MATLAB to find how each of these approximates $\cos(27°)$ and then find the actual error of each of these approximations.
(c) Find a general formula for $p_n(x)$.
(d) Using an appropriate MATLAB graph, estimate the length of the largest interval $[-a, a] = \{|x| \le a\}$ about $x = 0$ that $p_8(x)$ can be used to approximate $f(x)$ with an error always less than or equal to 0.2. What if we want the error to be less than or equal to 0.001?

SOLUTION: Part (a): We see from formula (2) or (3) that each Taylor polynomial is part of any higher-order Taylor polynomial. Since $a = 0$ in this example, formula (2) reduces to:

$$p_n(x) = f(0) + f'(0)x + \frac{1}{2} f''(0)x^2 + \frac{1}{3!} f'''(0)x^3 + \cdots$$
$$+ \frac{1}{n!} f^{(n)}(0)x^n = \sum_{k=0}^{n} \frac{1}{k!} f^{(k)}(0)x^k . \tag{4}$$

A systematic way to calculate these polynomials is by constructing a table for the derivatives:

n	$f^{(n)}(x)$	$f^{(n)}(0)$
0	$\cos(x)$	1
1	$-\sin(x)$	0
2	$-\cos(x)$	-1
3	$\sin(x)$	0
4	$\cos(x)$	1
5	$-\sin(x)$	0

We could continue on, but if one notices the repetitive pattern (when $n = 4$, the derivatives go back to where they began), this can save some time and help with finding a formula for the general Taylor polynomial $p_n(x)$. Using formula (4) in conjunction with the table (and indicated repetition pattern), we conclude that:

$$p_1(x) = 1, \ p_2(x) = 1 - \frac{x^2}{2} = p_3(x), \text{ and } p_8(x) = 1 - \frac{x^2}{2} + \frac{x^4}{4!} - \frac{x^6}{6!} + \frac{x^8}{8!}$$

Part (b): To use these polynomials to approximate $\cos(27°)$, we of course need to take x to be in radians, i.e., $x = 27° \left(\frac{\pi}{180°} \right) = .4712389.....$ Since two of these Taylor polynomials coincide, there are three different approximations at hand for $\cos(27°)$. In order to use MATLAB to make the desired calculations, we introduce a relevant MATLAB function:

```
To compute n! in MATLAB, use either:  factorial(n), or gamma(n+1)
```

Thus for example, to get 5! = 120, we could either type >>factorial(5) or >>gamma(6). Now we calculate:

```
>> x=27*pi/180;
>> format long
>> p1=1;
>> p2=1-x^2/2
→ 0.88896695048774
>> p8=p2+x^4/gamma(5)-x^6/gamma(7)+x^8/gamma(9)
→ 0.891006652433693
>> abs(p1-cos(x))    %abs, as you guessed, stands for absolute value
→ 0.10899347581163
>> abs(p2-cos(x))    → 0.00203957370062
>> abs(p8-cos(x))    → 1.485654932409375e-010
```

Transcribing these into usual mathematics notation, we have the approximations for $\cos(27°)$:

$$p_1(27°) = 1, \ p_2(27°) = p_3(27°) = .888967..., \ p_8(27°) = .89100694.... .$$

which have the corresponding errors:

$$|p_1(27°) - \cos(27°)| = 0.1089...,$$
$$|p_2(27°) - \cos(27°)| = |p_3(27°) - \cos(27°)| = 0.002039..., \text{ and}$$
$$|p_8(27°) - \cos(27°)| = 1.4856 \times 10^{-10}.$$

This demonstrates quite clearly how nicely these Taylor polynomials serve as approximating tools. As expected, the higher degree Taylor polynomials do a better job approximating but take more work to compute.

Part (c): Finding the general formula for the nth-order. Taylor polynomial $p_n(x)$ can be a daunting task, if it is even possible. It will be possible if some pattern can be discovered with the derivatives of the corresponding function at $x = a$. In this case, we have already discovered the pattern in part (b), which is quite simple: We just need a nice way to write it down, as a formula in n. It is best to separate into cases where n is even or odd. If n is odd we see $f^{(n)}(0) = 0$, end of story. When n is even $f^{(n)}(0)$ alternates between $+1$ and -1. To get a formula, we write an even n as $2k$, where k will alternate between even and odd integers. The trick is to use either $(-1)^k$ or $(-1)^{k+1}$ for $f^{(2k)}(0)$, which both also alternate between $+1$ and -1. To see which of the two to use, we need only check the starting values at $k = 0$ (corresponding to $n = 0$). Since $f^{(0)}(0) = 1$, we must use $(-1)^k$. Since any odd integer can be written as $n = 2k + 1$, in summary we have arrived at the following formula for $f^{(n)}(0)$:

$$f^{(n)}(0) = \begin{cases} (-1)^k, & \text{if } n = 2k \text{ is even,} \\ 0, & \text{if } n = 2k+1 \text{ is odd} \end{cases}.$$

Plugging this into equation (4) yields the formulas:

$$p_n(x) = 1 - \frac{x^2}{2!} + \frac{x^4}{4!} - \cdots + (-1)^k \frac{x^{2k}}{(2k)!} = \sum_{j=0}^{k} (-1)^j \frac{x^{2j}}{(2j)!}.$$
$$(\text{for } n = 2k \text{ or } n = 2k+1)$$

Part (d): In order to get a rough idea of how well $p_8(x)$ approximates $\cos(x)$, we will first need to try out a few plots. Let us first plot these two functions together on the domain: $-10 \le x \le 10$. This can be done as follows:

```
>> x=-10:.0001:10;
>> y=cos(x);
>> p8=1-x.^2/2+x.^4/gamma(5)-x.^6/gamma(7)+x.^8/gamma(9);
>> plot(x,y,x,p8,'r-.')
```

Notice that we were able to produce both plots (after having constructed the needed vectors) by a single line, without the `hold on/hold off` method. We have instructed MATLAB to plot the original function $y = \cos(x)$ in the default color and style (blue, solid line) and the approximating function $y = p_8(x)$ as a red plot with the dash/dot style. The resulting plot is the first one shown in Figure 2.2.

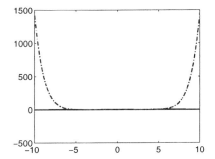 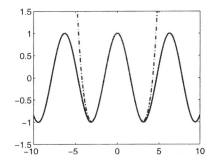

FIGURE 2.2: Graphs of $y = \cos(x)$ (solid) together with the eighth order Taylor approximating polynomial $y = p_8(x)$ (dash-dot) shown with two different y-ranges.

To answer (even just) the first question, this plot is not going to help much, owing to the fact that the scale on the y-axis is so large (increments are in 200 units and we need our error to be < 0.2). MATLAB always will choose the scales to accommodate all of the points in any given plot. The eighth-degree polynomial $y = p_8(x)$ gets so large at $x = \pm 10$ that the original function $y = \cos(x)$ is dwarfed in comparison so its graph will appear as a flat line (the x-axis). We could redo the plot trying out different ranges for the x-values and eventually arrive at a more satisfactory illustration[1]. Alternatively and more simply, we can work with the existing plot and get MATLAB to manually change the range of the x- and/or y-axes that appear in the plot. The way to do this is with the command:

`axis([xmin xmax ymin ymax])` $\rightarrow$	changes the range of a plot to : $x\min \le x \le x\max, \text{ and } y\min \le y \le y\max$

Thus to keep the x-range the same $[-10, 10]$, but to change the y-range to be $[-1.5, 1.5]$, we would enter the command to create the second plot of Figure 2.2.

```
>> axis([-10 10 -1.5 1.5])
```

We can see now from the second plot above that (certainly) for $-3 \le x \le 3$ we have $|\cos(x) - p_8(x)| \le 0.2$. This graph is, however, unsatisfactory in regards to the second question of the determination of an x-interval for which $|\cos(x) - p_8(x)| \le 0.001$. To answer this latter question and also to get a more satisfactory answer for the first question, we need only look at plots of the actual error $y = |\cos(x) - p_8(x)|$. We do this for two different y-ranges. There is a nice way to get MATLAB to partition its plot window into several (in fact, a matrix of) smaller subwindows.

`subplot(m,n,i)` $\rightarrow$	causes the plot window to partition into an $m \times n$ matrix of proportionally smaller subwindows, with the next plot going into the ith subwindow (listed in the usual "reading order"—left to right, then top to bottom)

The two error plots in Figure 2.3 were obtained with the following commands:

```
>> subplot(2,1,1)
>> plot(x,abs(y-p8)), axis([-10 10 -.1 .3])
>> subplot(2,1,2)
>> plot(x,abs(y-p8)), axis([-5 5 -.0005 .0015])
```

Notice that the ranges for the axes were appropriately set for each plot so as to make each more suitable to answer each of the corresponding questions.

---

[1] The zoom button ⊕ on the graphics window can save some time here. To use it, simply left click on this button with your mouse, then move the mouse to the desired center of the plot at which to zoom and left click (repeatedly). The zoom-out key ⊖ works in the analogous fashion.

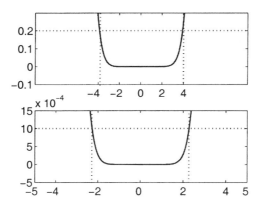

FIGURE 2.3: Plots of the error $y = |\cos(x) - p_8(x)|$ on two different y-ranges. Reference lines were added to help answer the question in part (d) of Example 2.4.

From Figure 2.3, we can deduce that if we want to guarantee an error of at most 0.2, then we can use $p_8(x)$ to approximate $\cos(x)$ anywhere on the interval $[-3.8, 3.8]$, while if we would like the maximum error to be only 0.001, we must shrink the interval of approximation to about $[-2.2, 2.2]$. In Figure 2.4 we give a MATLAB-generated plot of the function $y = \cos(x)$ along with several of its Taylor polynomials.

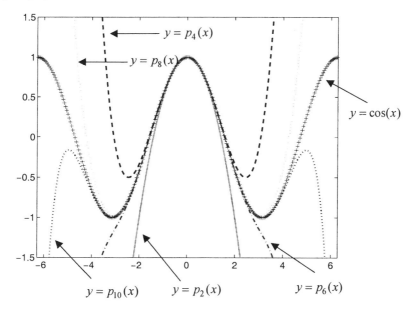

FIGURE 2.4: Some Taylor polynomials for $y = \cos(x)$.

EXERCISE FOR THE READER 2.1: Use MATLAB to produce the plot in Figure 2.4 (without the arrow labels).

It is a rare situation indeed in numerical analysis where we can actually compute the exact errors explicitly. In the next section, we will give Taylor's theorem, which gives us usable estimates for the error even in cases where it cannot be explicitly computed.

EXERCISES 2.2:

1. Find the second- and third-order Taylor polynomials $p_2(x)$ and $p_3(x)$, centered at $x = a$, for the each of the following functions.
   (a) $f(x) = \sin(x)$, $a = 0$ (b) $f(x) = \tan(x)$, $a = 0$
   (c) $f(x) = e^x$, $a = 1$ (d) $f(x) = x^{1/3}$, $a = 8$

2. Repeat Exercise 1 for each of the following:
   (a) $f(x) = \cos(x)$, $a = \pi/2$ (b) $f(x) = \arctan(x)$, $a = 0$
   (c) $f(x) = \ln x$, $a = 1$ (d) $f(x) = \cos(x^2)$, $a = 0$

3. (a) Approximate $\sqrt{65}$ by using the first-order Taylor polynomial of $f(x) = \sqrt{x}$ centered at $x = 64$ (this is tangent line approximation discussed in first-semester calculus) and find the error and the relative error of this approximation. (b) Repeat part (a) using instead the second-order Taylor polynomial to do the approximation.
   (c) Repeat part (a) once again, now using the fourth-order Taylor polynomial.

4. (a) Approximate $\sin(92°)$ by using the first-order Taylor polynomial of $f(x) = \sin(x)$ centered at $x = \pi/2$ (tangent line approximation) and find the error and the relative error of this approximation.
   (b) Repeat part (a) using instead the second-order Taylor polynomial to do the approximation.
   (c) Repeat part (a) using the fourth-order Taylor polynomial.

5. Find a general formula for the order n Taylor polynomial $p_n(x)$ centered at $x = 0$ for each of the following functions:
   (a) $y = \sin(x)$ (b) $y = \ln(1 + x)$
   (c) $y = e^x$ (d) $y = \sqrt{x + 1}$

6. Find a general formula for the order n Taylor polynomial $p_n(x)$ centered at $x = 0$ for each of the following functions:
   (a) $y = \tan(x)$ (b) $y = 1/(1 + x)$
   (c) $y = \arctan(x)$ (d) $y = x\sin(x)$

7. (a) Compute the following Taylor polynomials, centered at $x = 0$, of $y = \cos(x^2)$:

   $$p_1(x), \ p_2(x), \ p_6(x), \ p_{10}(x).$$

   (b) Next, use the general formula obtained in Example 2.4 for the general Taylor polynomials of $y = \cos(x)$ to write down the order 0, 1, 3 and 5 Taylor polynomials. Replace x with x^2 in each of these polynomials. Compare these with the Taylor polynomials in part (a).

8. Consider the function $f(x) = \sin(3x)$. All of the plots in this problem are to be done withMATLAB on the interval $[-3, 3]$. The Taylor polynomials refer to those of $f(x)$ centered at $x = 0$. Each graph of $f(x)$ should be done with the usual plot settings, while each graph of a Taylor polynomial should be done with the dot style.
    (a) Use the subplot command to create a graphic with 3×2 entries as follows:

The simultaneous graphs of $f(x)$ along with the first-order Taylor polynomial (= tangent line).	A graph of the error $\left|f(x) - p_1(x)\right|$
The simultaneous graphs of $f(x)$ along with the 3rd-order Taylor polynomial.	A graph of the error $\left|f(x) - p_3(x)\right|$
The simultaneous graphs of $f(x)$ along with the 9th-order Taylor polynomial.	A graph of the error $\left|f(x) - p_9(x)\right|$

    (b) By looking at your graphs in part (a), estimate on how large an interval $[-a, a]$ about $x = 0$ that the first-order Taylor polynomial would provide an approximation to $f(x)$ with error < 0.25. Answer the same question for p_3 and p_9.

9. (a) Let $f(x) = \ln(1 + x^2)$. Find formulas for the following Taylor polynomials of $f(x)$ centered at $x = 0$: $p_2(x)$, $p_3(x)$, $p_6(x)$. Next, using the subplot command, create a graphic window split in two sides (left and right). On the left, plot (together) the four functions $f(x)$, $p_2(x)$, $p_3(x)$, $p_6(x)$. In the right-side subwindow, plot (together) the corresponding graphs of the three errors: $\left|f(x) - p_2(x)\right|$, $\left|f(x) - p_3(x)\right|$, and $\left|f(x) - p_6(x)\right|$. For the error plot adjust the y- range so as to make it simple to answer the question in part (b). Use different styles/colors to code different functions in a given plot.
    (b) By looking at your graphs in part (a), estimate how large an interval $[-a, a]$ about $x = 0$ on which the second-order Taylor polynomial would provide an approximation to $f(x)$ with error < 0.25. Answer the same question for p_3 and p_6.

10. (a) Let $f(x) = x^2 \sin(x)$. Find formulas for the following Taylor polynomials of $f(x)$ centered at $x = 0$: $p_1(x)$, $p_4(x)$, $p_9(x)$. Next, using the subplot command, get MATLAB to create a graphic window split in two sides (left, and right). On the left, plot (together) the four functions $f(x)$, $p_1(x)$, $p_4(x)$, $p_9(x)$. In the right-side subwindow, plot (together) the corresponding graphs of the three errors: $\left|f(x) - p_1(x)\right|$, $\left|f(x) - p_4(x)\right|$, and $\left|f(x) - p_9(x)\right|$. For the error plot adjust the y- range so as to make it simple to answer the question in part (b). Use different styles/colors to code different functions in a given plot.
    (b) By looking at your graphs in part (a), estimate on how large an interval $[-a, a]$ about $x = 0$ the first-order Taylor polynomial would provide an approximation to $f(x)$ with error < 0.05. Answer the same question for p_4 and p_9.

11. In Example 2.3, we derived the general formula (2) for the zero- and first-order Taylor polynomial.
    (a) Do the same for the second-order Taylor polynomial, i.e., use the definition of the Taylor polynomial $p_2(x)$ to show that (2) is valid when $n = 2$.
    (b) Prove that formula (4) for the Taylor polynomials centered at $x = 0$ is valid for any n.
    (c) Prove that formula (2) is valid for all n.
    Suggestion: For part (c), consider the function $g(x) = f(x + a)$, and apply the result of part

(b) to this function. How are the Taylor polynomials of $g(x)$ at $x = 0$ related to those of $f(x)$ at $x = a$?

12.   (*Another Kind of Polynomial Interpolation*) In this problem we compare the fourth-order Taylor polynomial $p_3(x)$ of $y = \cos(x)$ at $x = 0$ (which is actually $p_4(x)$) with the third-order polynomial $p(x) = a_0 + a_1x + a_2x^2 + a_3x^3$, which has the same values and derivative as $\cos(x)$ at the points $x = 0$ and $x = \pi$. This means that $p(x)$ satisfies these four conditions:

$$p(0) = 1 \qquad p'(0) = 0 \qquad p(\pi) = -1 \qquad p'(\pi) = 0 .$$

Find the coefficients: $a_0, a_1, a_2,$ and a_3 of $p(x)$, and then plot all three functions together. Discuss the errors of the two different approximating polynomials.

2.3: TAYLOR'S THEOREM

In the examples and problems of previous section we introduced Taylor poly-

nomials $p_n(x)$ of a function $y = f(x)$ (appropriately differentiable) at $x = a$, and we saw that they appear to often serve as great tools for approximating the function near $x = a$. We also have seen that as the order n of the Taylor polynomial increases, so does its effectiveness in approximating $f(x)$. This of course needs to be reconciled with the fact that for larger values of n it is more work to form and compute $p_n(x)$. Additionally, the approximations seemed to improve, in general, when x gets closer to a . This latter observation seems plausible since $p_n(x)$ was constructed using only information about $f(x)$ at

FIGURE 2.5: Brook Taylor[2] (1685–1731), English mathematician.

$x = a$. Taylor's theorem provides precise quantitative estimates for the error

---

[2] Taylor was born in Middlesex, England, and his parents were quite well-rounded people of society. His father was rather strict but instilled in Taylor a love for music and painting. His parents had him educated at home by private tutors until he entered St. John's College in Cambridge when he was 18. He graduated in 1709 after having written his first important mathematical paper a year earlier (this paper was published in 1714).   He was elected rather early in his life (1712) to the Royal Society, the election being based more on his potential and perceived mathematical powers rather than on his published works, and two years later he was appointed to the prestigious post of Secretary of the Royal Society.   In this same year he was appointed to an important committee that was to settle the issue of "who invented calculus" since both Newton and Leibniz claimed to be the founders. Between 1712 and 1724 Taylor published 13 important mathematical papers on a wide range of subjects including magnetism, logarithms, and capillary action.

Taylor suffered some tragic personal events. His father objected to his marriage (claiming the bride's family was not a "good" one) and after the marriage Taylor and his father cut off all contact until 1723, when his wife died giving birth to what would have been Taylor's first child. Two years later he remarried (this time with his father's blessings) but the following year his new wife also died during childbirth, although this time his daughter survived.

$|f(x)-p_n(x)|$, which can be very useful in choosing an appropriate order n so that $p_n(x)$ will give an approximation within the desired error bounds. We now present Taylor's theorem. For its proof we refer the reader to any decent calculus textbook.

THEOREM 2.1: (*Taylor's Theorem*) Suppose that for a positive integer n, a function $f(x)$ has the property that its $(n+1)$st derivative is continuous on some interval I on the x-axis that contains the value $x=a$. Then the **nth-order remainder** $R_n(x) \equiv f(x) - p_n(x)$ resulting from approximating $f(x)$ by $p_n(x)$ is given by

$$R_n(x) = \frac{f^{(n+1)}(c)}{(n+1)!}(x-a)^{n+1} \quad (x \in I), \tag{5}$$

for some number c, lying between a and x (inclusive); see Figure 2.6.

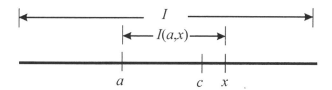

FIGURE 2.6: One possible arrangement of the three points relevant to Taylor's theorem.

REMARKS: (1) Note that like many such theorems in calculus, Taylor's theorem asserts the existence of the number c, but it does not tell us what it is.
(2) By its definition, (the absolute value of) $R_n(x)$ is just the actual error of the approximation of $f(x)$ by its nth-order Taylor polynomial $p_n(x)$, i.e.,

$$\text{error} = |f(x)-p_n(x)| = |R_n(x)|.$$

Since we do not know the exact value of c, we will not be able to calculate the error precisely; however, since we know that c must lie somewhere between a and x on I, call this interval $I(a,x)$, we can estimate that the unknown quantity $|f^{(n+1)}(c)|$ that appears $R_n(x)$, can be no more than the maximum value of this $(n+1)$st derivative function $|f^{(n+1)}(z)|$ as z runs through the interval $I(a,x)$. In mathematical symbols, this is expressed as:

$$|f^{(n+1)}(c)| \le \max\{|f^{(n+1)}(z)| : z \in I(a,x)\}. \tag{6}$$

EXAMPLE 2.5: Suppose that we wish to use Taylor polynomials (at $x=0$) to approximate $e^{0.7}$ with an error less than 0.0001.

(a) Apply Taylor's theorem to find out what order n of a Taylor polynomial we could use for the approximation to guarantee the desired accuracy.
(b) Perform this approximation and then check that the actual error is less than the maximum tolerated error.

SOLUTION: Part (a): Here $f(x) = e^x$, so, since $f(x)$ is its own derivative, we have $f^{(n)}(x) = e^x$ for any n, and so $f^{(n)}(0) = e^0 = 1$. From (4) (or (2) or (3) with $a = 0$), we can therefore write the general Taylor polynomial for $f(x)$ centered at $x = 0$ as

$$p_n(x) = 1 + x + \frac{x^2}{2!} + \frac{x^3}{3!} + \cdots + \frac{x^n}{n!} = \sum_{k=0}^{n} \frac{x^k}{k!},$$

and from (5) (again with $a = 0$), $R_n(0.7) = \dfrac{e^c}{(n+1)!}(0.7)^{n+1}$.

How big can e^c be? For $f(x) = e^x$, this is just the question of finding out the right side of (6). In this case the answer is easy: Since c lies between 0 and 0.7, and e^x is an increasing function, the largest value that e^c can be is $e^{0.7}$. To honestly use Taylor's theorem here (since "we do not know" what $e^{0.7}$ is—that's what we are trying to approximate) let's use the conservative upper bound: $e^c \le e^{0.7} \le e^1 = e < 3$.

Now Taylor's theorem tells us that

$$\text{error} = |e^{0.7} - p_n(0.7)| = |R_n(0.7)| = \frac{e^c}{(n+1)!}(0.7)^{n+1}.$$

(Since all numbers on the right side are nonnegative, we are able to drop absolute value signs.) As was seen above, we can replace e^c with 3 in the right side above, to get

$$\text{something larger than the error} = \frac{3}{(n+1)!}(0.7)^{n+1}.$$

The rest of the plan is simple: We find an n large enough to make the "something larger than actual error" to be less than the desired accuracy 0.0001. Then it will certainly follow that the actual error will also be less than 0.0001. We can continue to compute $3(0.7)^{n+1}/(n+1)!$ until it gets smaller than 0.0001. Better yet, let's use a while loop to get MATLAB to do the work for us; this will also provide us with a good occasion to introduce the remaining relational operators that can be used in any while loops (or subsequent programming). (See Table 2.1)

TABLE 2.1: Dictionary of MATLAB's relational operators.

Mathematical Relation	MATLAB Code
$>, <$	$>, <$
$\geq, \leq$	$>=, \quad <=$
$=, \neq$	$==, \quad \sim=$

We have already used one of the first pair. For the last one, we reiterate again that the single equal sign in MATLAB is reserved for "assignment." Since it gets used much less often (in MATLAB codes), the ordinary equals in mathematics got stuck with the more cumbersome MATLAB notation.

Now, back to our problem. A simple way to figure out that smallest feasible value of n would be to run the following code:

```
>> n=1;
>> while 3*(0.7)^(n+1)/gamma(n+2) >= 0.0001
      n=n+1;
end
```

This code has no output, but what it does is to keep making n bigger, one-by-one, until that "something larger than the actual error" gets less than 0.0001. The magic value of n that will work is now (by the way the while loop was constructed) simply the last stored value of n:

```
>>n    → 6
```

Part (b): The desired approximation is now: $e^{0.7} \approx p_6(0.7) = \sum_{k=0}^{6} \left. \frac{(0.7)^k}{k!} \right|_{x=0.7}$.

We can do the rest on MATLAB:

```
>> x=0.7;
>> n=0;
>> p6=0;   % we initialize the sum for the Taylor polynomial p6
>> while n<=6
 p6=p6+x^n/gamma(n+1);
 n=n+1;
end
>> p6
→ 2.0137 (approximation)
>> abs(p6-exp(0.7)) %we now check the actual error
→ 1.7889e-005 (this is less than 0.0001, as we knew from Taylor's theorem.)
```

EXERCISE FOR THE READER 2.2: If we use Taylor polynomials of $f(x) = \sqrt{x}$ centered at $x = 16$ to approximate $\sqrt{17} = f(16+1)$, what order Taylor polynomial should be used to ensure that the error of the approximation is less than 10^{-10}? Perform this approximation and then look at the actual error.

For any function $f(x)$, which has infinitely many derivatives at $x = a$, we can form the **Taylor series (centered)** at $x = a$:

$$f(a) + f'(a)(x-a) + \frac{f''(a)}{2}(x-a)^2 + \frac{f'''(a)}{3!}(x-a)^3 + \cdots$$
$$+ \frac{f^{(n)}(a)}{n!}(x-a)^n + \cdots = \sum_{k=0}^{\infty} \frac{f^{(k)}(a)}{k!}(x-a)^k. \tag{7}$$

Comparing this with (2) and (3), the formulas for the nth Taylor polynomial $p_n(x)$ at $x = a$, we see that the Taylor series is just the infinite series whose first n terms are exactly the same as those of $p_n(x)$. The Taylor series may or may not converge, but if Taylor's theorem shows that the errors $|p_n(x) - f(x)|$ go to zero, then it will follow that the Taylor series above converges to $f(x)$. When $a = 0$ (the most common situation), the series is called the **Maclaurin series** (Figure 2.7).

It is useful to have some Maclaurin series for reference. Anytime we are able to figure out a formula for the general Taylor polynomial at $x = 0$, we can write down the corresponding Maclaurin series. The previous examples we have done yield the Maclaurin series for $\cos(x)$ and e^x. We list these in Table 2.2, as well as a few other examples whose derivations will be left to the exercises.

TABLE 2.2: Some Maclaurin series expansions.

Function	Maclaurin Series	
e^x	$1 + x + \dfrac{x^2}{2!} + \dfrac{x^3}{3!} + \cdots + \dfrac{x^k}{k!} + \cdots$	(8)
$\cos(x)$	$1 - \dfrac{x^2}{2!} + \dfrac{x^4}{4!} + \cdots + \dfrac{(-1)^k x^{2k}}{(2k)!} + \cdots$	(9)
$\sin(x)$	$x - \dfrac{x^3}{3!} + \dfrac{x^5}{5!} \cdots + \dfrac{(-1)^k x^{2k+1}}{(2k+1)!} + \cdots$	(10)
$\arctan(x)$	$x - \dfrac{x^3}{3} + \dfrac{x^5}{5} - \cdots + \dfrac{(-1)^{2k+1} x^{2k+1}}{2k+1} + \cdots$	(11)
$\dfrac{1}{1-x}$	$1 + x + x^2 + x^3 + \cdots + x^k + \cdots$	(12)

One very useful aspect of Maclaurin and Taylor series is that they can be formally combined to obtain new Maclaurin series by all of the algebraic operations (addition, subtraction, multiplication, and division) as well as with substitutions, derivatives, and integrations. These informally obtained expansions are proved to be legitimate in calculus books. The word "formal" here means that all of the above operations on an infinite series should be done as if it were a finite sum. This method is illustrated in the next example.

EXAMPLE 2.6: Using formal manipulations, obtain the Maclaurin series of the functions (a) $x\sin(x^2)$ and (b) $\ln(1+x^2)$.

SOLUTION: Part (a): In (10) simply replace x with x^2 and formally multiply by x (we use the symbol $\sim$ to mean "has the Maclaurin series"):

FIGURE 2.7: Colin Maclaurin[3] (1698–1746), Scottish mathematician.

$$x\sin(x^2) \sim$$
$$x\left(x^2 - \frac{(x^2)^3}{3!} + \frac{(x^2)^5}{5!} \cdots + \frac{(-1)^k (x^2)^{2k+1}}{(2k+1)!} + \cdots \right)$$
$$= x^3 - \frac{x^7}{3!} + \frac{x^{11}}{5!} \cdots + \frac{(-1)^k x^{4k+3}}{(2k+1)!} + \cdots.$$

NOTE: This would have been a lot more work to do by using the definition and looking for patterns.

Part (b): We first formally integrate (12): $-\ln(1-x)$

$$\sim \int (1 + x + x^2 + x^3 + \cdots + x^n + \cdots) dx = C + x + \frac{x^2}{2} + \frac{x^3}{3} + \cdots + \frac{x^{n+1}}{n+1} + \cdots.$$

---

[3] Maclaurin was born in a small village on the river Ruel in Scotland. His father, who was the village minister, died when Colin was only six weeks old. His mother wanted Colin and his brother John to have good education so she moved the family to Dumbarton, which had reputable schools. Colin's mother died when he was only nine years old and he subsequently was cared for by his uncle, also a minister. Colin began studies at the Glasgow University at 11 years old (it was more common during these times in Scotland for bright youngsters to begin their university studies early—in fact universities competed for them). He graduated at age 14 when he defended an impressive thesis extending Sir Isaac Newton's theory on gravity. He then went on to divinity school with the intention of becoming a minister, but he soon ended this career path and became a chaired mathematics professor at the University of Aberdeen in 1717 at age 19.

Two years later, Maclaurin met the illustrious Sir Isaac Newton and they became good friends. The latter was instrumental in Maclaurin's appointment in this same year as a Fellow of the Royal Society (the highest honor awarded to English academicians) and subsequently in 1725 being appointed to the faculty of the prestigious University of Edinburgh, where he remained for the rest of his career. Maclaurin wrote several important mathematical works, one of which was a joint work with the very famous Leonhard Euler and Daniel Bernoulli on the theory of tides, which was published in 1740 and won the three a coveted prize from the Académie des Sciences in Paris. Maclaurin was also known as an excellent and caring teacher. He married in 1733 and had seven children. He was also known for his kindness and had many friends, including members of the royalty. He was instrumental in his work at the Royal Society of Edinburgh, having it transformed from a purely medical society to a more comprehensive scientific society. During the 1745 invasion of the Jacobite army, Maclaurin was engaged in hard physical labor in the defense of Edinburgh. This work, coupled with an injury from falling off his horse, weakened him to such an extent that he died the following year.

By making $x = 0$, we get that the integration constant C equals zero. Now negate both sides and substitute x with $-x^2$ to obtain:

$$\ln(1 + x^2) \sim x^2 - \frac{(-x^2)^2}{2} - \frac{(-x^2)^3}{3} - \cdots - \frac{(-x^2)^{n+1}}{n+1} - \cdots$$

$$\sim x^2 - \frac{x^4}{2} + \frac{x^6}{3} - \cdots + \frac{(-1)^{k+1} x^{2k}}{k} + \cdots.$$

Our next example involves another approximation using Taylor's theorem. Unlike the preceding approximation examples, this one involves an integral where it is impossible to find the antiderivative.

EXAMPLE 2.7: Use Taylor's theorem to evaluate the integral $\int_0^1 \sin(t^2)\, dt$ with an error $< 10^{-7}$.

SOLUTION: Let us denote the nth-order Taylor polynomial of $\sin(x)$ centered at $x = 0$ by $p_n(x)$. The formulas for each $p_n(x)$ are easily obtained from the Maclaurin series (10).   We will estimate $\int_0^1 \sin(t^2)\, dt$ by $\int_0^1 p_n(t^2)\, dt$ for an appropriately large n. We can easily obtain an upper bound for the error of this approximation:

$$\text{error} = \left| \int_0^1 \sin(t^2)\, dt - \int_0^1 p_n(t^2)\, dt \right| \leq \int_0^1 |\sin(t^2) - p_n(t^2)|\, dt \leq \int_0^1 |R_n(t^2)|\, dt ,$$

where $R_n(x)$ denotes Taylor's remainder.   Since any derivative of $f(x) = \sin(x)$ is one of $\pm\sin(x)$ or $\pm\cos(x)$, it follows that for any x, $0 \leq x \leq 1$, we have

$$|R_n(x)| = \left| \frac{f^{(n+1)}(c)x^{n+1}}{(n+1)!} \right| \leq \frac{1}{(n+1)!}.$$

Since in the above integrals, t is running from $t = 0$ to $t = 1$, we can substitute $x = t^2$ in this estimate for $|R_n(x)|$. We can use this and continue with the error estimate for the integral to arrive at:

$$\text{error} = \left| \int_0^1 \sin(t^2)\, dt - \int_0^1 p_n(t^2)\, dt \right| \leq \int_0^1 |R_n(t^2)|\, dt \leq \int_0^1 \frac{dt}{(n+1)!} = \frac{1}{(n+1)!}.$$

As in the previous example, let's now get MATLAB to do the rest of the work for us. We first need to determine how large n must be so that the right side above (and hence the actual error) will be less than 10^{-7}.

```
>> n=1;
>> while 1/gamma (n+2)  >= 10^(-7)
        n=n+1;
end
>> n→ 10
```

So it will be enough to replace $\sin(t^2)$ 10th-order Taylor polynomial evaluated at t^2, $p_n(t^2)$. The simplest way to see the general form of this polynomial (and its integral) will be to replace x with t^2 in the Maclaurin series (10) and then formally integrate it (this will result in the Maclaurin series for $\int_0^x \sin(t^2)dt$):

$$\sin(t^2) \sim t^2 - \frac{t^6}{3!} + \frac{t^{10}}{5!} - \cdots + \frac{(-1)^k t^{4k+2}}{(2k+1)!} + \cdots \quad \Rightarrow$$

$$\int_0^x \sin(t^2)dt \sim \int_0^x \left(t^2 - \frac{t^6}{3!} + \frac{t^{10}}{5!} + \cdots + \frac{(-1)^k t^{4k+2}}{(2k+1)!} + \cdots \right) dt$$

$$\sim C + \frac{x^3}{3} - \frac{x^7}{7\cdot 3!} + \frac{x^{11}}{11\cdot 5!} + \cdots + \frac{(-1)^k x^{4k+3}}{(4k+3)\cdot(2k+1)!} + \cdots.$$

If we substitute $x = 0$, we see that the constant of integration C equals zero. Thus,

$$\int_0^x \sin(t^2)dt \sim \frac{x^3}{3} - \frac{x^7}{7\cdot 3!} + \frac{x^{11}}{11\cdot 5!} + \cdots + \frac{(-1)^k x^{4k+3}}{(4k+3)\cdot(2k+1)!} + \cdots \quad .$$

We point out that the formal manipulation here is not really necessary since we could have obtained from (10) an explicit formula for $p_{10}(t^2)$ and then directly integrated this function. In either case, integrating this function from $t = 0$ to $t = 1$ gives the partial sum of the last Maclaurin expansion (for $\int_0^x \sin(t^2)dt$) gotten by going up to the $k = 4$ term, since this corresponds to integrating up through the terms of $p_{10}(t^2)$.

```
>> p10=0;
>> k=0;
>> while k<=4
        p10=p10+(-1)^k/(4*k+3)/gamma(2*k+2);
        k=k+1;
end
>> format long
>> p10
→p10 = 0.31026830280668
```

In summary, we have proved the approximation

$$\int_0^1 \sin(t^2)\,dt \approx 0.31026830280668.$$

Taylor's theorem has guaranteed that this is accurate with an error less than 10^{-7}.

EXERCISE FOR THE READER 2.3: Using formal manipulations, find the 10th-order Taylor polynomial centered at $x = 0$ for each of the following functions: (a) $\sin(x^2) - \cos(x^3)$, (b) $\sin^2(x^2)$.

EXERCISES 2.3:

1.   In each part below, we give a function $f(x)$, a center a to use for Taylor polynomials, a value for x, and a positive number ε that will stand for the error. The task will be to (carefully) use Taylor's theorem to find a value of the order n of a Taylor polynomial so that the error of the approximation $f(x) \approx p_n(x)$ is less than ε. Afterward, perform this approximation and check that the actual error is really less than what it was desired to be.
   (a) $f(x) = \sin(x)$, $a = 0$, $x = 0.2$rad, $\varepsilon = 0.0001$
   (b) $f(x) = \tan(x)$, $a = 0$, $x = 5°$, $\varepsilon = 0.0001$
   (c) $f(x) = e^x$, $a = 0$, $x = -0.4$, $\varepsilon = 0.00001$
   (d) $f(x) = x^{1/3}$, $a = 27$, $x = 28$, $\varepsilon = 10^{-6}$

2.   Follow the directions in Exercise 1 for the following:
   (a) $f(x) = \cos(x)$, $a = \pi/2$, $x = 88°$, $\varepsilon = 0.0001$
   (b) $f(x) = \arctan(x)$, $a = 0$, $x = 1/239$, $\varepsilon = 10^{-8}$
   (c) $f(x) = \ln x$, $a = 1$, $x = 3$, $\varepsilon = 0.00001$
   (d) $f(x) = \cos(x^2)$, $a = 0$, $x = 2.2$, $\varepsilon = 10^{-6}$

3.   Using only the Maclaurin series developed in this section, along with formal manipulations, obtain Maclaurin series for the following functions.

   (a) $x^2 \arctan(x)$ (b) $\ln(1 + x)$ (c) $\dfrac{x^2 + 3x}{1 - x}$ (d) $\int_0^x \dfrac{1}{1 - t^3} dt$

4.   Using only the Maclaurin series developed in this section, along with formal manipulations, obtain Maclaurin series for the following functions.

   (a) $\ln(1 + x)$ (b) $1/(1 + x^2)$ (c) $\arctan(x^2) - \sin(x)$ (d) $\int_0^x \cos(t^5) dt$

5.   Find the Maclaurin series for $f(x) = \sqrt{1 + x}$.

6.   Find the Maclaurin series for $f(x) = (1 + x)^{1/3}$.

7.   (a) Use Taylor's theorem to approximate the integral $\int_0^1 \cos(t^5) dt$ with an error less than 10^{-8} .
   (First find a large enough order n for a Taylor polynomial that can be used from the theorem, then actually perform the approximation.)
   (b) How large would n have to be if we wanted the error to be less than 10^{-30} ?

8.   The **error function** is given by the formula: $\operatorname{erf}(x) = (2/\sqrt{\pi}) \int_0^x e^{-t^2} dt$. It is used extensively in probability theory, but unfortunately the integral cannot be evaluated exactly. Use Taylor's theorem to approximate $\operatorname{erf}(2)$ with an error less than 10^{-6} .

9.   Since $\tan(\pi/4) = 1$ we obtain $\pi = 4\arctan(1)$. Using the Taylor series for the arctangent, this gives us a scheme to approximate π.
   (a) Using Taylor polynomials of $\arctan(x)$ centered at $x = 0$, how large an order n Taylor polynomial would we need to use in order for $4p_n(1)$ to approximate π with an error less than 10^{-12} ?

(b) Perform this approximation.

(c) How large an order n would we need for Taylor's theorem to guarantee that $4p_n(1)$ approximates π with an error less than 10^{-50} ?[4]

(d) There are more efficient ways to compute π. One of these dates back to the early 1700s, when Scottish mathematician John Machin (1680–1751) developed the inverse trig identity:

$$\frac{\pi}{4} = \arctan\left(\frac{1}{5}\right) - \arctan\left(\frac{1}{239}\right). \tag{13}$$

to calculate the first 100 decimal places of π. There were no computers back then, so his work was all done by hand and it was important to do it in way where not so many terms needed to be computed. He did it by using Taylor polynomials to approximate each of the two arctangents on the right side of (13). What order Taylor polynomials would Machin have needed to use (according to Taylor's theorem) to attain his desired accuracy?

(e) Prove identity (13).

Suggestion: For part (d), use the trig identity:

$$\tan(A \pm B) = \frac{\tan A \pm \tan B}{1 \mp \tan A \tan B}$$

to calculate first $\tan(2\tan^{-1}\frac{1}{5})$, then $\tan(4\tan^{-1}\frac{1}{5})$, and finally $\tan(4\tan^{-1}\frac{1}{5} - \tan^{-1}\frac{1}{239})$.

HISTORICAL ASIDE: Since antiquity, the problem of figuring out π to more and more decimals has challenged the mathematical world, and in more recent times the computer world as well. Such tasks can test the powers of computers as well as the methods used to compute them. Even in the 1970s π had been calculated to over 1 million places, and this breakthrough was accomplished using an identity quite similar to (13). See [Bec-71] for an enlightening account of this very interesting history.

10. (*Numerical Differentiation*) (a) Use Taylor's theorem to establish the following *forward difference formula*:

$$f'(a) = \frac{f(a+h) - f(a)}{h} - \frac{h}{2}f''(c),$$

for some number c between a and $a+h$, provided that $f'(x)$ is continuous on $[a, a+h]$. This formula is often used as a means of numerically approximating the derivative $f'(a)$ by the simple difference quotient on the right; in this case the error of the approximation would be $|hf''(c)/2|$ and could be made arbitrarily small if we take h sufficiently small.

(b) With $f(x) = \sinh(x)$, and $a = 0$, how small would we need to take h for the approximation in part (a) to have error less than 10^{-5}? Do this by first estimating the error, and then (using your value of h) check the actual error using MATLAB. Repeat with an error goal of 10^{-10}.

(c) Use Taylor's theorem to establish the following *central difference formula*:

$$f''(a) = \frac{f(a+h) - 2f(a) + f(a+h)}{h^2} - \frac{h^2}{12}f^{(4)}(c),$$

for some number c between $a-h$ and $a+h$, provided that $f^{(4)}(x)$ is continuous on $[a-h, a+h]$. This formula is often used as a means of numerically approximating the derivative $f''(a)$ by the simple difference quotient on the right; in this case the error of the approximation would be $|h^2 f^{(4)}(c)/12|$ and could be made arbitrarily small if we take h sufficiently small.

---

[4] Of course, since MATLAB's compiler keeps track of only about 15 digits, such an accurate approximation could not be done without the help of the Symbolic Toolbox (see Appendix A).

(d) Repeat part (b) for the approximation of part (c). Why do the approximations of part (c) seem more efficient, in that they do not require as small an h to achieve the same accuracy?

(e) Can you derive (and prove using Taylor's theorem) an approximation for $f'(x)$ whose error is proportional to h^2 ?

Chapter 3: Introduction to M-Files

3.1: WHAT ARE M-FILES?

Up to now, all of our interactions with MATLAB have been directly through the command window. As we begin to work with more complicated algorithms, it will be preferable to develop standalone programs that can be separately stored and saved into files that can always be called on (by their name) in any MATLAB session. The vehicle for storing such a program in MATLAB is the so-called **M-file**. M-files are programs that are plain-text (ASCII) files written with any word processing program (e.g., Notepad or MS Word) and are called M-files because they will always be stored with the extension `<filename>.m`.[1] As you begin to use MATLAB more seriously, you will start to amass your own library of M-files (some of these you will have written and others you may have gotten from other sources such as the Internet) and you will need to store them in various places (e.g., certain folders on your own computer, or also on your portable disk for when you do work on another computer). If you wish to make use of (i.e., "call on") some of your M-files during a particular MATLAB session from the command window, you will need to make sure that MATLAB knows where to look for your M-files. This is done by including all possible directories where you have stored M-files in MATLAB's **path**.[2]

 M-files are of two types: **script M-files** and **function M-files**. A script M-file is simply a list of MATLAB commands typed out just as you would do on the command line. The script can have any number of lines in it and once it has been saved (to a directory in the path) it can be invoked simply by typing its name in the command window.   When this is done, the script is "run" and the effect will be the same as having actually typed the lines one by one in the command window.

---

[1] It is recommended that you use the default MATLAB M-file editor gotten from the "File" menu (on the top left of the command window) and selecting "New"→ "M-File." This editor is designed precisely for writing M-files and contains many features that are helpful in formatting and debugging. Some popular word processing programs (notably MS Word) will automatically attach a certain extension (e.g., ".doc") at the end of any filename you save a document as and it can be difficult to prevent such things. On a Windows/DOS-based PC, one way to change an M-file that you have created in this way to have the needed ".m" extension is to open the DOS command window, change to the directory you have stored your M-file in and rename the file using the DOS command `ren` `<filename>.m.doc`   `<filename>.m`   (the   format   is: `ren`   `<oldfilename>` `.oldextension <newfilename>.newextension`).

[2] Upon installation, MATLAB sets up the path to include a folder "Work" in its directory, which is the default location for storing M-files. To add other directories to your path, simply select "Set Path" from the "File Menu" and add on the desired path. If you are using a networked computer, you may need to consult with the system administrator on this.

45

EXAMPLE 3.1: Here is a simple script which assumes that numbers x0,y0, and $r > 0$ have been stored in the workspace (before the script is invoked) and that will graph the circle with center $(x0, y0)$ and radius r.

```
t=0:.001:2*pi;
x=x0+r*cos(t);
y=y0+r*sin(t);
plot(x,y)
axis('equal')
```

If the above lines are simply typed as is into a text file and saved as, say, circcdrw.m into some directory in the path, then at any time later on, if we wish to get MATLAB to draw a circle of radius 2 and center $(5, -2)$, we could simply enter the following in the command window:

```
>> r=2; x0=5; y0= -2;
>> circdrw
```

and *voilà!* the graphic window pops up with the circle we desired. Please remember that any variables created in a script are **global variables**, i.e., they will enter the current workspace when the script is invoked in the command window. One must be careful of this since the script may have been written a long time ago and when it is run the lines of the script are not displayed (only executed).

Function M-files are stored in the same way as script M-files but are quite different in the way they work. Function M-files accept any number of input variables and can output any number of output variables (or none at all). The variables introduced in a function M-file are **local variables**, meaning that they do not remain in the workspace after a function M-file is called in a MATLAB session. Also, the first line of a function M-file must be in the following format:

```
function [<output variables>] = <function_name>(<input variables>)
```

Another important format issue is that the <function_name> (which you are free to choose) should coincide exactly with the filename that you save the function M-file under.

EXAMPLE 3.2: We create a function M-file that will do essentially the same thing as the script in the preceding example. There will be three input variables: We will make the first two be the coordinates of the center $(x0, y0)$ of the circle that we wish MATLAB to draw, and the third be the radius. Since there will be no output variables here (only a graphic), our function M-file will look like this:

```
function [ ] = circdrwf(x0,y0,r)
t=0:.001:2*pi;
x=x0+r*cos(t);
y=y0+r*sin(t);
plot(x,y)
axis('equal')
```

In particular, the word `function` must be in lowercase. We then save this M-file as `circdrwf.m` in an appropriate directory in the path. Notice we gave this M-file a different name than the one in Example 3.1 (so they may lead a peaceful coexistence if we save them to the same directory). Once this function M-file has been stored we can call on it in any MATLAB session to draw the circle of center $(5, -2)$ and radius 2 by simply entering

```
>> circdrwf(5, -2, 2)
```

We reiterate that, unlike with the script of the first example, after we use a function M-file, none of the variables created in the file will remain in the workspace. As you gain more experience with MATLAB you will be writing a lot of function M-files (and probably very soon find them more useful than script M-files). They are analogous to "functions" in the C-language, "procedures" in PASCAL, and "programs" or "subroutines" in FORTRAN. The <filenames> of M-files can be up to 19 characters long (older versions of MATLAB accepted only length up to 8 characters), and the first character must be a letter. The remaining characters can be letters, digits, or underscore (_).

EXERCISE FOR THE READER 3.1: Write a MATLAB script, call it `listp2`, that assumes that a positive integer has been stored as n and that will find and output all powers of 2 that are less than or equal to n. Store this script as an M-file `listp2.m` somewhere in MATLAB's path and then run the script for each of these values of n: $n = 5$, $n = 264$, *and* $n = 2917$.

EXERCISE FOR THE READER 3.2: Write a function M-file, call it `fact`, having one input variable—a nonnegative integer n, and the output will be the factorial of n: $n!$.. Write this program from scratch (using a while loop) without using a built-in function like `gamma`. Store this M-file and then run the following evaluations: `fact(4), fact(10), fact(0)`.

Since MATLAB has numerous built-in functions it is often advisable to check first if a proposed M-file name that you are contemplating is already in use. Let's say you are thinking of naming a function M-file you have just written with the name `det.m`. To check first with MATLAB to see if the name is already in use you can type:

```
>>exist('det')   %possible outputs are 0, 1, 2, 3, 4, 5, 6, 7, 8
→5
```

The output 5 means (as does any positive integer) `det` is already in use. Let's try again (with a trick often seen on vanity license plates).

```
>>exist('det1')    → 0
```

The output zero means the filename `det1` is not yet spoken for so we can safely assign this filename to our new M-file.

EXERCISES 3.1:

1.   (a) Write a MATLAB function M-file, call it rectdrwf(l,w), that has two input variables:
        l, the length and w, the width and no output variables, but will produce a graphic of a rectangle
        with horizontal length = l and vertical width = w. Arrange it so that the rectangle sits well
        inside the graphic window and so that the axes are equally scaled. (b) Store this M-file and then
        run the function rectdrwf(5,3) and rectdrwf(4, 4.5).

2.   (a) Write a function M-file, call it segdrwf(x,y), that has two input **vectors**
        $x = [x_1 \ x_2 \ \cdots x_n]$ and $y = [y_1 \ y_2 \ \cdots y_n]$ of the same size and no output variables, but will
        produce the graphic gotten by connecting the points (x_1, y_1), (x_2, y_2), $\cdots, (x_n, y_n)$. You might
        wish to make use of the MATLAB built-in function size(A) that, for an input matrix A, will
        output its size. (b) Run this program with the inputs x = [1 3 5 7 9 1] and y=[1 4 1
        4 8 1]. (c) Determine two vectors x and y so that segdrwf(x,y) will produce an
        equilateral triangle.

3.   Redo Exercise 1, creating a script M-file—called rectdrw —rather than a function M-file.

4.   Redo Exercise 2, creating a script M-file—called segdrw—rather than a function M-file.

5.   (*Finance*) Write a function M-file, call it compintf(r,P,F), that has three input variables:
        r, the annual interest rate, P, the principal, and F, the future goal. Here is what the function
        should do: It assumes that we deposit P dollars (assumed positive) into an interest bearing
        account that pays $100r$% interest per year compounded annually. The investment goal is F
        dollars (F is assumed larger than P, otherwise the goal is reached automatically as soon as the
        account is opened). The output will be one variable consisting of the number of years it takes
        for the account balance to first equal or exceed F . Store this M-file and run the following:
        comintf(0.06,   1000,   100000), comintf(0.085,   1000,   100000),
        comintf(0.10, 1000, 1000000), and comintf(0.05, 100, 1000000).

        Note: The formula for the account balance after t years is P dollars is invested at $100r$%
        compounded annually is $P(1+r)^t$

6.   (*Finance*)   Write a function M-file, call it loanperf(r,L,PMT), that has three input
        variables: r, the annual interest rate, L, the loan amount, and PMT, the monthly payment. There
        will be one output variable n, the number of months needed to pay off a loan of L dollars made
        at an annual interest rate of $100r$% (on the unpaid balance) where at the end of each month a
        payment of PMT dollars is made. (Of course L and PMT are assumed positive.) You will need
        to use a while loop construction as in Example 1.1 (Sec. 1.3). After storing this M-file, run the
        following: loanperf(.0799, 15000, 500), loanperf(.019, 15000, 500),
        loan(0.99, 22000, 450). What could cause the program loanerf(r, L, PMT) to
        go into an infinite loop? In the next chapter we will show, among other things, ways to
        safeguard programs from getting into infinite loops.

7.   Redo Exercise 5 writing a script M-file (which assumes the relevant input variables have been
        assigned) rather than a function M-file.

8.   Redo Exercise 6 writing a script M-file (which assumes the relevant input variables have been
        assigned) rather than a function M-file.

9.   Write a function M-file, call it oddfact(n), that inputs a positive integer n and will output
        the product of all odd positive integers that are less than or equal to n . So, for example,
        oddfact(8) will be $1 \cdot 3 \cdot 5 \cdot 7 = 105$. Store it and then run this function for the following
        values: oddfact(5), oddfact(22), oddfact(29). Get MATLAB to find the first

value of n for which oddfact(n) exceeds or equals 1 million, and then 5 trillion.

10.   Write a function M-file, call it evenfact(n), that inputs a positive integer n and will output the product of all even positive integers that are less than or equal to n. So, for example, evenfact(8) will be $2 \cdot 4 \cdot 6 \cdot 8 = 384$. Store it and then run this function for the following values: evenfact(5), evenfact(22), evenfact(29). Get MATLAB to find the first value of n for which evenfact(n) exceeds or equals 1 million, and then 5 trillion. Can you write this M-file without using a while loop, using instead some of MATLAB's built-in functions?

11.   Use the error estimate from Example 2.5 (Sec. 2.3) to write a function M-file called expcal(x,err) that does the following: The input variable x is any real number and the other input variable err is any positive number. The output will be an approximation of e^x by a Taylor polynomial $p_n(x)$ based at $x = 0$, where n is the first nonnegative integer such that the error estimate based on Taylor's theorem that was obtained in Example 2.5 gives a guaranteed error less than err. There should be two output variables, n, the order of the Taylor polynomial used, and $y = p_n(x) =$ the approximation. Run this function with the following input data: (2, 0.001), (−6, 10^{-12}), (15, 0.000001), (−30, 10^{-25}). For each of these y-outputs, check with MATLAB's built-in function exp to see if the actual errors are as desired. Is it possible for this program to ever enter into an infinite loop?

12.   Write a function M-file, called coscal(x,err), that does exactly what the function in Exercise 11 does except now for the function $y = \cos(x)$. You will need to obtain a Taylor's theorem estimate for the (actual) error $|\cos(x) - p_n(x)|$. Run this function with the following input data: (0.5, 0.0000001), (−2, 0.0001), ($20°$, 10^{-9}), and ($360020°$, 10^{-9}) (for the last two you will need to convert the inputs to radians). For each of these y-outputs, check with MATLAB's built-in function $\cos(x)$ to see if the actual errors are as desired. Is it possible for this program to ever enter into an infinite loop? Although $\cos(360020°) = \cos(20°)$, the outputs you get will be different; explain this discrepancy.

3.2: CREATING AN M-FILE FOR A MATHEMATICAL FUNCTION

Function M-files can be easily created to store (complicated) mathematical functions that need to be used repeatedly. Another way to store mathematical functions without formally saving them as M-files is to create them as "in-line objects." Unlike M-files, in-line objects are stored only as variables in the current workspace. The following example will illustrate such an M-file construction; in-line objects will be introduced in Chapter 6.

EXAMPLE 3.3: Write a function M-file, with filename bumpy.m that will store the function given by the following formula:

$$y = \frac{1}{4\pi} \left[\frac{1}{(x-2)^2 + 1} + \frac{1}{(x+0.5)^4 + 32} + \frac{1}{(x+1)^2 + 2} \right].$$

Once this is done, call on this newly created M-file to evaluate y at $x = 3$ and to sketch a graph of the function from $x = -3$ to $x = 3$.

SOLUTION: After the first "function definition line," there will be only one other line required: the definition of the function above written in MATLAB's language. Just like in a command window, anything we type after the percent symbol (%) is considered as a comment and will be ignored by MATLAB's processor. Comment lines that are put in immediately following the function definition line are, however, somewhat more special. Once a function has been stored (somewhere in MATLAB's path), and you type help <function_name> on the command window, MATLAB displays any comments that you inserted in the M-file after the function definition line. Here is one possibility of a function M-file for the above function:

```
function y = bumpy(x)
% our first function M-file
% x could be a vector
% created by <yourname> on <date>
y=1/(4*pi)*(1./((x-2).^2+1)+1./((x+.5).^4+32)+1./((x+1).^2+2));
```

Some comments are in order. First, notice that there is only one output variable, y, but we have not enclosed it in square brackets. This is possible when there is only one output variable (it would still have been okay to type function [y] = bumpy(x)). In the last line where we typed the definition of y (the output variable), notice that we put a semicolon at the end. This will suppress any duplicate outputs since a function M-file is automatically set up to print the output variable when evaluated.   Also, please look carefully at the placement of parentheses and (especially) the dots when we wrote down the formula. If x is just a number, the dots are not needed, but often we will need to create plots of functions and x will need to be a vector. The placement of dots was explained in Chapter 1.

The above function M-file should now be saved with the name bumpy.m with the same filename appearing (without the extension) in the function definition line into some directory contained in MATLAB's path (as explained in the previous section). This being done, we can use it just like any of MATLAB's built-in functions, like cos. We now proceed to perform the indicated tasks.

```
>> bumpy(3)
→ 0.0446
>> y     %Remember all the variables in a MATLAB M-file are local only.
→ Undefined function or variable 'y'.

>> x=-3:.01:3;
>> plot(x,bumpy(x))
```

From the plot we see that the function bumpy(x) has two peaks (local maxima) and one valley (local minimum) on the interval $-3 \le x \le 3$. MATLAB has many built-in functions to analyze mathematical functions.   Three very important numerical problems are to integrate a function on a specified interval, to find maximums and minimums on a specified interval, and to find zeros or roots of a function.   The next example will illustrate how to perform such tasks with MATLAB.

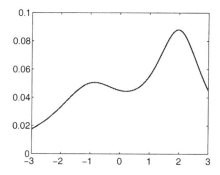

FIGURE 3.1: A graph of the function $y = \text{bumpy}(x)$ of Example 3.3.

EXAMPLE 3.4: For the function $\text{bumpy}(x)$ of the previous example, find "good" approximations to the following:

(a) $\int_{-3}^{3} \text{bumpy}(x)\,dx$

(b) The maximum and minimum values of $y = \text{bumpy}(x)$ on the interval $-1.2 \le x \le 1$ (i.e., the height of the left peak and that of the valley) and the corresponding x-coordinates of where these extreme values occur.

(c) A solution of the equation $\text{bumpy}(x) = 0.08$ on the interval $0 \le x \le 2$ (which can be seen to exist by examination of bumpy's graph in Figure 3.1).

SOLUTION: Part (a): The relevant built-in function in MATLAB for doing definite integrals is `quad` which is an abbreviation for *quadrature*, a synonym for integration.[3] The syntax is as follows:

quad('function', a, b, tol) →	approximates the integral $\int_{a}^{b} \text{function}(x)\,dx$ with the goal of the error being less than `tol`

The `function` must be stored as an M-file in MATLAB's path or the exact name of a built-in function, and the name must be enclosed in 'single quotes'. [4] If the whole last argument `tol` is omitted (along with the comma that precedes it), a maximum error goal of 10^{-3} is assumed. If this command is run and you just get

---

[3] MATLAB has another integrator, `quadl`, that gives more accurate results for well-behaved integrands. For most purposes, though, `quad` is quite satisfactory and versatile as a general quadrature tool.

[4] An alternative syntax (that avoids the single quotes) for this and other functions that call on M-files is `quad(@function, a, b, tol)`. Another way to create mathematical functions is to create them as so-called *inline functions* which are stored only in the workspace (as opposed to M-files) and get deleted when the MATLAB session is ended. Inline functions will be introduced in Chapter 6.

an answer (without any additional warnings or error messages), you can safely assume that the approximation is accurate within the sought-after tolerance.

```
>> quad('bumpy',-3,3)    → 0.3061
>> format long
>> ans                   → 0.30608471060690
```

As explained above, this answer should be accurate to at least three decimals. It is actually better than this since if we redo the calculation with a goal of six digits of accuracy, we obtain:

```
>> quad('bumpy',-3,3, 10^(-6))    → 0.30608514875582
```

This latter answer, which is accurate to at least six digits, agrees with the first answer (after rounding) to six digits, so the first answer is already quite accurate. There are limits to how accurate an answer we can get in this way. First of all, MATLAB works with only about 15 or so digits, so we cannot hope for an answer more accurate than this. But roundoff and other errors can occur in large-scale calculations and these can put even further restrictions on the possible accuracy, depending on the problem. We will address this issue in more detail in later chapters. For many practical purposes and applications the quad function and the others we discuss in this example will be perfectly satisfactory and in such cases there will be no need to write new MATLAB M-files to perform such tasks.

Part (b): To (approximately) solve the calculus problem of finding the minimum value of a function on a specified interval, the relevant MATLAB built-in function is fminbnd and the syntax is as follows:

`fminbnd('function', a, b, optimset('TolX',tol)) →`	approximates the x-coordinate of the minimum value of function(x) on [a, b] with a goal of the error being <tol

The usage and syntax comments for quad apply here as well. In particular, if the whole last argument optimset('TolX',tol) is omitted (along with the comma that precedes it), a maximum error goal of 10^{-3} is assumed. Note the syntax for changing the default tolerance goal is a bit different than for quad. This is due to the fact that fminbnd has more options and is capable of doing a lot more than we will have occasion to use it for in this text. For more information, enter help optimset.

```
>> xmin=fminbnd('bumpy',-1.2,1)   %We first find the x-coordinate of
>> %              the valley with a three digit accuracy (at least)
→0.21142776202687
>> xmin=fminbnd('bumpy',-1.2,1, optimset('TolX',1e-6))   %Next let's
>> %                           go for 6 digits of accuracy.
→0.21143721018793 (=x-coordinate of valley)
```

The corresponding y-coordinate (height of the valley) is now gotten by evaluating bumpy(x) at xmin.

```
>> ymin = bumpy(xmin)
```
→ 0.04436776267211 (= <u>y-coordinate of valley</u>)

Since we know that the x-coordinate is accurate to six decimals, a natural question is how accurate is the corresponding y-coordinate that we just obtained? One obvious thing to try would be to estimate this error by plotting bumpy(x) on the interval $x \min - 10^{-6} \le x \le x \min + 10^{-6}$ and then seeing how much the y-coordinates vary on this plot. This maximum variation will be an upper bound for the difference of ymax and the actual value of the y-coordinate for the valley. When we try and plot bumpy(x) on this interval as follows we get the following warning message:

```
>> x=(xmin-10^(-6)):10^(-9):(xmin+10^(-6));
>> plot(x,bumpy(x))
```
→Warning: Requested axes limit range too small; rendering with minimum range allowed by machine precision.

Also, the corresponding plot (which we do not bother reproducing) looks like that of a horizontal line, but the y-tick marks are all marked with the same number (.0444) and similarly for the x-tick marks. This shows that MATLAB's plotting precision works only up to a rather small number of significant digits. Instead we can look at the vector bumpy(x) with x still stored as the vector above, and look at the difference of the maximum less the minimum.

max(v) →	For a vector v, these MATLAB commands will return the maximum entry and
min(v) →	the minimum entry; e.g.: If $v = [2\ 8\ -5\ 0]$ then max(v) →8, and min(v) →
	−5

```
>> max(bumpy(x))-min(bumpy(x))
```
→ 1.785377401475330e-014

What this means is that, although xmin was guaranteed only to be accurate to six decimals, the corresponding y-coordinate seems to be accurate to MATLAB precision, which is about 15 digits!

EXERCISE FOR THE READER 3.3: Explain the possibility of such a huge discrepancy between the guaranteed accuracy of xmin (to the actual x-value of where the bottom of the valley occurs) being 10^{-6} and the incredibly smaller value 10^{-14} of the apparent accuracy of the corresponding ymin = bumpy(xmin). Make sure to use some calculus in your explanation.

EXERCISE FOR THE READER 3.4: Explain why the above vector argument does not necessarily guarantee that the error of ymin as an approximation to the actual y-coordinate of the valley is less than 1.8×10^{-14}.

We turn now to the sought-after maximum. Since there is no built-in MATLAB function (analogous to fminbnd) for finding maximums, we must make do with what we have. We can use fminbnd to locate maximums of functions as soon as we make the following observation: The maximum of a function $f(x)$ on an interval I, if it exists, will occur at the same x-value as the minimum value of the negative function $-f(x)$ on I. This is easy to see; just note that the graph of $y = -f(x)$ is obtained by the graph of $y = f(x)$ by turning the latter graph upside-down (more precisely, reflect it over the x-axis), and when a graph is turned upside-down, its peaks become valleys and its valleys become peaks. Let's initially go for six digits of accuracy:

```
>> xmax = fminbnd('-bumpy(x)', -1.2, 1, optimset('TolX',1e-6))
→ -0.86141835836638 (= x-coordinate of left peak)
```

The corresponding y-coordinate is now:

```
>> bumpy(xmax)    → 0.05055706241866 (= y-coordinate of left peak)
```

Part (c): One way to start would be to simultaneously graph $y = \text{bumpy}(x)$ together with the constant function $y = 0.08$ and continue to zoom in on the intersection point. As explained above, though, this graphical approach will limit the attainable accuracy to three or four digits. We must find the root (less than 2) of the equation $\text{bumpy}(x) = 0.08$. This is equivalent to finding a zero of the standard form equation $\text{bumpy}(x) - 0.08 = 0$. The relevant MATLAB function is `fzero` and its usage is as follows:

`fzero('function', a)` →	finds a zero of function(x) near the value $x = a$ (if one exists). Goal is machine precision (about 15 digits).

```
>> fzero('bumpy(x)-0.08', 1.5)
→Zero found in the interval: [1.38, 1.62].
→1.61904252091472 (=desired solution)
>> bumpy(ans) %as a check, let's see if this value of x does what we
>>%     want it to do.
→ 0.08000000000000 %Not bad!
```

EXERCISE FOR THE READER 3.5: Write a function M-file, with filename `wiggly.m` that will store the following function:

$$y = \sin\left(\exp\left[\frac{1}{(x^2 + .5)^2}\right]\right)\sin(x).$$

(a) Plot this function from $x = -2$ through $x = 2$.
(b) Integrate this function from $x = 0$ to $x = 2$ (use 10^{-5} as your accuracy goal).
(c) Approximate the x-coordinates of both the smallest positive local minimum (valley) and the smallest positive local maximum (peak) from $x = -2$ through $x = 2$.
(d) Approximate the smallest positive solution of $\text{wiggly}(x) = x/2$ (use 10^{-5} as your accuracy goal).

EXERCISES 3.2:

1. (a) Create a MATLAB function M-file for the function $y = f(x) = \exp\left(\sin[\pi/(x+0.001)^2]\right)$

    $+ (x-1)^2$ and then plot this function on the interval $0 \le x \le 3$. Do it first using 10 plotting points and then using 50 plotting points and finally using 500 points.

    (b) Compute the corresponding integral $\int_1^3 f(x)\,dx$.

    (c) What is the minimum value (y-coordinate) of $f(x)$ on the interval $[1,10]$? Make sure your answer is accurate to within the nearest $1/10{,}000$th.

2. (a) Create a MATLAB function M-file for the function $y = f(x) = \dfrac{1}{x}\sin(x^2) + \dfrac{x^2}{50}$ and then plot this function on the interval $0 \le x \le 10$. Do it first using 200 plotting points and then using 5000 plotting points.

    (b) Compute the corresponding integral $\int_1^{10} f(x)\,dx$.

    (c) What is the minimum value (y-coordinate) of $f(x)$ on the interval $[1,10]$? Make sure your answer is accurate to within the nearest $1/10{,}000$th. Find also the corresponding x-coordinate with the same accuracy.

3. Evaluate the integral $\int_0^1 \sin(t^2)\,dt$ (with an accuracy goal of 10^{-7}) and compare this with the answer obtained in Example 2.7 (Sec. 2.3).

4. (a) Find the smallest positive solution of the equation $\tan(x) = x$ using an accuracy goal of 10^{-12}. (b) Using calculus, obtain a bound for the actual error.

5. Find all zeros of the polynomial $x^3 + 6x^2 - 14x + 5$.

NOTE: We remind the reader about some facts on polynomials. A polynomial $p(x)$ of degree n can have at most n roots (that are the x-intercepts of the graph $y = p(x)$). If $x = r$ is a root and if the derivative $p'(r)$ is not zero, then we say $x = r$ is a root of **multiplicity 1**. If $p(r) = p'(r) = 0$ but $p''(r) \ne 0$, then we say the root $x = r$ has multiplicity 2. In general we say $z = r$ is a root of $p(x)$ of multiplicity a, if all of the first $a-1$ derivatives equal zero:

$$p(r) = p'(r) = p''(r) = \cdots p^{(a-1)}(r) = 0 \text{ but } p^{(a)}(r) \ne 0.$$

Algebraically $x = r$ is a root of multiplicity a means that we can factor $p(x)$ as $(x-r)^a q(x)$ where $q(x)$ is a polynomial of degree $n-a$. It follows that if we add up all of the multiplicities of all of the roots of a polynomial, we get the degree of the polynomial. This information is useful in finding all roots of a polynomial.

6. Find all zeros of the polynomial $2x^4 - 16x^3 - 2x^2 + 25$. For each one, attempt to ascertain its multiplicity.

7. Find all zeros of the polynomial

    $$x^6 - \frac{25}{4}x^5 + \frac{4369}{64}x^4 + \frac{8325}{32}x^3 + \frac{13655}{8}x^2 - \frac{325}{32}x + \frac{21125}{8}.$$

    For each one, attempt to ascertain its multiplicity.

8.   Find all zeros of the polynomial

$$x^8 + \frac{136}{5}x^7 + 210x^6 - \frac{165}{5}x^5 - 4094x^4 + \frac{4528}{5}x^3 + 17232x^2 + 320x + 5600 .$$

For each one, attempt to ascertain its multiplicity.

9.   Check that the value x = 2 is a zero of both of these polynomials:

$$P(x) = x^8 - 2x^7 + 6x^5 - 12x^4 + 2x^2 - 8$$
$$Q(x) = x^8 - 8x^7 + 28x^6 - 61x^5 + 95x^4 - 112x^3 + 136x^2 - 176x + 112$$

Next, use `fzero` to seek out this root for each polynomial using $a = 1$ (as a number near the root) and with accuracy goal 10^{-12}. Compare the outputs and try to explain why the approximation seemed to go better for one of these polynomials than for the other one.

Chapter 4: Programming in MATLAB

4.1: SOME BASIC LOGIC

Computers and their programs are designed to function very logically so that they always proceed by a well-defined set of rules. In order to write effective programs, we must first learn these rules so we can understand what a computer or MATLAB will do in different situations that may arise throughout the course of executing a program. The rules are set forth in the formal science of **logic**. Logic is actually an entire discipline that is considered to be part of both of the larger subjects of philosophy and mathematics. Thus there are whole courses (and even doctoral programs) in logic and any student who wishes to become adept in programming would do well to learn as much as possible about logic. Here in this introduction, we will touch only the surface of this subject, with the hope of supplying enough elements to give the student a working knowledge that will be useful in understanding and writing programs.

The basic element in logic is a **statement**, which is any declarative sentence or mathematical equation, inequality, etc. that has a **truth value** of either **true** or **false**.

EXAMPLE 4.1: For each of the English or mathematical expressions below, indicate which are statements, and for those that are, decide (if possible) the truth value.
(a) Al Gore was Bill Clinton's Vice President.
(b) $3 < 2$
(c) $x + 3 = 5$
(d) If $x = 6$ then $x^2 > 4x$.

SOLUTION: All but (c) are statements. In (c), depending on the value of the variable x, the equation could be either true (if $x = 2$) or false (if $x =$ any other number). The truth values of the other statements are as follows: (a) true, (b) false, and (d) true.

If you enter any mathematical relation (with one of the relational operators from Table 2.1), MATLAB will tell you if the statement is true or false in the following fashion:

Truth Value	MATLAB Code
True	1 (as output)
	Any nonzero number (as input)
False	0 (as input and output)

We shall shortly come to how the input truth values are relevant. For now, let us give some examples of the way MATLAB outputs truth values. In fact, let's use MATLAB to do parts (b) and (d) of the preceding example.

```
>> 3<2
→ 0  (MATLAB is telling us the statement is false.)
>> x=6; x^2>4*x
→ 1    (MATLAB is telling us the statement is true.)
```

Logical statements can be combined into more complicated compound statements using **logical operators**. We introduce the four basic logical operators in Table 4.1, along with their approximate English translations, MATLAB code symbols, and precise meanings.

TABLE 4.1: The basic logical operators. In the meaning explanation, it is assumed the p and q represent statements whose truth values are known.

Name of Operator	English Approxi-mation	MATLAB Code	Meaning
Negation	not p	~p	~p is true if p is false, and false if p is true.
Conjunction	p and q	p&q	p&q is true if both p and q are true, otherwise it's false.
Disjunction	p or q	p|q	p|q is true in all cases except if p and q are both false, in which case it is also false.
Exclusive Disjuntion	p or q (but not both)[1]	xor(p,q)	xor(p,q) is true if exactly one of p or q is true. If p and q are both true or both false then xor(p,q) is false.

EXAMPLE 4.2: Determine the truth value of each of the following compound statements.

(a) San Francisco is the capital of California and Egypt is in Africa.

(b) San Francisco is the capital of California or Egypt is in Africa.

(c) San Franciso is not the capital of California.

(d) not $(2 > -4)$

(e) letting $x = 2$, $z = 6$, and $y = -4$: $x^2 + y^2 > z^2/2$ or $zy < x$

(f) letting $x = 2$, $z = 6$, and $y = -4$: $x^2 + y^2 > z^2/2$ or $zy < x$ (but not both)

---

[1] Although most everyone understands the meaning of "and," in spoken English the word "or" is often ambiguous. Sometimes it is intended as the disjunction but other times as the exclusive disjunction. For example, if on a long airplane flight the flight attendant asks you, "Would you like chicken or beef?" Certainly here the exclusive disjunction is intended—indeed, if you were hungry and tried to ask for both, you would probably wind up with only one plus an unfriendly flight attendant. On the other hand, if you were to ask a friend about his/her plans for the coming weekend, he/she might reply, "Oh, I might go play some tennis or I may go to Janice's party on Saturday night." In this case the ordinary disjunction is intended. You would not be at all surprised if your friend wound up doing both activities. In logic (and mathematics and computer programming) there is no room for such ambiguity, so that is why we have two very precise versions of "or."

SOLUTION: To abbreviate parts (a) through (c) we introduce the symbols:
p = San Francisco is the capital of California.
q = Egypt is in Africa.

From basic geography, Sacremento is California's capital so p is false, and q is certainly true. Statements (a) through (c) can be written as: p and q, p or q, not p, respectively. From what was summarized in Table 4.1 we now see (a) is false, (b) is true, and (c) is true.

For part (d), since $2 > -4$ is true, the negation not $(2 > -4)$ must be false.

For parts (e) and (f), we note that substituting the values of x, y, and z the statements become:

(e) $20 > 18$ or $-24 < 2$ i.e., true or true, so true

(f) $20 > 18$ or $-24 < 2$ (but not both) i.e., true or true (but not both) so false.

MATLAB does not know geography but it could have certainly helped us with the mathematical questions (d) through (f) above. Here is how one could do these on MATLAB:

```
>> ~(2>-4)
→ 0 (=false)
>> x=2; z=6; y=-4; (x^2+y^2 > z^2/2) | (z*y < x)
→ 1 (=true)
>> x=2; z=6; y=-4; xor(x^2+y^2 > z^2/2, z*y < x)
→ 0 (=false)
```

EXERCISES 4.1:

1. For each of the English or mathematical expressions below, indicate which are statements, and for those that are statements, decide (if possible) the truth value.
   (a) Ulysses Grant served as president of the United States.
   (b) Who was Charlie Chaplin?
   (c) With $x = 2$ and $y = 3$ we have $\sqrt{x^2 + y^2} = x + y$.
   (d) What is the population of the United States?

2. For each of the English or mathematical statements below, determine the truth value.
   (a) George Harrison was a member of the Rolling Stones.
   (b) Canada borders the US or France is in the Pacific Ocean.
   (c) With $x = 2$ and $y = 3$ we have $x^x > y$ or $x^y > y^x$.
   (d) With $x = 2$ and $y = 3$ we have $x^x > y$ or $x^y > y^x$ (but not both).

3. Assume that we are in a MATLAB session in which the following variables have been stored: $x = 6$, $y = 12$, $z = -4$. What outputs would the following MATLAB commands produce? Of course, you should try to figure out these answers on your own and afterward use MATLAB to check your answers.
   (a) `>> x + y >= z`
   (b) `>> xor(z, x-2*y)`
   (c) `>> (x==2*z)|(x^2>50 & y^2>100)`

(d) >> (x==2*z)|(x^2>50 & y^2>100)

4.   The following while loops were separately entered in different MATLAB sessions. What will the resulting outputs be? Do this one carefully by hand and then use MATLAB to check your answers.

```
(a)
>> i = 1; x=-3;
>> while (i<3) & (x<35)
      x=-x*(i+1)
  end
(c)
>> i = 1; x=-3;
>> while xor(i<3, x<35)
      x=-x*(i+1)
  end
```

```
(b)
>> i = 1; x=-3;
>> while (i<3) | (x<35)
      x=-x*(i+1)
  end
```

5.   The following while loop was entered in a MATLAB session. What will the resulting output be? Do this one carefully by hand and then use MATLAB to check your answers.

```
>> i = 1; x=2; y =3;
>> while (i<5) | (x == y)
      x=x*2, y=y+x, i=i+1;
  end
```

4.2: LOGICAL CONTROL FLOW IN MATLAB

Up to this point, the reader has been given a reasonable amount of exposure to *while loops*. The while loop is quite universal and is particularly useful in those situations where it is not initially known how many iterations must be run in the loop. If we know ahead of time how many iterations we want to run through a certain recursion, it is more convenient to use a *for loop*. For loops are used to repeat (iterate) a statement or group of statements a specified number of times. The format is as follows:

```
>>for n=(start):(gap):(end)
    ...MATLAB commands...
end
```

The **counter** *n* (which could be any variable of your choice) gets automatically bumped up at each iteration by the "gap." At each iteration, the "...MATLAB commands..." are all executed in order (just like they would be if they were to be entered manually again and again). This continues until the counter meets or exceeds the "end" number.

EXAMPLE 4.3: To get the feel for how for loops function, we run through some MATLAB examples. In each case the reader is advised to try to guess the output of each new command before reading on (or hitting enter in MATLAB) as a check.

```
>> for n=1:5  % if "gap" is omitted it is assumed to be 1.
   x(n)=n^3;  % we will be creating a vector of cubes of successive
              % integers.
end  %all output has been suppressed, but a vector x has been
>>   %created.
```

```
>>x   %let's display x now
→ x = 1 8 27 64 125
```

Note that since a comma in MATLAB signifies a new line, we could also have written the above for loop in a single line. We do this in the next loop below:

```
>> for k=1:2:10,  x(k)=2; end
>> x   %we display x again.  Try to guess what it now looks like.
→ 2 8 2 64 2 0 2 0 2
```

Observe that there are now nine entries in the vector x. This loop overwrote some of the five entries in the previous vector x (which still remained in MATLAB's workspace). Let us carefully go through each iteration of this loop explaining exactly what went on at each stage:

$k = 1$ (start) → we redefine x(1) to be 2 (from its original value of 1)

$k = 1 + 2 = 3$ (augment k by gap = 2) → redefine x(3) to be 2 (x(2) was left to its original value of 8)

$k = 3 + 2 = 5$ → redefine x(5) to be 2

$k = 5 + 2 = 7$ → defines x(7) to be 2 (previously x was a length 5 vector, now it has 7 components), the skipped component x(6) is by default defined to be 0

$k = 7 + 2 = 9$ → defines x(9) to be 2 and the skipped x(8) to be 0

$k = 9 + 2 = 11$ (exceeds end = 10 so for loop is exited and thus completed).

The gap in a for loop can even be a negative number, as in the following example that creates a vector in backwards order. The semicolon is omitted to help the reader convince himself or herself how the loop progresses.

```
>> for i=3:-1:1, y(i)=i, end
→y = 0   0   3
→y = 0   2   3
→y = 1   2   3
```

A very useful tool in programming is the *if-branch*. In its basic form the syntax is as follows:

```
>>if <condition>
    …MATLAB commands…
end
```

The way such an if-branch works is that if the listed <condition> (which can be any MATLAB statement) is true (i.e., has a nonzero value), then all of the "…MATLAB commands…" listed are executed in order and upon completion the if-branch is then exited. If the <condition> is false then the "…MATLAB commands…" are ignored (i.e., they are bypassed) and the if-branch is immediately exited. As with loops in MATLAB, if-branches may be inserted within loops (or branches) to deal with particular situations that arise. Such loops/branches are said to be **nested**. Sometimes if-branches are used to "raise a flag" if a certain condition arises. The following MATLAB command is often useful for such tasks:

`fprintf('<any English /text phrase>')` →	causes MATLAB to print: <any English phrase>

Thus the output of the command `fprintf('Have a nice day!')` will simply be → Have a nice day! This command has a useful feature that allows one to print the values of variables that are currently stored within a text phrase. Here is how such a command would work: We assume that (previously in a MATLAB session) the values $w = 2$ and $h = 9$ have been calculated and stored and we enter:

```
>>fprintf('the width of the rectangle is %d,the length is %d.', w, h)
→the width of the rectangle is 2, the length is 9.»
```

Note that within the "text" each occurrence of %d was replaced, in order, by the (current) values of the variables listed at the end. They were printed as integers (without decimals); if we wanted them printed as floating point numbers, we would use %f in place of %d. Also note that MATLAB unfortunately put the prompt >> at the end of the output rather than on the next line as it usually does. To prevent this, simply add (at the end of your text but before the single right quote) \r—which stands for "carriage return." This carriage return is also useful for splitting up longer groups of text within an `fprintf`.

Sometimes in a nested loop we will want to exit from within an inner loop and at other times we will want exit from the **mother loop** (which is the outermost loop inside of which all the other loops/branches are a part of) and thus halt all operations relating to the mother loop. These two exits can be accomplished using the following useful MATLAB commands:

`break` (anywhere within a loop)	→ causes immediate exit only from the single loop in which `break` was typed.
`return` (anywhere within a nested loop)	→causes immediate exit from the mother loop, or within a function M-file, immediate exit from M-file (whether or not output has been assigned)

The next example illustrates some of the preceding concepts and commands.

EXAMPLE 4.4: Carefully analyze the following two nested loops, decide what exactly they cause MATLAB to do, and then predict the exact output. After you do this, read on (or use MATLAB) to confirm your predictions.

(a)
```
for n=1:5
      for k=1:3
       a=n+k
       if a>=4, break,   end
      end
   end
```

NOTE: We have inserted tabs to make the nested loops easier to distinguish. Always make certain that each loop/branch must be paired with its own end.

(b)
```
for n=1:5
   for k=1:3
```

```
     a=n+k
      if a>=4
       fprintf('We stop since a has reached the value %d  \r', a)
        return
       end
     end
   end
```

SOLUTION: Part (a): Both nested loops consist of two loops. The mother loop in each is, as usual, the outermost loop (with counter n). The first loop begins with the mother loop setting the counter n to be 1 then immediately moves to the second loop and sets k to be 1; now in the second loop a is assigned to be the value of $n+k=1+1=2$ and this is printed (since there is no semicolon). Since a = 2 now, the "if-condition" is not satisfied so the if-branch is bypassed and we now iterate the k-loop by bumping up k by 1 (= default gap). Note that the mother loop's n will not get bumped up again until the secondary k-loop runs its course. So now with $k=2$, a is reassigned as a = $n+k=1+2=3$ and printed, the if-branch is again bypassed and k gets bumped up to 3 (its ending value), a is now reassigned as a = $n+k=1+3=4$ (and printed). The if-branch condition is now met, so the commands within it (in this case only a single "break" command) are run. So we will break out of the k-loop (which is actually redundant at this point since k was at its ending value and the k-loop was about to end anyway). But we are still progressing within the mother n-loop. So now n gets bumped up by 1 to be 2 and we start afresh the k-loop again with $k=1$. The variable a is now assigned as a = $n+k=2+1=3$ (and printed), the if-branch is bypassed since the condition is not met and k gets bumped up to be 2. Next a gets reassigned as a = $n+k=2+2=4$, and printed. Now the if-branch condition is met so we exit the k-loop (this time prematurely) and n now gets bumped up to be 3. Next entering the k-loop with $k=1$, a gets set to be $n+k=3+1=4$, and printed, the "if branch condition" is immediately satisfied, and we exit the k-loop and n now gets bumped up to be 4. As in the last iteration, the k-loop will just reassign a to be 5 and print this, break and n will go to 5 (its final value). In the final stage, a gets assigned as 6, the if-branch breaks us out of the k-loop, and since n is at its ending value, the mother loop exits.

The actual output for part (a) is thus:
→a = 2 a = 3 a = 4 a = 3 a = 4 a = 4 a = 5 a = 6

Part (b): Apart from the fprintf command, the main difference here is the replacement of the break command with the return command. As soon as the if-branch condition is satisfied, the conditions within will be executed and the return will cause the whole nested loop to stop in its tracks. The output will be as follows:
→a =2 a = 3 a = 4 We stop since a has reached the value 4

EXERCISE FOR THE READER 4.1: Below are two nested loops. Carefully analyze each of them and try to predict resulting outputs and then use MATLAB to verify your predictions. For the second loop the output should be given in the default format short.

(a)
```
>>for i=1:5
        i
        if i>2,    fprintf('test'),   end
   end
```

(b)
```
>>for i=1:8, x(i)=0; end  %initialize vector
>>for i=6:-2:2
        for j=1:i
             x(i)=x(i)+1/j;
        end
   end
>>x
```

The basic form of the if-branch as explained above, allows one to have MATLAB perform a list of commands in the event that one certain condition is fulfilled. In its more advanced form, if-branches can be set up to perform different sets of commands depending on which situation might arise. In the fullest possible form, the syntax of an if-branch is as follows:

```
>>if <condition_1>
   ...MATLAB commands_1...
elseif <condition_2>
   ...MATLAB commands_2...
...
elseif <condition_n>
   ...MATLAB commands_n...
else
   ...MATLAB commands...
end
```

There can be any number of else if cases (with subsequent MATLAB commands) and the final else is optional. Here is how such an if-branch would function: The first thing to happen is that <condition_1> gets tested. If it tests true (nonzero), then the "...MATLAB commands_1..." are all executed in order, after which the if-branch is exited. If <condition_1> tests false (zero), then the if-branch moves on to the next <condition_2> (associated with the first elseif). If this condition tests true, then "...MATLAB commands_2..." are executed and the if-branch is exited, otherwise it moves on to test the next <condition_3>, and so on. If the final else is not present, then once the loop goes through testing all of the conditions, and if none were satisfied, the if-branch would exit without performing any tasks. If the else is present, then in such a situation the "...MATLAB commands..." after the else would be performed as a catch-all to all remaining situations not covered by the conditions listed above.

Our next example will illustrate the use of such an extended if-branch and will also bring forth a rather subtle but important point.

EXAMPLE 4.5: Create a function M-file for the mathematical function defined by the formula:

$$y = \begin{cases} -x^2 - 4x - 2, & \text{if } x < -1 \\ |x|, & \text{if } |x| \le 1, \\ 2 - e^{\sqrt{x-1}}, & \text{if } x > 1 \end{cases}$$

then store this M-file and get MATLAB to plot this function on the interval $-4 \le x \le 4$.

SOLUTION: The M-file can easily be written using an if-branch. If we use the filename ex4_5, here is one possible M-file:

```
function y = ex4_5(x)
if x<-1
     y = -x.^2-4*x-2;
elseif x>1
     y = 2-exp(sqrt(x-1));
else
     y=abs(x);
end
end
```

It is tempting to now obtain the desired plot using the usual command sequence:

```
>>   x=-4:.001:4; y=ex4_5(x); plot(x,y)
```

There is a subtle problem here, though. If we were to enter these commands, we would obtain the graph of (the last function in the formula) $y = |x|$ on the whole interval $[-4, 4]$. Before we explain this and show how to resolve this problem, it would behoove the reader to try to decide what is causing this to happen and to figure out a way to fix this problem.

If we carefully go on to see what went wrong with the last attempt at a plot, we observe that since x is a vector, the first condition $x < -1$ in the if-branch now becomes a bit ambiguous. When asked about such a vector inequality, MATLAB will return a vector of the same size as x that is made up of zeros and ones. In each slot, the vector is 1 if the corresponding component of x satisfies the inequality $(x < -1)$ and 0 if it does not satisfy this inequality. Here is an example:

```
>> [2 -5  3  -2  -1] < -1     %causes MATLAB to test each of the 5
>>%                            inequalities as true (1) or false (0)
 →0   1  0   1   0
```

Here is what MATLAB did in the above attempt to plot our function. Since x is a (large) vector, the first condition $x < -1$ produced another vector as the same size as x made up of (both) 0's and 1's. Since the vector was not all true (1's), the condition as a whole was not satisfied so it moved on to the next condition $x > 1$, which for the same reason was also not satisfied and so bypassed. This left us with the catch-all command y=abs(x) for the whole vector x, which, needless to say, is not what we had intended.

So now how can we fix this problem? One simple fix would be to use a for loop to construct the y-coordinate vector using ex4_5(x) with only scalar values for x. Here is one such way to construct y:

```
>> size(x)    %first we find out
>>%            the size of x
→ 1    8001
>> for n=1:8001
y(n)=ex4_5(x(n));
end
>> plot(x,y)  %now we can
>>%     get the desired plot
```

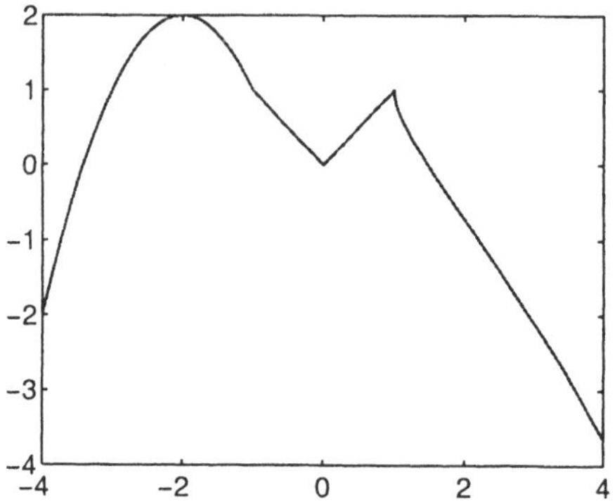

FIGURE 4.1: The plot of the function of Example 4.5.

A more satisfying solution would be to rebuild the M-file in such a way that when vectors are inputted for x, the if-branch testing is done separately for each component. The following program will do the job

```
function y = ex4_5v2(x)
for i = 1:length(x)
 if x(i)<-1
      y(i) = -x(i).^2-4*x(i)-2;
 elseif x(i)>1
      y(i) = 2-exp(sqrt(x(i)-1));
 else
      y(i)=abs(x(i));
end
```

With this M-file stored in our path the following commands would then indeed produce the desired plot of Figure 4.1:

```
>>  x=-4:.001:4; y=ex4_5v2(x); plot(x,y)
```

In dealing with questions involving integers, MATLAB has several number-theoretic functions available. We mention three here. They will be useful in the following exercise for the reader as well as in the exercises of this section.

floor(x) →	gives the greatest integer that is $\leq x$ (the **floor** of x)
ceil(x) →	gives the least integer that is $\geq x$ (the **ceiling** of x)
round(x) →	gives the nearest integer to x

For example, floor(2.5) = 2, ceil(2.5) = 3, ceil(−2.5) = −1, and round(−2.2) = −2. Observe that a real number x is an integer exactly when it equals its floor (or ceiling, or round(x) = x).

EXERCISE FOR THE READER 4.2: (a) Write a MATLAB function M-file, call it `sum2sq`, that will take as input a positive integer n and will produce the following output:

(i) In case n cannot be written as a sum of squares (i.e., if it is not possible to write $n = a^2 + b^2$ for some nonnegative integers a and b) then the output should be the statement: "the integer $<n>$ cannot be written as a sum of squares" (where $<n>$ will print as an actual numerical value).

(ii) If n can be written as a sum of squares (i.e., $n = a^2 + b^2$ can be solved for nonnegative integers a and b then the output should be "the integer $<n>$ can be written as the sum of the squares of $<a>$ and $<b>$ (here again, $<n>$ and also $<a>$ and $<b>$ will print as an actual numerical values) where a and b are actual solutions of the equation.

(b) Run your program with the following inputs: $n = 5$, $n = 25$, $n = 12,233$, $n = 100,000$.

(c) Write a MATLAB program that will determine the largest integer $< 100,000$ that cannot be written as a sum of squares. What is this integer?

(d) Write a MATLAB program that will determine the first integer > 1000 that cannot be written as a sum of squares. What is this integer?

(e) How many integers are there (strictly) between 1000 and 100,000 that cannot be expressed as a sum of the squares of two integers?

A useful MATLAB command syntax for writing interactive script M-files is the following:

`x = input('<Enter input` `phrase> : ') →`	When a script with this command is run, you will be prompted in command window by the same <Enter input phase> to enter an input for script after which your input will be stored as variable x and the script will be executed.

The command can, of course, also be invoked in a function M-file, or at any time in the MATLAB command window. The next example presents a way to use this command in a nice mathematical experiment.

EXAMPLE 4.6: (*Number Theory: The Collatz Problem*) Suppose we start with any positive integer a_1, and perform the following recursion to define the rest of the sequence $a_1, a_2, a_3, \cdots$:

$$a_{n+1} = \begin{cases} a_n/2, & \text{if } a_n \text{ is even} \\ 3a_n + 1, & \text{if } a_n \text{ is odd} \end{cases}.$$

We note that if a term a_n in this sequence ever reaches 1, then from this point on the sequence will *cycle* through the values 1, 4, 2. For example, if we start with $a_1 = 5$, the recursion formula gives $a_2 = 3 \cdot 5 + 1 = 16$, and then $a_3 = 16/2 = 8$, $a_4 = 8/2 = 4$, $a_5 = 4/2 = 2$, $a_6 = 2/2 = 1$, and so on (4,2,1,4, 2,1,...). Back in 1937, German mathematician Lothar Collatz conjectured that no matter what positive integer we start with for a_1, the above recursively defined sequence will always reach the 1,4,2 cycle. Collatz is an example of a mathematician who is more famous for a question he asked than for problems he solved or theorems he proved (although he did significant research in numerical differential equations). The *Collatz conjecture* remains an open problem to this day.[2] Our next example will give a MATLAB script that is useful in examining the Collatz conjecture. Some of the exercises will outline other ways to use MATLAB to run some illuminating experiments on the Collatz conjecture.

EXAMPLE 4.7: We write a script (and save it as `collatz`) that does the following. It will ask for an input for a positive integer to be the initial value a(1) of a Collatz experiment. The program will then run through the Collatz iteration scheme until the sequence reaches the value 1, and so begins to cycle (if ever). The script should output a sentence telling how many iterations were used for this Collatz experiment, and also give the sequence of numbers that were run through until reaching the value of one.

```
%Collatz script
a(1) = input('Enter a positive integer:   ');
n=1;
while a(n) ~= 1
    if ceil(a(n)/2)==a(n)/2   %tests if a(n) is even
        a(n+1)=a(n)/2;
    else
        a(n+1)=3*a(n)+1;
    end
    n=n+1;
end
fprintf('\r Collatz iteration with initial value a(1)= %d \r', a(1))
fprintf(' took %d iterations before reaching the value 1 and ',n-1)
fprintf(' beginning \r to cycle. The resulting pre-cycling')
fprintf(' sequence is as follows:')
a
clear a %lets us start with a fresh vector a on each run
```

With this script saved as an M-file `collatz`, here is a sample run using $a(1) = 5$:

```
>> collatz
```

---

[2] The Collatz problem has an interesting history; see, for example [Lag-85] for some details. Many mathematicians have proved interesting results that strongly support the truth of the conjecture. For example, in 1972, the famous Princeton number-theorist J. H. Conway [Con-72] proved that if a Collatz iteration enters into a cycle other than (1,4,2), the cycle must be of length at least 400 (i.e., the cycle itself must consist of at least 400 different numbers). Subsequently, J. C. Lagarias (in [Lag-85]) extended Conway's bound from 400 to 275,000! Recent high-speed computer experiments (in 1999, by T. Oliveira e Silvio [OeS-99]) have shown the Collatz conjecture to be true for all initial values of the sequence less than about 2.7×10^{16} . Despite all of these breakthroughs, the problem remains unsolved. P. Erdös, who was undoubtedly one of the most powerful problem-solving mathematicians of the twentieth century, was quoted once as saying "Mathematics is not yet ready for such problems," when talking about the Collatz conjecture. In 1996 a prize reward of £1,000 (approx. $2,000) was offered for settling the Collatz conjecture. Other math problems have (much) higher bounties. For example the *Clay Foundation* (URL: www.claymath.org/prizeproblems/statement.htm) has listed seven math problems and offered a prize of $1 million for each one.

Enter a positive integer: 5 (MATLAB gives the first message, we only enter 5, and enter to then get all of the informative output below.)
→Collatz iteration with initial value a(1) = 5 took
5 iterations before reaching the value 1 and beginning
to cycle. The resulting pre-cycling sequence is as follows:
a =5   16   8   4   2   1

EXERCISE FOR THE READER 4.3: (a) Try to understand this script, enter it, and run it with these values: $a(1) = 6, 9, 1, 12, 19, 88, 764$. Explain the purpose of the last command in the above script that cleared the vector a.

(b) Modify the M-file to a new one, called `collctr` (Collatz counter script), which will only give as output the total number of iterations needed for the sequence to reach 1. Make sure to streamline your program so that it does not create the whole vector a (which is not needed here) but rather overwrites new entries for the sequence over the previous ones.

EXERCISES 4.2:

1. Write a MATLAB function M-file, called `sumodsq(n)`, that does the following: The input is a positive integer n. Your function should compute the sum of the squares of all odd integers that do not exceed n:

$$1^2 + 3^2 + 5^2 + \cdots + k^2$$

where k is the largest odd integer that does not exceed n. If this sum is less than 1 million, the output will be the actual sum (a number); if this sum is greater than or equal to 1 million, the output will be the statement "$<n>$ is too big" where $<n>$ will appear as the actual number that was inputted.

2. Write a function M-file, call it `sevenpow(n)`, that inputs a positive integer n and that figures out how many factors of 7 n has (call this number k) and outputs the statement: "the largest power of 7 which $<n>$ contains as a factor is $<k>$."

   So for example, if you run sevenpow(98) your output should be the sentence "the largest power of 7 which   98 contains as a factor is 2." Run the commands: `sevenpow(36067)`, `sevenpow(671151153)`, and `sevenpow(3080641535629)`.

3. (a) Write a function M-file, call it `sumsq(n)`, that will input a positive integer n and output the sum of the squares of all positive integers that are less than or equal to n (sumsq(n) $= 1^2 + 2^2 + 3^2 + \cdots + n^2$). Check and debug this program with the results sumsq(1) = 1, sumsq(3) = 14.

   (b) Write a MATLAB loop to determine the <u>largest</u> integer n for which sumsq(n) does not exceed 5,000,000. The output of your code should be this largest integer but nothing else.

4. (a) Write a MATLAB function M-file, call it `sum2s(n)`, that will take as input a positive integer n and will produce for the output either of the following:

   (i) The sum of all of the positive powers of 2 $(2 + 4 + 8 + 16 + ...)$ that do not exceed n, provided this sum is less than 50 million.

   (ii) In case the sum in (i) is greater than or equal to 50 million, the output should simply be "overflow."

   (b) Run your function with the following inputs: $n = 1$, $n = 10$, $n = 265$, $n = 75,000$, $n = 65,000,000$.

(c) Write a short MATLAB code that will <u>determine the largest integer</u> n for which this program <u>does not</u> produce "overflow."

5.   (a) Write a MATLAB function M-file, called `bigpro(x)`, that does the following: The input is a real number x. The <u>only</u> output of your function should be the real number formed by the product $x(2x)(3x)(4x)\cdots(nx)$, where n is the first positive integer such that either nx is an integer or $\left|nx\right|$ exceeds x^2 (whichever comes first).

(b) Of course, after you write the program you need to debug it. What values should result if we were to use the (correctly created) program to find: bigpro(4), bigpro(2.5), bigpro(12.7)? Run your program for these values as well as for the values $x = -3677/9$, $x = 233.6461$, and $x = 125,456.789$.

(c) Find a negative number x that is not an integer such that bigpro(x) is negative.

6.   (*Probability: The Birthday Problem*) This famous problem in probability goes as follows: If there is a room with a party of people and everyone announces his or her birthday, how many people (at least) would there need to be in the room so that there is more than a 50% chance that at least two people have the same birthday?
To solve this problem, we let $P(n)$ = the probability of a common birthday if there are n people in the room. Of course $P(1) = 0$ (no chance of two people having the same birthday if there is only one person in the room), and $P(n) = 1$ when $n > 365$ (there is a 100% chance, i.e., guaranteed, two people will have the same birthday if there are more people in the room than days of the year; we ignore leap-year birthdays). We can get an expression for $P(n)$ by calculating the *complementary probability*, i.e., the probability that there will not be a common birthday among n different people. This must be

$$\frac{365}{365}\cdot\frac{364}{365}\cdot\frac{363}{365}\cdots\frac{366-n}{365}$$

This can be seen as follows: The first person can have any of the 365 possible birthdays, the second person can have only 364 possibilities (since he/she cannot have the same birthday as the first person), the third person is now restricted to only 363 possible birthdays and so on. We multiply the individual probabilities (fractions) to get the combined probability of no common birthday. Now this is the complementary probability of what we want (i.e., it must add to $P(n)$ to give 1 = 100% since it is guaranteed that either there is a common birthday or not). Thus

$$P(n) = 1 - \frac{365}{365}\cdot\frac{364}{365}\cdot\frac{363}{365}\cdots\frac{366-n}{365}$$

(a) Write a MATLAB function M-file for this function $P(n)$. Call the M-file bprob(n) and set it up so that it does the following: If n is a nonnegative integer, the function bprob(n) will output the sentence: "If a room contains $< n >$ people then the probability of a common birthday is $< P(n) >$" where $< n >$ and $< P(n) >$ should be the actual numerical values. If n is any other type of number (e.g., a negative number or 2.6) the output should be "input $< n >$ is not a natural number so the probability is undefined." Save your M-file and then run it for the following values: $n = 3$, $n = 6$, $n = 15$, $n = 90$, $n = 110.5$, and $n = 180$.
(b) Write a MATLAB code that uses your function in part (a) to solve the birthday problem, i.e., determine the smallest n for which $P(n) > .5$. More precisely, create a for loop whose <u>only</u> <u>output</u> will be, $n =$ the minimum number needed (for $P(n)$ to be $> .5$) and the associated probability $P(n)$.
(c) Get MATLAB to draw a neat plot of $P(n)$ vs. n (for all n between 1 and 365), and on the same plot, include the plots of the two horizontal lines with y-intercepts .5 and .9. Interpret the intersections.

7.   Write a function M-file, call it pythag(n) that inputs a positive integer n and determines

whether n is the hypotenuse of a right triangle with sides of integer lengths. Thus your program will determine whether the equation $n^2 = a^2 + b^2$ has a solution with a and b both being positive integers.

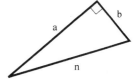

FIGURE 4.2: Pythagorean triples.

Such triples n, a, b are called Pythagorean triples (Figure 4.2). In case there is no solution (as, for example, if $n = 4$), your program should output the statement: "There are no Pythagorean triples with hypotenuse $< n >$." But if there is a solution your output should be a statement that actually gives a specific Pythagorean triple for your value of n. For example, if you type pythag(5), your output should be something like: "There are Pythagorean triples having 5 as the hypotenuse, for example: 3, 4, 5 is one such triple." Run this for several different values of n. Can you find a value of n larger than 1000 that has a Pythagorean triple? Can you find an n that has two different Pythagorean triples associated with it (of course not just by switching a and b)?

Historical Note: Since the ancient times of Pythagoras, mathematicians have tried long and hard to find integer triple solutions of the corresponding equation with exponent 3: $n^3 = a^3 + b^3$. No one has ever succeeded. In the 1700s the amateur French mathematician Pierre Fermat conjectured that no such triples can exist. He claimed to have a truly remarkable proof of this but there was not enough space in the margin of his notes to include it. There has been an incredible amount of research trying to come up with this proof. Just recently, more than 300 years since Fermat stated his conjecture, Princeton mathematician Andrew Wiles came up with a proof. He was subsequently awarded the Fields medal, the most prestigious award in mathematics.

8. (*Plane Geometry*) For an integer n that is at least equal to 3, a *regular n-gon* in the plane is the interior of a set whose boundary consists of n flat edges (sides) each having the same length (and such that the interior angles made by adjacent edges are all equal). When $n = 3$ we get an equilateral triangle, when $n = 4$ we get a square, and when $n = 8$ we get a regular octagon, which is the familiar stop-sign shape. There are regular *n*-gons for any such value of n; some are pictured in Figure 4.3.

FIGURE 4.3: Some regular polygons.

(a) Write a MATLAB function M-file, ngonper1(n, dia), that has two input variables, n = the number of sides of the regular *n*-gon, and dia = the diameter of the regular *n*-gons. The *diameter* of an *n*-gons is the length of the longest possible segment that can be drawn connecting two points on the boundary. When n is even, the diameter segment cuts the *n*-gons into two congruent (equal) pieces. Assuming that n is an even integer greater than 3 and dia is any positive number, your function should have a single output variable that equals the perimeter of the regular *n*-gons with diameter = dia. Your solution should include a handwritten mathematical derivation of the formula for this perimeter. This will be the hard part of this exercise, and it should be done, of course, before you write the program. Run your program for the following sets of input data: (i) $n = 4$, dia $= \sqrt{4}$, (ii) $n = 12$, dia $= 12$, (iii) $n = 1000$, dia $= 5000$.

(b) Remove the restriction that n is even from your program in part (a). The new function (call it now `ngonper(n, dia)` will now do everything that the one you constructed in part (a) did but it will be able to input and deal with any integer n greater than or equal to 3. Again, include with your solution a mathematical derivation of the perimeter formula you are using in your program. Run your program for these sets of values: (i) $n = 3$, dia = 2, (ii) $n = 5$, dia = 4, (iii) $n = 999$, dia = 500.

(c) For which values of n (if any) will your function in part (b) continue to give the correct perimeter of an n-gon that is no longer regular? An *irregular n-gon* is the interior of a set in the plane whose boundary consists of n flat edges whose interior angles are not all equal. Examples of irregular n-gons include any nonequilateral triangle $(n = 3)$, any quadrilateral that is not a square $(n = 4)$. For those n's for which you say things still work, a (handwritten mathematical) proof should be included and for those n's for which you say things no longer continue to work, a (handwritten) counterexample should be included.

9.   (*Plane Geometry*) This exercise consists of doing what is asked for in Exercise 8 (a)(b)(c) but with changing all occurrences of the word "perimeter" to "area." In parts (a) and (b) use the M-file names `ngonarl(n, dia)` and `ngonarea(n, dia)`.

10. (*Finance: Compound Interest*) Write a script file called `compints` that will compute (as output) the future value A in a savings account after prompting the user for the following inputs: the principal P (= amount deposited), the annual interest rate r (as a decimal), the number k of compoundings per year (so quarterly compounding means $k = 4$, monthly means $k = 12$, daily means $k = 365$, etc.), and the time t that the money is invested (measured in years).   The relevant formula from finance is $A = P(1 + r/k)^{kt}$. Run the script using the following sets of inputs: $P = \$10,000$, $r = 8\%$ (.08), $k = 4$, *and* $t = 10$, then changing t to 20, then also changing r to 11%.
Suggestion: You probably want to have four separate "input" lines in your script file. The first asking for the principal, etc. Also, to get the printout to look nice, you should switch to `format bank` inside the script and then (at the very end) switch back to `format short`.

11. (*Finance: Compound Interest*) Write a script file called `comintgs`, that takes the same inputs as in the previous exercise, but instead of producing the output of the future account balance, it should produce a graph of the future value A as a function of time as the time t ranges from zero (day money was invested) until the end of the time period that was entered. Run the script for the three sets of data in the previous problem.

12. (*Finance: Future Value Annuities*) Write a script file called `fvanns`, that will compute (as output) the future value FV in an annuity after prompting the user for the following inputs: the periodic payment PMT (= amount deposited in account per period), the annual interest rate r (as a decimal), the number k of periods per year, that is, the number of compoundings per year (so quarterly compoundings/deposits means $k = 4$, monthly means $k = 12$, bimonthly means $k = 24$, etc.), and the time t that the money is invested (measured in years). The relevant formula from finance is $FV = PMT((1 + r/k)^{kt} - 1)/(r/k)$. Run the script using the following sets of inputs: $PMT = 200$, $r = 7\%$ (.07), $k = 12$, and $t = 30$, then changing t to 40, then also changing r to 9%. Next change PMT to 400 on each of these three sets of inputs. Note, the first set of inputs could correspond to a worker who starts a supplemental retirement plan (say a 401(k)), deposits $200 each month starting at age 35, and continues until he/she plans to retire at age 65 ($t = 30$ years later). The FV will be his/her retirement nest egg at time of retirement. The next set of data could correspond to the same retirement plan but started at age 25 (10 years more time). In each case compare the future value with the total amount of contributions. To encourage such supplemental retirement plans, the federal government allows such contributions (with limits) to be done before taxation.
Suggestion: You probably want to have four separate "input" lines in your script file, the first

asking for the principal, etc. Also, to get the printout to look nice, you should switch to `format bank` inside the script and then (at the very end) switch back to `format short`.

13. (*Finance: Future Value Annuities*) In this exercise you will be writing a script file that will take the same inputs as in the previous exercise (interactively), but instead of just giving the future value at the end of the time period, this script will produce a graph of the growth of the annuity's value as a function of time.
   (a) Base your script on the formula given in the preceding exercise for future value annuities. Call this script `fvanngs`. Run the script for the same sets of inputs that were given in the previous exercise.
   (b) Rewrite the script file, this time constructing the vector of future values using a recursion formula rather than directly (as was asked in part (a)). Call this script `fvanng2s`. Run the script for the same sets of inputs that were given in the previous exercise.

14. (*Number Theory: The Collatz Problem*) Write a function M-file, call it `collctr`, that takes as input, a positive integer an (the first element for a Collatz experiment), and has as output the positive integer n, which equals the number of iterations required for the Collatz iteration to reach the value of 1. What is the first positive integer n for which this number of iterations exceeds 100? 200? 300?

4.3: WRITING GOOD PROGRAMS

Up to this point we have introduced the two ways that programs can be written and stored for MATLAB to use (function M-files and script M-files) and we have also introduced the basic elements of control flow and a few very useful built-in MATLAB functions. To write a good program for a specified task, we will need to put all of our skills together to come up with an M-file that, above all, does what it is supposed to do, is efficient, and is as eloquent as possible. In this section we present some detailed suggestions on how to systematically arrive at such programs. Programming is an art and the reader should not expect to master it in a short time or easily.

STEP 1: **Understand the problem, do some special cases by hand, and draw an outline.** Before you begin to actually type out a program, you should have a firm understanding of what the problem is (that the program will try to solve) and know how to solve it by hand (in theory, at least). Computers are not creative. They can do very well what they are told, but you will need to tell them exactly what to do, so you had better understand how to do what needs to be done. You should do several cases by hand and record the answers. This data will be useful later when you test your program and debug it if necessary. Draw pictures (a flowchart), write in plain English an explanation of the program, trying to be efficient and avoiding unnecessary tasks that will use up computer time.

STEP 2: **Break up larger programs into smaller module programs.** Larger programs can usually be split up into smaller independent programs. In this way the main program can be considerably reduced in size since it can call on the smaller module programs to perform secondary tasks. Such a strategy has numerous advantages. Smaller programs are easier to write (and debug) than larger ones and they may be used to create other large or improved programs later on down the road.

STEP 3: **Test and debug every program.** This is not an option. You should always test your programs with a variety of inputs (that you have collected output data for in Step 1) to make sure all of the branches and loops function appropriately. Novice and experienced programmers alike are often shocked at how rarely a program works after it is first written. It may take many attempts and changes to finally arrive at a fully functional program, but a lot of valuable experience can be gained in this step. It is one thing to look at a nice program and think one understands it well, but the true test of understanding programming is to be able to create and write good programs. Before saving your program for the first time, always make sure that every "for," "while," or "if" has a matching "end." One useful scheme when debugging is to temporarily remove all semicolons from the code, perhaps add in some auxiliary output to display, and then run your program on the special cases that you went through by hand in Step 1. You can see first hand if things are proceeding along the lines that you intended.

STEP 4: **After it finally works, try to make the program as efficient and easy to read as possible.** Look carefully for redundant calculations. Also, try to find ways to perform certain required tasks that use minimal amounts of MATLAB's time. Put in plenty of comments that explain various elements of the program. While writing a complicated program, your mind becomes full of the crucial and delicate details. If you read the same program a few months (or years) later (say, to help you to write a program for a related task), you might find it very difficult to understand without a very time-consuming analysis. Comments you inserted at the time of writing can make such tasks easier and less time consuming. The same applies even more so for other individuals who may need to read and understand your program.

The efficiency mentioned in Step 4 will become a serious issue with certain problems whose programs (even good ones) will push the computer to its limits. We will come up with many examples of such problems this book. We mention here two useful tools in testing efficiency of programs or particular tasks. A **flop** (abbreviation for **floating point operation**) is roughly equivalent to a single addition, subtraction, multiplication, or division of two numbers in full MATLAB precision (rather than a faster addition of two single-digit integers, say). Counting flops is a common way of comparing and evaluating efficiency of various programs and parts thereof. MATLAB has convenient ways of counting flops[3] or elapsed time(`tic/toc`):

---

[3] The `flop` commands in MATLAB are actually no longer available since Version 5 (until further notice). This is due to the fact that, starting with Version 6, the core programs in MATLAB got substantially revised to be much more efficient in performing matrix operations. It was unfortunate that the `flop` counting features could no longer be made to perform in this newer platform (collateral damage). Nonetheless, we will, on occasion, use this function in cases where flop counts will help to

flops(0)	the flops(0) resets the flop counter at zero. The flops tells
...MATLAB commands...	the number of flops used to execute the "MATLAB commands" in
flops	between

tic	this tic resets the stopwatch to zero. The toc will tell the
...MATLAB commands...	elapsed time used to execute the "MATLAB commands"
toc	

The results of tic/toc depend not just on the MATLAB program but on the speed of the computer being used, as well as other factors, such as the number of other tasks concurrently being executed on the same computer. Thus the same MATLAB routines will take varying amounts of time on different computers (or even on the same computer under different circumstances). So, unlike flop comparisons, tic/toc comparisons cannot be absolute.

EXAMPLE 4.8: Use the tic/toc commands to compare two different ways of creating the following large vector: $x = [1 \quad 2 \quad 3 \quad \cdots \quad 10,000]$. First use the non-loop construction and then use a for loop. The results will be quite shocking, and since we will need to work with such large single vectors quite often, there will be an important lesson to be learned from this example. **When creating large vectors in MATLAB, avoid, if possible, using "loops."**

SOLUTION:

```
>> tic, for n=1:10000, x(n)=n; end, toc
 →elapsed_time =8.9530 (time is measured in seconds)
>> tic, y=1:10000; toc
 →elapsed_time = 0
```

The loop took nearly nine seconds, but the non-loop construction of the same vector went so quickly that the timer did not detect any elapsed time. Let's try to build a bigger vector:

```
>> tic, y=1:100000; toc
 →elapsed_time = 0.0100
```

This gives some basis for comparison. We see that the non-loop technique built a vector 10 times as large in about 1/1000th of the time that it took the loop construction to build the smaller vector! The flop-counting comparison method would not apply here since no flops were done in these constructions.

Our next example will deal with a concept from linear algebra called the **determinant** of a square matrix, which is a certain important number associated

illustrate important points. Readers that do not have access to older versions of MATLAB will not be able to mimic these calculations.

with the matrix. We now give the definition of the determinant of a square $n \times n$ matrix[4]

$$A = \begin{bmatrix} a_{11} & a_{12} & a_{13} & \cdots & a_{1n} \\ a_{21} & a_{22} & a_{23} & \cdots & a_{2n} \\ a_{31} & a_{32} & a_{33} & \cdots & a_{3n} \\ \vdots & \vdots & \vdots & \ddots & \vdots \\ a_{n1} & a_{n2} & a_{n3} & \cdots & a_{nn} \end{bmatrix}.$$

If $n = 1$, so $A = [a_{11}]$, then the determinant of A is simply the number a_{11}. If $n = 2$, so $A = \begin{bmatrix} a_{11} & a_{12} \\ a_{21} & a_{22} \end{bmatrix}$, the determinant of A is defined to be the number $a_{11}a_{22} - a_{12}a_{21}$ that is just the product of the main diagonal entries (top left to bottom right) less the product of the off diagonal entries (top right to bottom left).

For $n = 3$, $A = \begin{bmatrix} a_{11} & a_{12} & a_{13} \\ a_{21} & a_{22} & a_{23} \\ a_{31} & a_{32} & a_{33} \end{bmatrix}$ and the determinant can be defined using the $n = 2$ definition by the so-called cofactor expansion (on the first row). For any entry a_{ij} of the 3×3 matrix A, we define the corresponding **submatrix** A_{ij} to be the 2×2 matrix obtained from A by deleting the row and column of A that contain the entry a_{ij}. Thus, for example,

$$A_{13} = \begin{bmatrix} a_{11} & a_{12} & a_{13} \\ a_{21} & a_{22} & a_{23} \\ a_{31} & a_{32} & a_{33} \end{bmatrix} = \begin{bmatrix} a_{21} & a_{22} \\ a_{31} & a_{32} \end{bmatrix}.$$

Abbreviating the determinant of the matrix A by det(A), the determinant of the 3×3 matrix A is given by the following formula:

$$\det(A) = a_{11} \det(A_{11}) - a_{12} \det(A_{12}) + a_{13} \det(A_{13}).$$

Since we have already shown how to compute the determinant of a 2×2 matrix, the right side can be thus computed. For a general $n \times n$ matrix A, we can compute it with a similar formula in terms of some of its $(n-1) \times (n-1)$ submatrices:

---

[4] The way we define the determinant here is different from what is usually presented as the definition. One can find the formal definition in books on linear algebra such as [HoKu-71]. What we use as our definition is often called *cofactor expansion on the first row*. See [HoKu-71] for a proof that this is equivalent to the formal definition. The latter is actually more complicated and harder to compute and this is why we chose cofactor expansion.

$$\det(A) = a_{11}\det(A_{11}) - a_{12}\det(A_{12}) + a_{13}\det(A_{13}) - \cdots + (-1)^{n+1}a_{1n}\det(A_{1n})$$

It is proved in linear algebra books (e.g., see [HoKu-71]) that one could instead take the corresponding (cofactor) expansion along any row or column of A, using the following rule to choose the alternating signs: The sign of $\det(A_{ij})$ is $(-1)^{i+j}$.

Below are two MATLAB commands that are used to work with entries and submatrices of a given general matrix A.

`A(i,j)` →	represents the entry a_{ij} located in the ith row and the jth column of the matrix A
`A([i1 i2 … imax],` `[j1 j2 …jmax])` →	represents the submatrix of the matrix A formed using the rows $i1, i2, ..., imax$ and columns $j1, j2, ..., jmax$
`A([i1 i2 … imax], :)` →	represents the submatrix of the matrix A formed using the rows $i1, i2, ..., imax$ and all columns

EXAMPLE 4.9: (a) Write a MATLAB function file, called `mydet2(A)`, that calculates the determinant of a 2×2 matrix A.

(b) Using your function `mydet2` of part (a), build a new MATLAB function file, called `mydet3(A)` that computes the determinant of a 3×3 matrix A (by performing cofactor expansion along the first row).

(c) Write a program `mydet(A)` that will compute the determinant of a square matrix of <u>any</u> size. Test it on the matrices shown below. MATLAB has a built-in function `det` for computing determinants. Compare the results, flop counts (if available) and times using your function `mydet` versus MATLAB's program `det`. Perform this comparison also for a randomly generated 8×8 matrix. Use the following command to generate random matrices:

`rand(n,m)` → NOTE: `rand(n)` is equivalent to `rand(n,n)`, and `rand` to `rand(1)`.	generates an $n\times m$ matrix whose entries are randomly selected from $0 \le x \le 1$ [5]

$$A = \begin{bmatrix} 2 & 7 & 8 & 10 \\ 0 & -1 & 4 & -9 \\ 0 & 0 & 3 & 6 \\ 0 & 0 & 0 & 5 \end{bmatrix}, \quad A = \begin{bmatrix} 1 & 2 & -1 & -2 & 1 & 2 \\ 0 & 3 & 0 & 2 & 0 & 1 \\ 1 & 0 & 2 & 0 & 3 & 0 \\ 1 & 1 & 1 & 1 & 1 & 1 \\ -2 & -1 & 0 & 1 & 2 & 3 \\ 1 & 2 & 3 & 1 & 2 & 3 \end{bmatrix}$$

---

[5] Actually, the `rand` function, like any computer algorithm, uses a deterministic program to generate random numbers that is based on a certain seed number (starting value). The numbers generated meet statistical standards for being truly random, but there is a serious drawback that at each fresh start of a MATLAB session, the sequence of numbers generated by successive applications of `rand` will always result in the same sequence. This problem can be corrected by entering `rand('state',sum(100*clock))`, which resets the seed number in a somewhat random fashion based on the computer's internal clock. This is useful for creating simulation trials.

SOLUTION: The programs in parts (a) and (b) are quite straightforward:

```
function y = mydet2(A)
y=A(1,1)*A(2,2)-A(1,2)*A(2,1);

function y = mydet3(A)
y=A(1,1)*mydet2(A(2:3,2:3))-A(1,2)*mydet2(A(2:3,[1...
 3]))+A(1,3)*mydet2(A(2:3,1:2));
```

NOTE: The three dots (. . .) at the end of a line within the second function indicate (in MATLAB) a continuation of the command. This prevents the carriage return from executing a command that did not fit on a single line.

The program for part (c) is not quite so obvious. The reader is strongly urged to try and write one now before reading on.

Without having the mydet program call on itself, the code would have to be an extremely inelegant and long jumble. Since MATLAB allows its functions to (recursively) call on themselves, the program can be elegantly accomplished as follows:

```
function y = mydet(A)
y=0;  %initialize y
[n, n] = size(A);  %record the size of the square matrix A
if n ==2
   y=mydet2(A);
   return
end
for i=1:n
   y=y+(-1)^(i+1)*A(1,i)*mydet(A(2:n, [1:(i-1) (i+1):n]));
end
```

   Let's now run this program side by side with MATLAB's det to compute the requested determinants.

```
>> A=[2 7 8 10; 0 -1 4 -9; 0 0 3 6; 0 0 0 5];
>> A1=[1 2 -1 -2 1 2;0 3 0 2 0 1;1 0 2 0 3 0;1 1 1 1 1;...
-2 -1 0 1 2 3; 1 2 3 1 2 3];

>> flops(0), tic,  mydet(A), toc, flops
→ans = -30(=determinant), elapsed_time =  0.0600, ans = 182 (=flop count)
>> flops(0), tic,  mydet(A1), toc, flops
→ans = 324, elapsed_time = 0.1600, ans =5226

>> flops(0), tic,  det(A), toc, flops
→ans =-30, elapsed_time = 0, ans =52

>> flops(0), tic,  det(A1), toc, flops
→ans =324, elapsed_time = 0, ans = 117
```

   So far we can see that MATLAB's built-in det works quicker and with a lot less flops than our mydet does. Although mydet still performs reasonably well, check out the flop-count ratios and how they increased as we went from the 4×4 matrix A to the 6×6 matrix $A1$. The ratio of flops `mydet` used to the number that `det` used rose from about a factor of 3.5 to a factor of nearly 50. For larger matrices, the situation quickly gets even more extreme and it becomes no longer practical to use `mydet`. This is evidenced by our next computation with an 8×8 matrix.

```
>> Atest= rand(8); %we suppress output here.
>> flops(0), tic,  det(Atest), toc, flops
→ans = -0.0033, elapsed_time = 0, ans = 326
>> flops(0), tic,  mydet(Atest), toc, flops
→ans = -0.0033, elapsed_time =8.8400, ans = 292178
```

MATLAB's `det` still works with lightning speed (elapsed time was still undetectable) but now `mydet` took a molasses-slow nearly 9 seconds, and the ratio of the flop count went up to nearly 900! If we were to go to a 20×20 matrix, at this pace, our `mydet` would take over 24 years to do! (See Exercise 5 below.) Suprisingly though, MATLAB's `det` can find the determinant of such a matrix in less than 1/100th of a second (on the author's computer) with a flop count of only about 5000. This shows that there are more practical ways of computing (large matrix determinants) than by the definition or by cofactor expansion. In Chapter 7 such a method will be presented.

Each time when the `rand` command is invoked, MATLAB uses a program to generate a random numbers so that in any given MATLAB session, the sequence of "random numbers" generated will always be the same. Random numbers are crucial in the important subject of *simulation*, where trials of certain events that depend on chance (like flipping a coin) need to be tested. In order to assure that the random sequences are different at each start of a MATLAB session, the following command should be issued before starting to use `rand`:

`rand('state',sum(100*clock))` →	This sets the "state" of MATLAB's random number generator in a way that depends in a complicated fashion on the current computer time. It will be different each time MATLAB is started.

Our next exercise for the reader will require the ability to store strings of text into rows of numerical matrices, and then later retrieve them. The following basic example will illustrate how such data transformations can be accomplished:
We first create text string `T` and a numerical vector `v`:

```
>> T = 'Test', v = [1 2 3 4 5 6]
→T = Test,   v = 1   2   3   4   5   6
```

If we examine how MATLAB has stored each of these two objects, we learn that both are "arrays," but `T` is a "character array" and `v` is a "double array" (meaning a matrix of numbers):

```
>> whos T  v  →Name    Size           Bytes  Class
                 T      1x4                8  char array
                 v      1x6               48  double array
```

If we redefine the first four entries of the vector v to be the vector T, we will see that the characters in T get transformed into numbers:

```
>> v(1:4)=T
→v = 84   101   115   116   5   6
```

MATLAB does this with an internal dictionary that translates all letters, numbers and symbols on the keyboard into integers (between 1 and 256, in a case-sensitive fashion).   To transform back to the original characters, we use the char command, as follows:

```
>> U=char(v(1:4))
→U =Test
```

Finally, to call on a stored character (string) array within an fprintf statement, the symbol %s is used as below:

```
>> fprintf('The %s has been performed.', U)
→The Test has been performed.
```

EXERCISE FOR THE READER 4.4: (*Electronic Raffle Drawing Program*)
(a) Create a script M-file, raffledraw, that will randomly choose the winner of a raffle as follows: When run, the first thing the program will do is prompt the user to enter the number of players (this can be any positive integer). Next it will successively ask the user to input the names of the players (in single quotes, as text strings are usually inputted) along with the corresponding weight of each player. The weight of a player can be any positive integer and corresponds to the number of raffle tickets that the player is holding. Then the program will randomly select one of these tickets as the winner and output a phrase indicating the name of the winner.
(b) Run your M-file with the following data on four players: Alfredo has four tickets, Denise has two tickets, Sylvester has two tickets and Laurie has four tickets. Run it again with the same data.

EXERCISES 4.3:

1.   Write a MATLAB function M-file, call it sum3sq(n), that takes as input a positive integer n and as output will do the following. If n can be expressed as a sum of three squares (of positive integers), i.e., if the equation:

$$n = a^2 + b^2 + c^2$$

has a solution with a, b, c all positive integers, then the program should output the sentence, "the number $< n >$ can be written as the sum of the squares of the three positive integers $< a >$, $< b >$, and $< c >$." Each of the numbers in brackets must be actual integers that solve the equation. In case the equation has no solution (for a, b, c), the output should be the sentence: "The number $< n >$ cannot be expressed as a sum of the squares of three positive integers." Run your program with the numbers $n = 3$, $n = 7$, $n = 43$, $n = 167$, $n = 994$, $n = 2783$, $n = 25,261$. Do you see a pattern for those integers n for which the equation does/does not have a solution?

2. Repeat Exercise 1 with "three squares" being replaced by "four squares," so the equation becomes:

$$n = a^2 + b^2 + c^2 + d^2 .$$

Call your function sum4sq. In each of these problems feel free to run your programs for a larger set of inputs so as to better understand any patterns that you may perceive.

3 (*Number Theory: Perfect Numbers*) (a) Write a MATLAB function M-file, call it divsum(n), that takes as input a positive integer n and gives as output the sum of all of the proper divisors of n. For example, the proper divisors of 10 are 1, 2, and 5 so the output of divsum(10) should be 8 (=1+2+5). Similarly, divsum(6) should equal 6 since the proper divisors of 6 are 1, 2, and 3. Run your program for the following values of n: $n = 10$, $n = 224$, $n = 1410$ (and give the outputs).

(b) In number theory, a **perfect number** is a positive integer n that equals the sum of its proper divisors, i.e., $n = $ divsum(n). Thus from above we see that 6 is a perfect number but 10 is not. In ancient times perfect numbers were thought to carry special magical properties. Write a program that uses your function in part (a) to get MATLAB to find and print all of the perfect numbers that are less than 1000. Many questions about perfect numbers still remain perfect mysteries even today. For example, it is not known if the list of perfect numbers goes on forever.

4 (*Number Theory: Prime Numbers*) Recall that a positive integer n is called a **prime** number if the only positive integers that divide evenly into n are 1 and itself. Thus 4 is not a prime since it factors as 2×2. The first few primes are as follows: 2, 3, 5, 7, 11, 13, 17, 19, 23, ... (1 is not considered a prime for some technical reasons). There has been a tremendous amount of research done on primes, and there still remain many unanswered questions about them that are the subject of contemporary research. One of the first questions that comes up about primes is whether there are infinitely many of them (i.e., does our list go on forever?). This was answered by an ancient Greek mathematician, Euclid, who proved that there are infinitely many primes. It is a very time-consuming task to determine if a given (large) number is prime or not (unless it is even or ends in 5).

(a) Write a MATLAB function M-file, call it primeck(n), that will input a positive integer n > 1, and will output either the statement: "the number < n > is prime," if indeed, n is prime, or the statement "the number < n > is not prime, its smallest prime factor is < k >," if n is not prime, and here k will be the actual smallest prime factor of n.

Test (and debug) your program for effectiveness with the following inputs for n:

$$n = 51, n = 53, n = 827, n = 829.$$

Next test your program for efficiency with the following inputs (depending on how you wrote your program and also how much memory your computer has, it may take a very long time or not finish with these tasks)

$$n = 8237, n = 38877, n = 92173, n = 1,875,247, n = 2038074747, n = 22801763489,$$
$$n = 1689243484681, n = 7563374525281.$$

In your solution, make sure to give exactly what the MATLAB printout was; also, next to each of these larger numbers, write down how much time it took MATLAB to perform the calculation.

(b) Given enough time (and assuming you are working on a computer that will not run out of memory) will this MATLAB program always work correctly no matter how large n is? Recall that MATLAB has an accuracy of about 15 significant digits.

5. We saw in Example 4.7 that by calculating a determinant by using cofactor expansion, the number of flops (additions, subtractions, multiplications, and divisions) increases dramatically. For a 2×2 matrix, the number (worst-case scenario, assuming no zero entries) is 3; for a 3×3 matrix it is 14. What would this number be for a 5×5 matrix?, For a 9×9 matrix? Can you determine a general formula for an $n \times n$ matrix?

6. (*Probability and Statistics*) Write a program, called `cointoss(n)` that will have one input variable *n* = a positive integer and will simulate n coin tosses, by (internally) generating a sequence of n random numbers (in the range $0 \le x \le 1$) and will count each such number that is less than 0.5 as a "HEAD" and each such number that is greater than 0.5 as a "TAIL". If a number in the generated sequence turns out to be exactly = 0.5, another simulated coin toss should be made (perhaps repeatedly) until a "HEAD" or a "TAIL" comes up. There will be only one output variable: *P* = the ratio of the total number of "HEADS" divided by *n* . But the program should also cause the following sentence to be printed: "In a trial of < *n* > coin tosses, we had <H> flips resulting in "HEAD" and <T> flips resulting in "TAIL," so "HEADS" came up <100P>% of the time." Here, <H> and <T> are to denote the actual numbers of "HEAD" and "TAIL" results. Run your program for the following values of *n*: 2, 4, 6, 10, 50, 100, 1000, 5000, 50,000. Is it possible for this program to enter into an infinite loop? Explain!

7. (*Probability and Statistics*) Write a program similar to the one in the previous exercise except that it will not print the sentence, and it will have three output variables: P (as before), H = the number of heads, and T = the number of tails. Set up a loop to run this program with *n* = 1000 fixed for *k* = 100 times. Collect the outcomes of the variable *H* as a vector: $[h_0, h_1, h_2, \cdots h_{n+1}]$ (with *n*+1 = 1001 entries) where each h_i denotes the number of times that the experiment resulted in having exactly h_i heads (so $H = h_i$) and then plot the graph of this vector (on the x-axis n runs from 0 to 1001 and on the y-axis we have the h_i -values). Repeat this exercise for *k* = 200 and then *k* = 500 times.

8. (*Probability: Random Integers*) Write a MATLAB function M-file `randint(n,k)`, that has two input variables n and k being positive integers. There will be one output variable R, a vector with *k* components $R = [r_1, r_2, \cdots, r_k]$, each of whose entries is a positive integer randomly selected from the list {1, 2, ..., *n* }. (Each integer in this list has an equal chance of being generated at any time.)

9. (*Probability: Random Walks*) Create a MATLAB M-file, called `ran2walk(n)`, that simulates a random walk in the plane. The input *n* is the number of steps in the walk. The starting point of the walk is at the origin (0,0). At each step, random numbers are chosen (with uniform distribution) in the interval $[-1/2, 1/2]$ and are added to the present *x*- and *y*-coordinates to get the next *x*- and *y*-coordinates. The MATLAB command `rand` generates a random number in the interval [0,1], so we must subtract 0.5 from these to get the desired distributions. There will be no output variables, but MATLAB will produce a plot of the generated random walk.

    Run this function for the values *n* = 8, 25, 75, 250 and (using the subplot option) put them all into a single figure. Repeat once again with the same values. In three dimensions, these random walks simulate the chaotic motion of a dust particle that makes many microscopic collisions and produces such strange motions. This is because the microscopic particles that collide with our particle are also in constant motion. We could easily modify our program by adding a third *z*-coordinate (and using `plot3(x,y,z)` instead of `plot(x,y)`) to make a program to simulate such three dimensional random walks. Interestingly, each time you run the `ran2walk` function for a fixed value of *n*, the paths will be different. Try it out a few times.

    Do you notice any sort of qualitative properties about this motion? What are the chances (for a fixed *n*) that the path generated will cross itself? How about in three dimensions? Does the motion tend to move the particle away from where it started as *n* gets large? For these latter questions do not worry about proofs, but try to do enough experiments to lead you to make some educated hypotheses.

10. (*Probability Estimates by Simulation*) In each part, run a large number of simulations of the following experiments and take averages to estimate the indicated quantities.
    (a) Continue to generate random numbers in (0,1) using `rand` until the accumulated sum

exceeds 1. Let N denote the number that of such random numbers that get added up when this sum first exceeds 1. Estimate the *expected value* of N, which can be thought of as the theoretical (long-run) average value of N if the experiment gets repeated indefinitely.

(b) Number a set of cards from 1 to 20, and shuffle them. Turn the cards over one by one and record the number of times K that card number i ($1 \le i \le 20$) occurs at (exactly) the ith draw. Estimate the expected value of K.

Note: Simulation is a very useful tool for obtaining estimates for quantities that can be impossible to estimate analytically; see [Ros-02] for a well-written introduction to this interesting subject. In it the reader can also find a rigorous definition of the expectation of a random variable associated with a (random) experiment. The quantities K and N above are examples of random variables. Their outcomes are numerical quantities associated with the outcomes of (random) experiments. Although the outcomes of random variables are somewhat unpredictable, their long-term averages do exhibit patterns that can be nicely characterized. For the above two problems, the exact expectations are obtainable using methods of probability; they are $N = e$ and $K = 1$.

The next four exercises will revisit the Collatz conjecture that was introduced in the preceding section.

11. (*Number Theory: The Collatz Problem*) Write a function M-file, call it `collsz`, that takes as input a positive integer an (the first element for a Collatz experiment), and has as output a positive integer `size` equaling the size of the largest number in the Collatz iteration sequence before it reaches the value of 1. What is the first positive integer an for which this maximum size exceeds the value 100? 1000? 100,000? 1,000,000?

12. (*Number Theory: The Collatz Problem*) Modify the script file, `collatz`, of Example 4.7 in the text to a new one, `collatzg`, that will interactively take the same input and internally construct the same vector a, but instead of producing output on the command window, it should produce a graphic of the vector a's values versus the index of the vector. Arrange the plot to be done using blue pentagrams connected with lines. Run the script using the following inputs: 7, 15, 27, 137, 444, 657.
Note: The syntax for this plot style would be `plot(index, a, bp-)`.

13. (*Number Theory: The Collatz Problem*) If a Collatz experiment is started using a negative integer for $a(1)$, all experiments so far done by researchers have shown that the sequence will eventually cycle. However, in this case, there is more than one possible cycle. Write a script, `collatz2`, that will take an input for a(1) in the same way as the script `collatz` in Example 4.7 did, and the script will continue to do the Collatz iteration until it detects a cycle. The output should include the number of iterations done before detecting a cycle as well as the actual cycle vector. Run your script using the following inputs: $-2, -4, -8, -10, -56, -88, -129$.
Suggestion: A cycle can be detected as soon as the same number $a(n)$ has appeared previously in the sequence. So your script will need to store the whole Collatz sequence. For example, each time it has constructed a new sequence element, say $a(20)$, the script should compare with the previous vector elements $a(1)$, $a(20)$, ..., $a(19)$ to see if this new element has previously appeared. If not, the iteration goes on, but if there is a duplication, say, $a(20) = a(15)$, then there will be a cycle and the cycle vector would be ($a(15)$, $a(16)$, $a(17)$, $a(18)$, $a(19)$).

14. (*Number Theory: The Collatz Problem*) Read first the preceding exercise. We consider two cycles as *equivalent* in a Collatz experiment if they contain the same numbers (but not necessarily in the same order). Thus the cycle (1,4,2) has the equivalent forms (4,2,1), and (2,1,4). The program in the previous exercise, if encountering a certain cycle, may output any of the possible equivalent forms, depending on the first duplication encountered. We say that two cycles are *essentially different* if they are not equivalent cycles. In this exercise, you are to use MATLAB to help you figure out the number of essentially different Collatz cycles that

come up from using negative integers for $a(1)$ ranging from -1 to $-20,000$.

Note: The Collatz conjecture can be paraphrased as saying that all Collatz iterations starting with a positive integer must eventually cycle and the resulting cycles are all equivalent to (4,2,1). The essentially different Collatz cycles for negative integer inputs in this problem will cover all that are known to this date. It is also conjectured that there are no more.

Chapter 5: Floating Point Arithmetic and Error Analysis

5.1: FLOATING POINT NUMBERS

We have already mentioned that the data contained in just a single irrational real number such as π has more information in its digits than all the computers in the world could possibly ever store. Then again, it would probably take all the scientific surveyors in the world to look for and not be able to find any scientist who vitally needed, say, the 534th digit of this number. What is usually required in scientific work is to maintain accuracy with a certain number of so-called **significant digits**, which constitutes the portion of a numerical answer to a problem that is trusted to be correct. For example, if we want π to three significant digits, we could use 3.14. A computer can only work with a finite set of numbers; these computer numbers for a given system are usually called floating point numbers. Since there are infinitely many real numbers, what has to happen is that big (infinite) sets of real numbers must get identified with single computer numbers. Floating point numbers are best understood by their relations with numbers in scientific notation, such as 3.14159×10^0, although they need not be written in this form.

A **floating point number system** is determined by a **base** β (any positive integer greater than one), a **precision** s (any positive integer; this will be the number of significant digits), and two integers m (negative) and M (positive) that determine the exponent range. In such a system, a **floating point number** can always be expressed in the form:

$$\pm . d_1 d_2 \cdots d_s \times \beta^e , \tag{1}$$

where,

$$d_i = 0, 1, 2, \cdots, \text{or } \beta - 1 \text{ but } d_1 \neq 0 \text{ and } m \leq e \leq M .$$

The number zero is represented as $.00 \cdots 0 \times \beta^{-m}$. In a computer, each of the three parts (the sign $\pm$, mantissa $d_1 d_2 \cdots d_s$, and the exponent β) of a floating point number is stored in its own separate fixed width field. Most contemporary computers and software on the market today (MATLAB included) use **binary arithmetic** ($\beta = 2$). Hand-held calculators use decimal base $\beta = 10$. In the past, other computers have used different bases that were usually powers of two, such as $\beta = 16$ (hexadecimal arithmetic). Of course, such arithmetic (different from base 10) is done in internal calculations only. When the number is displayed, it is converted to decimal form. An important quantity for determining the precision of a given computing system is known as the **unit roundoff** u (or the **machine epsilon**[1]), which is the maximum relative error that can occur when a real number is approximated by a floating point number[2]. For example, one Texas Instruments graphing calculator uses the floating point parameters $\beta = 10$, $s = 12$, $m = -99$, and $M = 99$, which means that this calculator can effectively handle numbers whose absolute values lie (approximately) between 10^{-99} and 10^{99}, and the unit roundoff is $u = 10^{-12}$. MATLAB's arithmetic uses the parameters: $\beta = 2$, $s = 53$, $m = -1074$, and $M = 1023$, which conforms to the **IEEE double precision standard**.[3] This means that MATLAB can effectively handle numbers with absolute values from $2^{-1074} \approx 10^{-324}$ to $2^{1023} \approx 10^{308}$; also the unit roundoff is $u = 2^{-53} \approx 10^{-16}$.

5.2: FLOATING POINT ARITHMETIC: THE BASICS

Many students have gotten accustomed to the reliability and logical precision of exact mathematical arithmetic. When we get the computer to perform calculations for us, we must be aware that floating point arithmetic compounded with roundoff errors can lead to unexpected and undesirable results. Most large-scale numerical algorithms are not exact algorithms, and when such methods are used, attention must be paid to the error estimates. We saw this at a basic level in Chapter 2, and it will reappear again later on several occasions. Here we will talk about different sorts of errors, namely, those that arise and are compounded by the computer's floating point arithmetic. We stress the distinction with the first type of errors. Even if an algorithm is mathematically guaranteed to work, floating point errors may arise and spoil its success. All of our illustrations below will be in base 10 floating point arithmetic, since all of the concepts can be covered and better understood in this familiar setting; changing to a different base is merely a technical issue. To get a feel for the structure of a floating point number system, we begin with an example of a very small system.

---

[1] The rationale for this terminology is that the Greek letter epsilon (ε) is usually used in mathematical analysis to represent a very small number.

[2] There is another characterization of the unit roundoff as the gap between the floating point number 1 and the next floating point number to the right. These two definitions are close, but not quite equivalent; see Example 5.4 and Exercise for the Reader 5.3 for more details on how these two quantities are related, as well as explicit formulas for the unit roundoff.

[3] The IEEE (Eye-triple-E) is a nonprofit, technical professional association of more than 350,000 individual members in 150 countries. The full name is the Institute of Electrical and Electronics Engineers, Inc. The IEEE single-precision (SP) and double-precision (DP) standards have become the international standard for computers and numerical software. The standards were carefully developed to help avoid some problems and incompatibilities with previous floating point systems. In our notation, the IEEE SP standard specifies $\beta = 2$, $s = 24$, $m = -126$, $= -126$, and $M = 127$ and the IEEE DP standard has $\beta = 2$, $s = 53$, $m = -1022$, and $M = 1023$.

EXAMPLE 5.1: Find all floating point numbers in the system with $\beta = 10$, $s = 1$, $m = -1$, and $M = 1$.

SOLUTION: In this case, it is a simple matter to write down all of the floating point numbers in the system:

$\pm .1 \times 10^{-1} = \pm .01$ $\pm .1 \times 10^{0} = \pm .1$ $\pm .1 \times 10^{1} = \pm 1$

$\pm .2 \times 10^{-1} = \pm .02$ $\pm .2 \times 10^{0} = \pm .2$ $\pm .2 \times 10^{1} = \pm 2$

$\vdots$ $\vdots$ $\vdots$

$\pm .9 \times 10^{-1} = \pm .09$ $\pm .9 \times 10^{0} = \pm .9$ $\pm .9 \times 10^{1} = \pm 9$.

Apart from these, there is only $0 = .0 \times 10^{-1}$. Of these 55 numbers, the nonnegative ones are pictured on the number line in Figure 5.1. We stress that the gaps between adjacent floating point numbers are not always the same; in general, these gaps become smaller near zero and more spread out far away from zero (larger numbers).

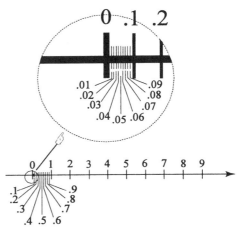

FIGURE 5.1: The nonnegative floating point numbers of Example 5.1. The omitted negative floating point numbers are just (in any floating point number system) the opposites of the positive numbers shown. The situation is typical in that as we approach zero, the density of floating point numbers increases.

Let us now talk about how real numbers get converted to floating point numbers. Any real number x can be expressed in the form

$$x = \pm .d_1 d_2 \cdots d_s d_{s+1} \cdots \times 10^{e} \tag{2}$$

where there are infinitely many digits (this is the only difference from the floating point representation (1) with $\beta = 10$) and there is no restriction on the exponent's range. The part $.d_1 d_2 \cdots d_s d_{s+1} \cdots$ is called the **mantissa** of x. If x has a finite decimal expansion, we can trail it with an infinite string of zeros to conform with (2). In fact, the representation (2) is unique for any real number (i.e., there is only one way to represent any x in this way) provided that we adopt the convention that an infinite string of 9's not be allowed; such expansions should just be rounded up. (For example, the real number .37999999999... is the same as .38.)

At first glance, it may seem straightforward how to represent a real number x in form (2) by a floating point number of form (1) with $\beta = 10$; simply either chop off or round off any of the digits past the allowed number. But there are serious problems that may arise, stemming from the fact that the exponent 'e' of the real number may be outside the permitted range. Firstly, if $e > M$, this means that x is (probably much) larger than any floating point number and so cannot be represented by one. If such a number were to come up in a computation, the computation is said to have **overflowed**. For example, in the simple setting of Example 5.1, any number $x \geq 10$ would overflow this simple floating point system. Depending on the computer system, overflows will usually result in termination of calculation or a warning. For example, most graphing calculators, when asked to evaluate an expression like e^{5000}, will either produce an error message like "OVERFLOW", and freeze up or perhaps give '∞' as an output. MATLAB behaves similarly to the latter pattern for overflows:

```
>> exp(5000)
→ans = Inf  %MATLAB's abbreviation for infinity.
```

'Inf' (or 'inf') is MATLAB's way of saying the number is too large to continue to do any more number crunching, except for calculations where the answer will be either 'Inf' or '−Inf' (a very large negative number). Here are some examples:

```
>> exp(5000)
→ans = Inf  % MATLAB tells us we have a very big positive number here
```

```
>> 2*exp(5000)
→ans = Inf  %No new information
```

```
>> exp(5000)/-5
→ans = -Inf  %OK now we have a very big negative number.
```

```
>> 2*exp(5000)-exp(5000)
→ans = NaN   % "NaN" stands for "not a number"
```

The last calculation is more interesting. Obviously, the expression being evaluated is just e^{5000}, which, when evaluated separately, is outputted as Inf. What happens is that MATLAB tries instead to do inf−inf, which is undefined (once large numbers are converted to inf, their relative sizes are lost and it is no longer possible for MATLAB to compare them).[4]

A very different situation occurs if the exponent e of the real number x in (2) is less than m (too small). This means that the real number x has absolute value (usually much) smaller than that of any nonzero floating point number. In this case, a computation is said to have **underflowed**. Most systems will represent an underflow by zero without any warning, and this is what MATLAB does.

---

[4] We mention that the optional "Symbolic Toolbox" for MATLAB allows, among other things, the symbolic manipulation of such expressions. The Symbolic Toolbox does come with the student version of MATLAB. Some of its features are explained in Appendix A.

Underflows, although less serious than overflows, can be a great source of problems in large-scale numerical calculations. Here is a simple example. We know from basic rules of exponents that $e^p e^{-p} = e^{p-p} = e^0 = 1$, but consider the following calculation:

```
>> exp(-5000)
→ans = 0  %this very small number has underflowed to zero

>> exp(5000)*exp(-5000)
→ans = NaN
```

The latter calculation had both underflows and overflows and resulted in 0*Inf ($= 0 \cdot \infty$), which is undefined. We will give another example shortly of some of the tragedies that can occur as the result of underflows. But now we show two simple ways to convert a real number to a floating point number in case there is no overflow or underflow. So we assume the real number x in (2) has exponent e satisfying $m \le e \le M$. The two methods for converting the real number x to its floating point representative $fl(x)$ are as follows:

(i) Chopped (or Truncated) Arithmetic: With this system we simply drop all digits after d_s :

$$fl(x) = fl(\pm.d_1 d_2 \cdots d_s d_{s+1} \cdots \times 10^e) = \pm.d_1 d_2 \cdots d_s \times 10^e .$$

(ii) Rounded Arithmetic: Here we do the usual rounding scheme for the first s significant digits. If $d_{s+1} < 5$ we simply chop as in method (i), but if $d_{s+1} \ge 5$, we need to round up. This may change several digits depending if there is a string of 9's or not. For example, with $s = 4$, ...2456823... would round to .2457 (one-digit changed), but .2999823 would round to .3000 (four digits changed). So a nice formula as in (i) is not possible. There is, however, an elegant way to describe rounded arithmetic in terms of chopped arithmetic using two steps.
Step 1: Add $5 \times 10^{-(s+1)}$ to the mantissa $.d_1 d_2 \cdots d_s d_{s+1} \cdots$ of x.
Step 2: Now chop as in (i) and retain the sign of x.

EXAMPLE 5.2: The following example parallels some calculations in exact arithmetic with the same calculations in 3-digit floating point arithmetic with $m = -8$ and $M = 8$. The reader is encouraged to go through both sets of calculations, using either MATLAB or a calculator. Note that at each point in a floating point calculation, the numbers need to be chopped accordingly before any math operations can be done.

Exact Arithmetic	Floating Point Arithmetic
$x = \sqrt{3}$	$fl(x) = 1.73 \ (\equiv .173 \times 10^1)$
$x^2 = 3$	$fl(x)^2 = 2.99$

Thus, in floating point arithmetic, we get that $\sqrt{3}^2 = 2.99$. This error is small but understandable.

Exact Arithmetic	Floating Point Arithmetic
$x = \sqrt{1000}$	$fl(x) = 31.6 \ (\equiv .316 \times 10^2)$
$x^2 = 1000$	$fl(x)^2 = 998$

The same calculation with larger numbers, of course, results in a larger error; but relatively it is not much different. A series of small errors can pile up and amount to more catastrophic results, as the next calculations show.

Exact Arithmetic	Floating Point Arithmetic
$x = 1000$	$fl(x) = 1000$
$y = 1/x = .001$	$fl(y) = .001$
$z = 1 + y = 1.001$	$fl(z) = 1$
$w = (z - 1) \cdot x^2$	$fl(w) = (1 - 1) \cdot 1000^2$
$\quad = y \cdot x^2$	$\quad = 0 \cdot 1000^2$
$\quad = \dfrac{1}{x} \cdot x^2$	$\quad = 0$
$\quad = x = 1000$	

The floating point answer of 0 is a ridiculous approximation to the exact answer of 1000!   The reason for this tragedy was the conversion of an underflow to zero. By themselves, such conversions are rather innocuous, but when coupled with a sequence of other operations, problematic results can sometimes occur.

When we do not make explicit mention of the exponent range $m \le e \le M$, we assume that the numbers that come up have their exponents in the appropriate range and so there will be no underflows or overflows.

EXERCISE FOR THE READER 5.1: Perform the following calculations in two-digit rounded arithmetic, and compare with the exact answers.

(a) $(.15)^2$

(b) $365,346 \times .4516$

(c) $8001 \div 123$

Our next example should alert the reader that one needs to be cautious on many different fronts when using floating point arithmetic. Many arithmetic rules that we have become accustomed to take for granted sometimes need to be paid careful attention when using floating point arithmetic.

EXAMPLE 5.3: Working in three-digit chopped floating point arithmetic with the exponent e restricted to the range $-8 \le e \le 8$, perform the following tasks:

(a) Compute the infinite series: $\sum_{n=1}^{\infty} \frac{1}{n^2} = 1 + \frac{1}{4} + \frac{1}{9} + \cdots$

(b) In each part below an equation is given and your task will be to decide how many solutions it will have in this floating point arithmetic. For each part you should give one of these four answers: **NO SOLUTION**, **EXACTLY ONE SOLUTION**, **BETWEEN 2 AND 10 SOLUTIONS**, or **MORE THAN 10 SOLUTIONS** (Work here only with real numbers; take all underflows as zero.)

(i) $3x = 5$

(ii) $x^3 = 0$

SOLUTION: Part (a): Unlike with exact arithmetic, when we sum this infinite series in floating point arithmetic, it is really going to be a finite summation since eventually the terms will be getting too small to have any effect on the accumulated sum. We use the notation $S_N = \sum_{n=1}^{N} \frac{1}{n^2} = 1 + \frac{1}{2^2} + \frac{1}{3^2} + \cdots + \frac{1}{N^2}$ for the partial sum (a finite sum). To find the infinite sum, we need to calculate (in order) in floating point arithmetic $S_1, S_2, S_3, \cdots$ and continue until these partial sums no longer change. Here are the step-by-step details:

$S_1 = 1$

$S_2 = S_1 + 1/4 = 1 + .25 = 1.25$

$S_3 = S_2 + 1/9 = 1.25 + .111 = 1.36$

$S_4 = S_3 + 1/16 = 1.36 + .0625 = 1.42$

$S_5 = S_4 + 1/25 = 1.42 + .040 = 1.46$

$S_6 = S_5 + 1/36 = 1.46 + .0277 = 1.48$

$S_7 = S_6 + 1/49 = 1.48 + .0204 = 1.50$

$S_8 = S_7 + 1/64 = 1.50 + .0156 = 1.51$

$S_9 = S_8 + 1/81 = 1.51 + .0123 = 1.52$

$S_{10} = S_9 + 1/100 = 1.52 + .010 = 1.53$

$S_{11} = S_{10} + 1/121 = 1.53 + .00826 = 1.53$

We can now stop this infinite process since the terms being added are small enough that when added to the existing partial sum 1.53, their contributions will just get chopped. Thus in the floating point arithmetic of this example, we have computed $\sum_{n=1}^{\infty} \frac{1}{n^2} = 1.53$, or more correctly we should write $\text{fl}\left(\sum_{n=1}^{\infty} \frac{1}{n^2}\right) = 1.53$.

Compare this result with the result from exact arithmetic $\sum_{n=1}^{\infty} \frac{1}{n^2} = \frac{\pi^2}{6} = 1.64\ldots$.

Thus in this calculation we were left with only one significant digit of accuracy!

Part (b): (i) The equation $3x = 5$ has, in exact arithmetic, only one solution, $x = 5/3 = 1.666....$ Let's look at the candidates for floating point arithmetic solutions that are in our system. This exact solution has floating point representative 1.66. Checking this in the equation (now working in floating point arithmetic) leads to: $3 \cdot 1.66 = 4.98 \neq 5$. So this will not be a floating point solution. Let's try making the number a bit bigger to 1.67 (this would be the smallest possible jump to the next floating point number in our system). We have (in floating point arithmetic) $3 \cdot 1.67 = 5.01 \neq 5$, so here $3x$ is too large. If these two numbers do not work, no other floating point numbers can (since for other floating point numbers $3x$ would be either less than or equal to 4.98 or greater than or equal to 5.01). Thus we have "NO SOLUTION" to this equation in floating point arithmetic![5]

(ii) As in (i), the equation $x^3 = 0$ has exactly one real number solution, namely $x = 0$. This solution is also a floating point solution. But there are many, many others. The lower bound on the exponent range $-8 \leq e$ is relevant here. Indeed, take any floating point number whose magnitude is less than 10^{-3}, for example, $x = .0006385$. Then $x^3 = (.0006385)^3 = 2.60305... \times 10^{-10} = .260305 \times 10^{-9}$ (in exact arithmetic). In floating point arithmetic, this computation would underflow and hence produce the result $x^3 = 0$. We conclude that in floating point arithmetic, this equation has "MORE THAN 10 SOLUTIONS" (see also Exercise 10 of this section).

EXERCISE FOR THE READER 5.2: Working two-digit rounded floating point arithmetic with the exponent e restricted to the range $-8 \leq e \leq 8$, perform the following tasks:

(a) Compute the infinite series: $\sum_{n=1}^{\infty} \frac{1}{n} = 1 + \frac{1}{2} + \frac{1}{3} + \cdots$

(b) In each part below an equation is given and your task will be to decide how many solutions it will have in this floating point arithmetic. For each part you should give one of these four answers: **NO SOLUTION, EXACTLY ONE SOLUTION, BETWEEN 2 AND 10 SOLUTIONS,** or **MORE THAN 10 SOLUTIONS**. (Work here only with real numbers; take all underflows as zero.)

(i) $x^2 = 100$

(ii) $8x^2 = x^5$

---

[5] We point out that when asked to (numerically) solve this equation in floating point arithmetic, we would simply use the usual (pure) mathematical method but work in floating point arithmetic, i.e., divide both sides by 3. The question of how many solutions there are in floating point arithmetic is a more academic one to help highlight the differences between exact and floating point arithmetic. Indeed, any time one uses a calculator or any floating point arithmetic software to solve any sort of mathematical problem with an exact mathematical method we should be mindful of the fact that the calculation will be done in floating point arithmetic.

EXERCISES 5.2:

NOTE: Unless otherwise specified, assume that all floating point arithmetic in these exercises is done in base 10.

1. In three-digit chopped floating point arithmetic, perform the following operations with these numbers: $a = 10000$, $b = .05$, and $c = 1/3$.
    (a) Write c as a floating point number, i.e., find $fl(c)$.
    (b) Find $a + b$.
    (c) Solve the equation $ax = c$ for x.

2. In three-digit rounded floating point arithmetic, perform the following tasks:
    (a) Find $1.23 + .456$ (b) Find $110,000 - 999$ (c) Find $(.055)^2$

3. In three-digit chopped floating point arithmetic, perform the following tasks:
    (a) Solve the equation $5x + 8 = 0$.
    (b) Use the quadratic formula to solve $1.12x^2 + 88x + 1 = 0$.
    (c) Compute $\displaystyle\sum_{n=1}^{\infty} \frac{1}{n^4} = \frac{1}{1^4} + \frac{1}{2^4} + \frac{1}{3^4} + \cdots$.

4. In three-digit rounded floating point arithmetic, perform the following tasks:
    (a) Solve the equation $5x + 4 = 17$.
    (b) Use the quadratic formula to solve $x^2 - 2.2x + 3 = 0$.
    (c) Compute $\displaystyle\sum_{n=1}^{\infty} \frac{(-1)^n \cdot 10}{n^4 + 2} = \frac{-10}{1^4 + 2} + \frac{10}{2^4 + 2} - \frac{10}{3^4 + 2} + \cdots$.

5. In each part below an equation is given and your task will be to decide how many solutions it will have in 3-digit chopped floating point arithmetic. For each part you should give one of these four answers: **NO SOLUTION**, **EXACTLY ONE SOLUTION**, **BETWEEN 2 AND 10 SOLUTIONS**, or **MORE THAN 10 SOLUTIONS**. (Work here only with real numbers with exponent e restricted to the range $-8 \le e \le 8$, and take all underflows as zero.)
    (a) $2x + 7 = 16$
    (b) $(x + 5)^2 (x + 1/3) = 0$
    (c) $2^x = 20$

6. Repeat the directions of Exercise 5, for the following equations, this time using 3-digit rounded floating point arithmetic with exponent e restricted to the range $-8 \le e \le 8$.
    (a) $2x + 7 = 16$ (b) $x^2 - x = 6$ (c) $\sin(x^2) = 0$

7. Using three-digit chopped floating point arithmetic (in base 10), do the following:
    (a) Compute the sum: $1 + 8 + 27 + 64 + 125 + 216 + 343 + 512 + 729 + 1000 + 1331$, then find the relative error of this floating point answer with the exact arithmetic answer.
    (b) Compute the sum in part (a) in the reverse order, and again find the relative answer of this floating point answer with the exact arithmetic answer.
    (c) If you got different answers in parts (a) and (b), can you explain the discrepancy?

8. Working in two-digit chopped floating point arithmetic, compute the infinite series $\displaystyle\sum_{n=1}^{\infty} \frac{1}{n}$.

9.   Working in two-digit rounded floating point arithmetic, compute the infinite series

$$\sum_{n=2}^{\infty} \frac{1}{n^{3/2} \ln n} \ .$$

10.   In the setting of Example 5.3(b)(ii), exactly how many floating point solutions are there for the equation $x^3 = 0$?

11.   (a) Write a MATLAB function M-file $z=rfloatadd(x,y,s)$, that has inputs x and y being any two real numbers, a positive integer s and the output z will be the sum $x+y$ using s-digit rounded floating point arithmetic. The integer s should not be more than 14 so as not to transcend MATLAB's default floating point accuracy.
      (b) Use this program (perhaps in conjunction with loops) to redo Exercise for the Reader 5.2, and Exercise 9.

12.   (a) Write a MATLAB function M-file $z=cfloatadd(x,y,s)$ that has inputs x and y being any two real numbers, a positive integer s, and the output z will be the sum $x+y$ using s-digit chopped floating point arithmetic. The integer s should not be more than 14 so as not to transcend MATLAB's default floating point accuracy.
      (b) Use this program (perhaps in conjunction with loops) to redo Example 5.3(a), and Exercise 7.

13.   (a) How many floating point numbers are there in the system with $\beta = 10$, $s = 2$, $m = -2$, $M = 2$? What is the smallest real number that would cause an overflow in this system?
      (b) How many floating point numbers are there in the system with $\beta = 10$, $s = 3$, $m = -3$, $M = 3$? What is the smallest real number that would cause an overflow in this system?
      (c) Find a formula that depends on s, m, and M that gives the number of floating point numbers in a general base 10 floating point number system ($\beta = 10$). What is the smallest real number that would cause an overflow in this system?

NOTE: In the same fashion as we had with base 10, for any base $\beta > 1$, any nonzero real number x can be expressed in the form:

$$x = \pm .d_1 d_2 \cdots d_s d_{s+1} \cdots \times \beta^e \ ,$$

where there are infinitely many digits $d_i = 0, 1, \cdots, \beta - 1$, and $d_1 \neq 0$. This notation means the following infinite series:

$$x = \pm (d_1 \times \beta^{-1} + d_2 \times \beta^{-2} + \cdots + d_s \times \beta^{-s} + d_{s+1} \times \beta^{-s-1} + \cdots) \times \beta^e$$

To represent any nonzero real number with its **base** β expansion, we first would determine the exponent e so that the inequality $1/\beta \le |x|/\beta^e < 1$ is valid. Next we construct the "digits" in order to be as large as possible so that the cumulative sum multiplied by β^e does not exceed $|x|$. As an example, we show here how to get the binary expansions ($\beta = 2$) of each of the numbers $x = 3$ and $x = 1/3$.   For $x = 3$, we get first the exponent $e = 2$, since $1/2 \le |3|/2^2 < 1$.   Since $(1 \times 2^{-1}) \times 2^2 = 2 < 3$, the first digit d_1 is 1 (in binary arithmetic, the digits can only be zeros or ones). The second digit d_2 is also 1 since the cumulative sum is now

$$(1 \times 2^{-1} + 1 \times 2^{-2}) \times 2^2 = 2 + 1 = 3 \ .$$

Since the cumulative sum has now reached $x = 3$, all remaining digits are zero, and we have the binary expansion of $x = 3$:

$$3 = .1100 \cdots 00 \cdots \times 2^2 \ .$$

Proceeding in the same fashion for $x = 1/3$, we first determine the exponent e to be -1 (since $1/3/2^{-1} = 2/3$ lies in $[1/2, 1)$). We then find the first digit $d_1 = 1$, and cumulative sum is $(1 \times 2^{-1}) \times 2^{-1} = 1/4 < 1/3$. Since $(1 \times 2^{-1} + 1 \times 2^{-2}) \times 2^{-1} = 3/8 > 1/3$, we see that the second digit $d_2 = 0$. Moving along, we get that $d_3 = 1$ and the cumulative sum is

$$(1 \times 2^{-1} + 0 \times 2^{-2} + 1 \times 2^{-3}) \times 2^{-1} = 5/16 < 1/3.$$

Continuing in this fashion, we will find that $d_4 = d_6 = \cdots = d_{2n} = 0$, and $d_5 = d_7 = \cdots = d_{2n+1} = 1$ and so we obtain the binary expansion:

$$1/3 = .101010 \cdots 1010 \cdots \times 2^{-1}.$$

If we require that there be no infinite string of ones (the construction process given above will guarantee this), then these expansions are unique. Exercises 14–19 deal with such representations in nondecimal bases ($\beta \neq 10$).

14. (a) Find the binary expansions of the following real numbers: $x = 1000$, $x = -2$, $x = 2.5$.
   (b) Find the binary expansions of the following real numbers: $x = 5/32$, $x = 2/3$, $x = 1/5$, $x = -0.3$, $x = 1/7$.
   (c) Find the exponent e and the first 5 digits of the binary expansion of π.
   (d) Find the real numbers with the following (terminating) binary expansions: $.1010 \cdots 00 \cdots \times 2^8$, $.1110 \cdots 00 \cdots \times 2^{-3}$.

15. (a) Use geometric series to verify the binary expansion of 1/3 that was obtained in the previous note.
   (b) Use geometric series to find the real numbers having the following binary expansions: $.10010101 \cdots \cdots \times 2^1$, $.11011011 \cdots \cdots \times 2^0$, $.1100011011011 \cdots \cdots \times 2^1$
   (c) What sort of real numbers will have binary expansions that either end in a sequence of zeros, or repeat, like the one for 1/3 obtained in the note preceding Exercise 14?

16. (a) Write down all floating point numbers in the system with $\beta = 2$, $s = 1$, $m = -1$, $M = 1$. What is the smallest real number that would cause an overflow in this system?
   (b) Write down all floating point numbers in the system with $\beta = 2$, $s = 2$, $m = -1$, $M = 1$. What is the smallest real number that would cause an overflow in this system?
   (c) Write down all floating point numbers in the system with $\beta = 3$, $s = 2$, $m = -1$, $M = 1$. What is the smallest real number that would cause an overflow in this system?

17. (a) How many floating point numbers are there in the system with $\beta = 2$, $s = 3$, $m = -2$, $M = 2$? What is the smallest real number that would cause an overflow in this system?
   (b) How many floating point numbers are there in the system with $\beta = 2$, $s = 2$, $m = -3$, $M = 3$? What is the smallest real number that would cause an overflow in this system?
   (c) Find a formula that depends on s, m, and M that gives the number of floating point numbers in a general binary floating point number system ($\beta = 2$). What is the smallest real number that would cause an overflow in this system?

18. Repeat each part of Exercise 17, this time using base $\beta = 3$.

19. Chopped arithmetic is defined in arbitrary bases exactly the same as was explained for decimal bases in the text. Real numbers must first be converted to their expansion in base β. For rounded floating point arithmetic using s-digits with base β, we simply add $\beta^{-s}/2$ to the mantissa and then chop. Perform the following floating point additions by first converting the numbers to floating point numbers in base $\beta = 2$, doing the operation in two-digit chopped

arithmetic, and then converting back to real numbers. Note that your real numbers may have more digits in them than the number s used in base 2 arithmetic, after conversion.
(a) $2 + 6$ (b) $22 + 7$ (c) $120 + 66$

5.3: FLOATING POINT ARITHMETIC: FURTHER EXAMPLES AND DETAILS

In order to facilitate further discussion on the differences in floating point and exact arithmetic, we introduce the following notation for operations in floating point arithmetic:

$$
\begin{aligned}
x \oplus y &\equiv \mathrm{fl}(x + y) \\
x \ominus y &\equiv \mathrm{fl}(x - y) \\
x \otimes y &\equiv \mathrm{fl}(x \cdot y) \\
x \oslash y &\equiv \mathrm{fl}(x \div y) ,
\end{aligned}
\tag{3}
$$

i.e., we put circles around the standard arithmetic operators to represent the corresponding floating point operations. To better illustrate concepts and subtleties of floating point arithmetic without getting into technicalities with different bases, we continue to work only in base $\beta = 10$.

  In general, as we have seen, floating point operations can lead to different answers than exact arithmetic operations. In order to track and predict such errors, we first look, in the next example, at the relative error introduced when a real number is approximated by its floating point number representative.

EXAMPLE 5.4: Show that in s-digit chopped floating point arithmetic, the unit roundoff u is 10^{1-s}, and that this number equals the distance from one to the next (larger) floating point number. We recall that the unit roundoff is defined to be the maximum relative error that can occur when a real number is approximated by a floating point number.

SOLUTION: Since $\mathrm{fl}(0)=0$, we may assume that $x \neq 0$.   Using the representations (1) and (2) for the floating point and exact numbers, we can estimate the relative error as follows:

$$
\begin{aligned}
\left| \frac{x - \mathrm{fl}(x)}{x} \right| &= \left| \frac{.d_1 d_2 \cdots d_s d_{s+1} \cdots \times 10^e - .d_1 d_2 \cdots d_s \times 10^e}{.d_1 d_2 \cdots d_s d_{s+1} \cdots \times 10^e} \right| \quad \begin{array}{l} (s)\ \text{slot} \\ (s+1)\ \text{slot} \end{array} \\
&= \left| \frac{.00 \cdots 0 d_{s+1} d_{s+2} \cdots \times 10^e}{.d_1 d_2 \cdots d_{s+1} d_{s+2} \cdots \times 10^e} \right| \leq \frac{.00 \cdots 099 \cdots}{.10 \cdots 000 \cdots} \\
&= \frac{.00 \cdots 100 \cdots}{.10 \cdots 000 \cdots} = \frac{10^{-s}}{10^{-1}} = 10^{1-s}
\end{aligned}
$$

Since equality can occur, this proves that the number on the right side is the unit roundoff. To see that this number u coincides with the gap between the floating point number 1 and the next (larger) floating point number on the right, we write the number 1 in the form (1):

$$1 = .10 \cdots 00 \times 10^1,$$

(note there are s digits total on the right, $d_1 = 1$, and $d_2 = d_3 = \cdots = d_s = 0$); we see that the next larger floating point number of this form will be:

$$1 + \text{gap} = .10 \cdots 01 \times 10^1$$

Subtracting gives us the unit roundoff:

$$\text{gap} = .00 \cdots 01 \times 10^1 = 10^{-s} \times 10^1 = 10^{1-s},$$

as was claimed.

EXERCISE FOR THE READER 5.3: (a) Show that in s-digit rounded floating point arithmetic the unit roundoff is $u = \frac{1}{2} 10^{1-s}$, but that the gap from 1 to the next floating point number is still 10^{1-s} .
(b) Show also that in any floating point arithmetic system and for any real number x, we can write

$$\text{fl}(x) = x(1+\delta), \quad \text{where } |\delta| \le u. \tag{4}$$

In relation to (4), we also assume that for any single floating point arithmetic operation: $x \circledcirc y$, with "$\circledcirc$" representing any of the floating point arithmetic operations from (3), we can write

$$x \circledcirc y = (x \circ y)(1+\delta), \quad \text{where } |\delta| \le u \tag{5}$$

and where "$\circ$" denotes the exact arithmetic operation corresponding to "$\circledcirc$" . This assumption turns out to be valid for IEEE (and hence, MATLAB's) arithmetic but for other computing environments may require that the bound on δ be replaced by a small multiple of u . We point out that IEEE standards require that $x \circledcirc y = \text{fl}(x \circ y)$.

In scientific computing, we will often need to do a large number of calculations and the resulting roundoff (or floating point) errors can accumulate. Before we can trust the outputs we get from a computer on an extensive computation, we have to have some confidence of its accuracy to the true answer. There are two major types of errors that can arise: roundoff errors and algorithmic errors. The first results from propagation of floating point errors, and the second arises from mathematical errors in the model used to approximate the true answer to a problem. To decrease mathematical errors, we will need to do more computations, but more computations will increase computer time and roundoff errors. This is a major dilemma of scientific computing! The best strategy and ultimate goal is to try to find efficient algorithms; this point will be applied and reemphasized frequently in the sequel.

To illustrate roundoff errors, we first look at the problem of numerically adding up a set of positive numbers. Our next example will illustrate the following general principle:

A General Principle of Floating Point Arithmetic: When numerically computing a large sum of positive numbers, it is best to start with the smallest number and add in increasing order of magnitude.

Roughly, the reason the principle is valid is that if we start adding the large numbers first, we could build up a rather large cumulative sum. Thus, when we get to adding to this sum some of the smaller numbers, there are much better chances that all or parts of these smaller numbers will have decimals beyond the number of significant digits supported and hence will be lost or corrupted.

EXAMPLE 5.5:   (a) In   exact   mathematics,   addition   is   associative: $(x+y)+z = x+(y+z)$. Show that in floating point arithmetic, addition is no longer associative.
(b) Show that for adding up a finite sum $S_N = a_1 + a_2 + \cdots + a_N$ of positive numbers (in the order shown), in floating point arithmetic, the error of the floating point answer $\mathrm{fl}(S_N)$ in approximating the exact answer S_N can be estimated as follows:

$$| \mathrm{fl}(S_N) - S_N | \le u[(N-1)a_1 + (N-1)a_2 + (N-2)a_3 + \cdots + 2a_{N-1} + a_N], \tag{6}$$

where u is the unit roundoff.

REMARK: Formula (6), although quite complicated, can be seen to demonstrate the above principle. Remember that u is an extremely small number so that the error on the right will normally be small as long as there are not an inordinate number of terms being added. In any case, the formula makes it clear that the relative contribution of the first term being added is the largest since the error estimate (right side of (6)) is a sum of terms corresponding to each a_i multiplied by the proportionality factor $(N-i)u$. Thus if we are adding $N = 1,000,001$ terms then this proportionality factor is $1,000,000u$ (worst) for a_1 but only u (best) for a_N, and these factors decrease linearly for intermediate terms. Thus it is clear that we should start adding the smaller terms first, and save the larger ones for the end.

SOLUTION: Part (a): We need to find (in some floating point arithmetic system) three numbers x, y and z such that $(x \oplus y) \oplus z \ne x \oplus (y \oplus z)$. Here is a simple example that also demonstrates the above principle: We use 2-digit chopped arithmetic with $x = 1$, $y = z = .05$. Then,

$$(x \oplus y) \oplus z = (1 \oplus .05) \oplus .05 = 1 \oplus .05 = 1,$$

but

$$x \oplus (y \oplus z) = 1 \oplus (.05 \oplus .05) = 1 \oplus .1 = 1.1.$$

This not only provides a counterexample, but since the latter computation (gotten by adding the smaller numbers first) gave the correct answer it also demonstrates the above principle.

Part (b): We continue to use the notation for partial sums that was employed in Example 5.3: i.e., $S_1 = a_1$, $S_2 = a_1 + a_2$, $S_3 = a_1 + a_2 + a_3$, etc. By using identity (5) repeatedly, we have:

$$\mathrm{fl}(S_2) = a_1 \oplus a_2 = (a_1 + a_2)(1 + \delta_2) = S_2 + (a_1 + a_2)\delta_2, \quad \text{where } |\delta_2| \leq u \text{, and so}$$

$$\begin{aligned} \mathrm{fl}(S_3) &= \mathrm{fl}(S_2) \oplus a_3 = (\mathrm{fl}(S_2) + a_3)(1 + \delta_3) \quad \text{where } |\delta_3| \leq u \\ &= (S_2 + (a_1 + a_2)\delta_2 + a_3)(1 + \delta_3) \\ &= S_3 + (a_1 + a_2)\delta_2 + (a_1 + a_2 + a_3)\delta_3 + (a_1 + a_2)\delta_2\delta_3 \\ &\approx S_3 + (a_1 + a_2)\delta_2 + (a_1 + a_2 + a_3)\delta_3. \end{aligned}$$

To get to the last estimate, we ignored the higher-order (last) term of the second-to-last expression. Continuing to estimate in the same fashion leads us to

$$\mathrm{fl}(S_N) \approx S_N + u \begin{cases} a_1(\delta_2 + \delta_3 + \delta_4 + \cdots + \delta_N) \\ +a_2(\delta_2 + \delta_3 + \delta_4 + \cdots + \delta_N) \\ +a_3(\quad\quad \delta_3 + \delta_4 + \cdots + \delta_N) \\ +a_4(\quad\quad\quad\quad \delta_4 + \cdots + \delta_N) \\ \vdots \\ +a_N(\quad\quad\quad\quad\quad\quad\quad \delta_N) \end{cases},$$

where each of the δ_i's arise from application of (5) and thus satisfy $|\delta_i| \leq u$. Bounding each of the $|\delta_i|$'s above with u and using the triangle inequality produces the asserted error bound in (6).

EXERCISE FOR THE READER 5.4: From estimate (6) deduce the following (cleaner but weaker) error estimates for the roundoff error in performing a finite summation of positive numbers $S_N = a_1 + a_2 + \cdots + a_N$ in floating point arithmetic:

(a) Error $= |\mathrm{fl}(S_N) - S_N| \leq Nu \sum_{n=1}^{N} a_n$.

(b) Relative error $= \left| \dfrac{\mathrm{fl}(S_N) - S_N}{S_N} \right| \leq Nu$.

Next, we give some specific examples that will compare these estimates against the actual roundoff errors. Recall from calculus that an (infinite) p-series

$$\sum_{n=1}^{\infty} \frac{1}{n^p} = 1 + \frac{1}{2^p} + \frac{1}{3^p} + \cdots \text{converges (i.e., adds up to a finite number) exactly}$$

when $p > 1$, otherwise it diverges (i.e., adds up to infinity). If we ask MATLAB (or any floating point computing system) to add up the terms in any p-series (or any series with positive terms that decrease to zero), eventually the terms will get too small to make any difference when they are added to the accumulated sum, so it will eventually appear that the series has converged (if the computer is given enough time to finish its task).[6] Thus, it is not possible to detect divergence of such a series by asking the computer to perform the summation. Once a series is determined to converge, however, it is possible to get MATLAB to help us estimate the actual sum of the infinite series. The key question in such a problem is to determine how many terms need to be summed in order for the partial sums to approximate the actual sum within the desired tolerance for error. We begin with an example for which the actual sum of the infinite series is known. This will allow us to verify if our accuracy goal is met.

EXAMPLE 5.6: Consider the infinite p-series $\sum_{n=1}^{\infty} \frac{1}{n^2} = 1 + \frac{1}{2^2} + \frac{1}{3^2} + \cdots$. Since $p = 2 > 1$, the series converges to a finite sum S.

(a) How many terms N of this infinite sum would we need to sum up to so that the corresponding partial sum $\sum_{n=1}^{N} \frac{1}{n^2} = 1 + \frac{1}{2^2} + \frac{1}{3^2} + \cdots + \frac{1}{N^2}$ is within an error of 10^{-7} of the actual sum S, i.e., Error $= |S - S_N| \le 10^{-7}$?

(b) Use MATLAB to perform this summation and compare the result with the exact answer $S = \pi^2/6$ to see if the error goal has been met. Discuss roundoff errors as well.

SOLUTION: Part (a): The mathematical error analysis needed here involves a nice geometric estimation method for the error that is usually taught in calculus courses under the name of the integral test. In estimating the infinite sum $S = \sum_{n=1}^{\infty} \frac{1}{n^2}$ with the finite partial sum $S_N = \sum_{n=1}^{N} \frac{1}{n^2}$, the error is simply the tail of the series:

$$\text{Error} = \left| \sum_{n=1}^{\infty} \frac{1}{n^2} - \sum_{n=1}^{N} \frac{1}{n^2} \right| = \sum_{n=N+1}^{\infty} \frac{1}{n^2} = \frac{1}{(N+1)^2} + \frac{1}{(N+2)^2} + \frac{1}{(N+3)^2} + \cdots$$

---

[6] We point out, however, that many symbolic calculus rules, and in particular, abilities to detect convergence of infinite series, are features available in MATLAB's Symbolic Toolbox (included in the Student Version). See Appendix A.

The problem is to find out how large N must be so that this error is $\leq 10^{-7}$; of course, as is usually the case, we have no way of figuring out this error exactly (if we could, then we could determine S exactly). But we can estimate this "Error" with something larger, let's call it ErrorCap, that we CAN compute. Each term in the "Error" is represented by an area of a shaded rectangle (with base = 1) in Figure 5.2. Since the totality of the shaded rectangles lies under the graph of $y = 1/x^2$, from $x = N$ to $x = \infty$, we have

$$\text{Error} < \text{Error Cap} \equiv \int_N^\infty \frac{dx}{x^2} = \frac{x^{-1}}{-1}\Bigg]_{x=N}^{x=\infty} = 1/N ,$$

and we conclude that our error will be less than 10^{-7}, provided that Error Cap $\leq 10^{-7}$, or $1/N \leq 10^{-7}$, or $N \geq 10^7$.

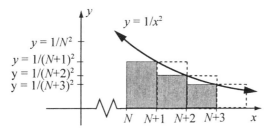

FIGURE 5.2: The areas of the shaded rectangles (that continue on indefinitely to the right) add up to precisely the Error = $\sum_{n=N+1}^\infty \frac{1}{n^2}$ of Example 5.6. But since they lie directly under the curve $y = 1/x^2$ from $x = N$ to $x = \infty$, we have Error < Error Cap $\equiv \int_N^\infty \frac{dx}{x^2}$.

Let us now use MATLAB to perform this summation, following the principle of adding up the numbers in order of increasing size.

```
>> Sum=0;   %initialize sum
>> for n=10000000:-1:1
Sum=Sum +1/n^2;
end
>> Sum    →Sum = 1.64493396684823
```

This took about 30 seconds on the author's PC. Since we know that the exact infinite sum is $\pi^2/6$, we can look now at the actual error.

```
>> pi^2/6-Sum
→ans = 9.999999495136080e-008   %This is indeed (just a wee bit) less than the desired
```
tolerance 10^{-7}.

Let us look briefly at (6) (with $a_n = 1/n^2$) to see what kind of a bound on roundoff errors it gives for this computation. At each step, we are adding to the accumulated sum a quotient. Each of these quotients ($1/n^2$) gives a round off error (by (5)) at most $a_n u$. Combining this estimate with (6) gives the bound

$$|\,\mathrm{fl}(S_N) - S_N\,| \le 2u \left\{ \begin{bmatrix} \dfrac{9999999}{(10000000)^2} \end{bmatrix} + \begin{bmatrix} \dfrac{9999999}{(9999999)^2} \end{bmatrix} + \begin{bmatrix} \dfrac{9999998}{(9999998)^2} \end{bmatrix} + \\ \cdots + \begin{bmatrix} \dfrac{2}{(2)^2} \end{bmatrix} + \begin{bmatrix} \dfrac{1}{(1)^2} \end{bmatrix} \end{bmatrix} \right\}$$

$$\le 2u \left\{ \dfrac{1}{10000000} + \dfrac{1}{9999999} + \dfrac{1}{9999998} + \cdots + \dfrac{1}{2} + 1 \right\}.$$

The sum in braces is $\displaystyle\sum_{n=1}^{10^7} \dfrac{1}{n}$. By a picture similar to the one of Figure 5.2, we can estimate this sum as follows: $\displaystyle\sum_{n=1}^{10^7} \dfrac{1}{n} \le 1 + \int_1^N \ln(x)\,dx = 1 + \ln(N)$. Since the unit roundoff for MATLAB is 2^{-53}, we can conclude the following upper bound on the roundoff error of the preceding computation:

$$|\,\mathrm{fl}(S_N) - S_N\,| \le 2u(1 + \ln(10^7)) \approx 3.8 \times 10^{-15}.$$

We thus have confirmation that roundoff error did not play a large role in this computation. Let's now see what happens if we perform the same summation in the opposite order.

```
>> Sum=0;
>> for n=1:10000000
Sum=Sum +1/n^2;
end
>> pi^2/6-Sum          →ans = 1.000009668405966e-007
```

Now the actual error is a bit worse than the desired tolerance of 10^{-7}. The error estimate (6) would also be a lot larger if we reversed the order of summation. Actually for both of these floating point sums the roundoff errors were not very significant; such problems are called *well-conditioned*. The reason for this is that the numbers being added were getting small very fast. In the exercises and later in this book, we will encounter problems that are *ill-conditioned*, in that the roundoff errors can get quite large relative to the amount of arithmetic being done. The main difficulty in the last example was not roundoff error, but the computing time. If we instead wanted our accuracy to be 10^{-10}, the corresponding calculation would take over 8 hours on the same computer! A careful look at the strategy we used, however, will allow us to modify it slightly to get a much more efficient method.

A Better Approach: Referring again to Figure 5.2, we see that by sliding all of the shaded rectangles one unit to the right, the resulting set of rectangles will completely cover the region under the graph of $y = 1/x^2$ from $x = N + 1$ to $x = \infty$. This gives the inequality:

$$\text{Error} > \int_{N+1}^{\infty} \frac{dx}{x^2} = \frac{x^{-1}}{-1}\Bigg]_{x=N}^{x=\infty} = 1/(N+1).$$

In conjunction with the previous inequality, we have in summary that:

$$\frac{1}{N+1} < \text{Error} < \frac{1}{N}.$$

If we add to our approximation S_N the average value of these two upper and lower bounds, we will obtain the following much better approximation for S:

$$\tilde{S}_N \equiv S_N + \frac{1}{2}\left[\frac{1}{N} + \frac{1}{N+1}\right].$$

The new error will be at most one-half the length of the interval from $1/(N+1)$ to $1/N$:

$$|S - \tilde{S}_N| \equiv \text{New Error} \le \frac{1}{2}\left[\frac{1}{N} - \frac{1}{N+1}\right] = \frac{1}{2N(N+1)} < \frac{1}{2N^2}.$$

(The elementary last inequality was written so as to make the new error bound easier to use, as we will now see.) With this new scheme much less work will be required to attain the same degree of accuracy. Indeed, if we wanted the error to be less than 10^{-7}, this new scheme would require that the number of terms N needed to sum should satisfy $1/2N^2 \le 10^{-7}$ or $N \ge \sqrt{10^7/2} = 2236.07...$, a far cry less than the 10 million terms needed with the original method! By the same token, to get an error less than 10^{-10}, we would need only take $N = 70,711$! Let us now verify, on MATLAB, this 10-significant-digit approximation:

```
>>   Sum=0; N=70711;
>> for n=N:-1:1
Sum=Sum +1/n^2;
end
>> format long
>> Sum=Sum+(1/N + 1/(N+1))/2       →Sum = 1.64493406684823
>> abs(Sum-pi^2/6)       →ans =8.881784197001252e-016
```

The actual error here is even better than expected; our approximation is actually as good as machine precision! This is a good example where the (worst-case) error guarantee (New Error) is actually a lot larger than the true error. A careful examination of Figure 5.2 once again should help to make this plausible.

We close this section with an exercise for the reader that deals with the approximation of an alternating series. This one is rather special in that it can be used to approximate the number π and, in particular, we will be able to check the accuracy of our numerical calculations against the theory. We recall from calculus that an alternating series is an infinite series of the form $\sum (-1)^n a_n$, where $a_n > 0$ for each n. Leibniz's Theorem states that if the a_n's decrease, $a_n \geq a_{n+1}$ for each n (sufficiently large), and converge to zero, $a_n \to 0$ as $n \to \infty$, then the infinite series converges to a sum S. It furthermore states that if $S_N = \sum^{N} (-1)^n a_n$ denotes a partial sum (the lower index is left out since it can be any integer), then we have the error estimate: $|S - S_N| \leq a_{N+1}$.

EXERCISE FOR THE READER 5.5: Use the infinite series expansion:

$$ 1 - \frac{1}{3} + \frac{1}{5} - \frac{1}{7} + \cdots = \frac{\pi}{4} , $$

to estimate π with an error less than 10^{-7} .

The series in the above exercise required the summation of a very large number of terms to get the required degree of accuracy. Such series are said to converge "slowly." Exercise 12 will deal with a better (faster-converging) series to use for approximating π .

EXERCISES 5.3:

NOTE: Unless otherwise specified, assume that all floating point arithmetic in these exercises is done in base 10.

1. (a) In chopped floating point arithmetic with s digits and exponent range $m \leq e \leq M$, write down (in terms of these parameters s, m, and M) the largest positive floating point number and the smallest positive floating point number.
    (b) Would these answers change if we were to use rounded arithmetic instead?

2. Recall that for two real numbers x and y, the average value $(x + y)/2$ of x and y lies between the values of x and y .
    (a) Working in a chopped floating point arithmetic system, find an example where $fl((x+y)/2)$ is strictly less than $fl(x)$ and $fl(y)$.
    (b) Repeat part (a) in rounded arithmetic.

3. (a) In chopped floating point arithmetic with base β, with s digits and exponent range $m \leq e \leq M$, write down (in terms of these parameters β, s, m, and M) the largest positive floating point number and the smallest positive floating point number.
    (b) Would these answers change if we were to use rounded arithmetic instead?

4. For each of the following arithmetic properties, either explain why the analog will be true in floating point arithmetic, or give an example where it fails. If possible, provide counterexamples that do not involve overflows, but take underflows to be zero.

(a)   (*Commutativity of Addition*) $x + y = y + x$

(b)   (*Commutativity of Multiplication*) $x \cdot y = y \cdot x$

(c)   (*Associativity of Addition*) $x \cdot (y \cdot z) = (x \cdot y) \cdot z$

(d)   (*Distributive Law*) $x \cdot (y + z) = x \cdot y + x \cdot z$

(e)   (*Zero Divisors*) $x \cdot y = 0 \Rightarrow x = 0 \text{ or } y = 0$

5.   Consider the infinite series: $1 + \dfrac{1}{8} + \dfrac{1}{27} + \dfrac{1}{64} + \cdots \dfrac{1}{n^3} + \cdots$

    (a)   Does it converge? If it does not, stop here; otherwise continue.

    (b)   How many terms would we have to sum to get an approximation to the infinite sum with an absolute error $< 10^{-7}$?

    (c)   Obtain such an approximation.

    (d)   Use an approach similar to what was done after Example 5.6; add an extra term to the partial sums so as to obtain an improved approximation for the infinite series. How many terms would be required with this improved scheme? Perform this approximation and compare the answer with that obtained in part (c).

6.   Consider the infinite series: $\displaystyle\sum_{n=1}^{\infty} \dfrac{1}{n\sqrt{n}}$

    (a) Does it converge? If it does not, stop here, otherwise continue.

    (b) How many terms would we have to sum to get an approximation to the infinite sum with an absolute error $1/500$?

    (c) Obtain such an approximation.

    (d) Use an approach similar to what was done after Example 5.6; add an extra term to the partial sums so as to obtain an improved approximation for the infinite series. How many terms would be required with this improved scheme? Perform this approximation and compare the answer with that obtained in part (c).

7.   Consider the infinite series: $\displaystyle\sum_{n=1}^{\infty} \dfrac{(-1)^{n+1}}{n} = 1 - \dfrac{1}{2} + \dfrac{1}{3} - \dfrac{1}{4} + \cdots$

    (a)   Show that this series satisfies the hypothesis of Leibniz's theorem (for n sufficiently large) so that from the theorem, we know the series will converge to a sum S.

    (b)   Use Leibniz's theorem to find an integer N so that summing up to the first N terms only will give approximation to the sum with an error less than .0001.

    (c)   Obtain such an approximation.

8.   Repeat all parts of Exercise 7 for the series: $\displaystyle\sum_{n=1}^{\infty} (-1)^n \dfrac{\ln n}{n}$.

9.   (a) In an analogous fashion to what was done in Example 5.5, establish the following estimate for floating point multiplications of a set of N positive real numbers, $P_N = a_1 \cdot a_2 \cdots a_N$:

$$|\mathrm{fl}(P_N) - P_N| \le Nu. \qquad\qquad (7)$$

We have, as before, ignored higher-order error terms. Thus, as far as minimizing errors is concerned, unlike for addition, the roundoff errors do not depend substantially on the order of multiplication.

(b) In forming a product of positive real numbers, is there a good order to multiply so as to minimize the chance of encountering overflows or underflows? Explain your answer with some examples.

10.   (a) Using an argument similar to that employed in Example 5.4, show that in base β chopped

floating point arithmetic the unit roundoff is given by $u = \beta^{1-s}$.

(b) Show that in rounded arithmetic, the unit roundoff is given by $u = \beta^{1-s}/2$.

11.   Compare and contrast a two-digit ($s = 2$) floating point number system in base $\beta = 4$ and a four-digit ($s = 4$) binary ($\beta = 2$) floating point number system.

12.   In Exercise for the Reader 5.5, we used the following infinite series to approximate π :

$$\frac{\pi}{4} = 1 - \frac{1}{3} + \frac{1}{5} - \frac{1}{7} \cdots \;\Rightarrow\; \pi = 4 - \frac{4}{3} + \frac{4}{5} - \frac{4}{7} \cdots.$$

This alternating series was not a very efficient way to compute π, since it converges very slowly. Even to get an approximation with an accuracy of 10^{-7}, we would need to sum about 20 million terms. In this problem we will give a much more efficient (faster-converging) series for computing π by using Machin's identity ((13) of Chapter 2):

$$\frac{\pi}{4} = 4\tan^{-1}\!\left(\frac{1}{5}\right) - \tan^{-1}\!\left(\frac{1}{239}\right).$$

(a) Use this identity along with the arctangent's MacLaurin series (see equation (11) of Chapter 2) to express π either as a difference of two alternating series, or as a single alternating series. Write your series both in explicit notation (as above) and in sigma notation.

(b) Perform an error analysis to see how many terms you would need to sum in the series (or difference of series) to get an approximation to π with error $< 10^{-7}$. Get MATLAB to perform this summation (in a "good" order) to thus obtain an approximation to π .

(c) How many terms would we need to sum so that the (exact mathematical) error would be less than 10^{-30} ? Of course, MATLAB only uses 16-digit floating point arithmetic so we could not directly use it to get such an approximation to π (nnless we used the symbolic toolbox; see Appendix A).

13.   Here is π, accurate to 30 decimal places:

$$\pi = 3.141592653589793238462643383279\ldots$$

Can you figure out a way to get MATLAB to compute π to 30 decimals of accuracy without using its symbolic capabilities?
Suggestions: What is required here is to build some MATLAB functions that will perform certain mathematical operations with more significant digits (over 30) than what MATLAB usually guarantees (about 15). In order to use the series of Exercise 12, you will need to build functions that will add/subtract, and multiply and divide (at least). Here is an example of a possible syntax for such new function: `z=highaccuracyadd(x,y)` where x and y are vectors containing the mantissa (with over 30 digits) as well as the exponent and sign (can be stored as 1 for plus, 2 for negative in one of the slots). The output z will be another vector of the same size that represents a 30 + significant digit approximation to the sum $x + y$. This is actually quite a difficult problem, but it is fun to try.

Chapter 6: Rootfinding

6.1: A BRIEF ACCOUNT OF THE HISTORY OF ROOTFINDING

The mathematical problems and applications of solving equations can be found in the oldest mathematical documents that are known to exist. The *Rhind Mathematical Papyrus*, named after Scotsman A. H. Rhind (1833–1863), who purchased it in a Nile resort town in 1858, was copied in 1650 B.C. from an original that was about 200 years older. It is about 13 inches high and 18 feet long, and it currently rests in a museum in England. This Papyrus contains 84 problems and solutions, many of them linear equations of the form (in modern algebra notation): $ax + b = 0$. It is fortunate that the mild Egyptian climate has so well preserved this document. A typical problem in this papyrus runs as follows: "A heap and its 1/7 part become 19. What is the heap?" In modern notation, this problem amounts to solving the equation $x + (1/7)x = 19$, and is easily solved by basic algebra. The arithmetic during these times was not very well developed and algebra was not yet discovered, so the Egyptians solved this equation with an intricate procedure where they made an initial guess, corrected it and used some complicated arithmetic to arrive at the answer of 16 5/8. The exact origin and dates of the methods in this work are not well documented and it is even possible that many of these methods may have been handed down by Imhotep, who supervised the construction of the pyramids around 3000 B.C.

Algebra derives from the Latin translation of the Arabic word, *al-jabr*, which means "restoring" as it refers to manipulating equations by performing the same operation on both sides. One of the earliest known algebra texts was written by the Islamic mathematician Al-Khwarizmi (c. 780–850), and in this book the quadratic equation $ax^2 + bx + c = 0$ is solved. The Islamic mathematicians did not deal with negative numbers so they had to separate the equation into 6 cases. This important work was translated into Latin, which was the language of scholars and universities in all of the western world during this era. After algebra came into common use in the western world, mathematicians set their sights on solving the next natural equation to look at: the general **cubic** equation $ax^3 + bx^2 + cx + d = 0$. The solution came quite a bit later in the Renaissance era in 16th century and the history after this point gets quite interesting.

The Italian mathematician Niccolo Fontana (better known by his nickname Tartaglia; see Figure 6.1) was the first to find the solution of the general cubic equation. It is quite a complicated formula and this is why it is rarely seen in textbooks.

FIGURE 6.1:
Niccolo Fontana ("Tartaglia") (1491–1557), Italian mathematician.

A few years later, Tartaglia's contemporary, Girolamo Cardano[1] (sometimes the English translation "Cardan" is used; see Figure 6.2), had obtained an even more complicated formula (involving radicals of the coefficients) for the solution of the general **quartic** equation $ax^4 + bx^3 + cx^2 + dx + e = 0$. With each extra degree of the polynomial equations solved thus far, the general solution was getting inordinately more complicated, and it became apparent that a general formula for the solution of an nth-degree polynomial equation would be very unlikely and that the best mathematics could hope for was to keep working at obtaining general solutions to higher-order polynomials at one-degree increments.

FIGURE 6.2:
Girolamo Cardano (1501–1576), Italian mathematician.

Three centuries later in 1821, a brilliant, young yet short-lived Norwegian mathematician named Niels Henrik Abel (Figure 6.3) believed he had solved the general **quintic** (5th-degree polynomial) equation, and submitted his work to the Royal Society of Copenhagen for publication. The editor contacted Abel to ask for a numerical example. In his efforts to construct examples, Abel found that his method was flawed, but in doing so he was able to prove that no formula could possibly exist (in terms of radicals and algebraic combinations of the coefficients) for the solution of a general quintic. Such nonexistence results are very deep and this one had ended generations of efforts in this area of rootfinding.[2]

---

[1] When Tartaglia was only 12 years old, he was nearly killed by French soldiers invading his town. He suffered a massive sword cut to his jaw and palate and was left for dead. He managed to survive, but he always wore a beard to hide the disfiguring scar left by his attackers; also his speech was impaired by the sword injury and he developed a very noticeable stutter. (His nickname Tartaglia means stammerer.) He taught mathematics in Venice and became famous in 1535 when he demonstrated publicly his ability of solving cubics, although he did not release his "secret formula". Other Italian mathematicians had publicly stated that such a solution was impossible. The more famous Cardano, located in Milan, was intrigued by Tartaglia's discovery and tried to get the latter to share it with him. At first Tartaglia refused but after Cardano tempted Tartaglia with his connections to the governor, Tartalgia finally acquiesced, but he made Cardano promise never to reveal the formula to anyone and never to even write it down, except in code (so no one could find it after he died). Tartaglia presented his solution to Cardano as a poem, again, so there would be no written record. With his newly acquired knowledge, Cardano was eventually able to solve the general quartic.

[2] Abel lived during a very difficult era in Norway and despite his mathematical wizardry, he was never able to obtain a permanent mathematics professorship. His short life was marked with constant poverty. When he proved his impossibility result for the quintic, he published it immediately on his own but in order to save printing costs, he trimmed down his proof to very bare details and as such it was difficult to read and did not give him the recognition that he was due. He later became close friends with the eminent German mathematician and publisher Leopold Crelle who recognized Abel's genius and published much of his work. Crelle had even found a suitable professorship for Abel in

FIGURE 6.3: Niels Henrik Abel (1802 -1829), Norwegian mathematician.

FIGURE 6.4: Evariste Galois[3] (1811–1832), French mathematician.

At roughly the same time, across the continent in France, another young mathematician, Evariste Galois (Figure 6.4), had worked on the same problems. Galois' life was also tragically cut short, and his brilliant and deep mathematical achievements were not recognized or even published until after his death. Galois invented a whole new area in mathematical group theory and he was able to use his development to show the impossibility of having a general formula (involving radicals and the four basic mathematical operations) for solving the general polynomial equation of degree 5 or more and, furthermore, he obtained results that developed special conditions on polynomial equations under which such formulas could exist.

The work of Abel and Galois had a considerable impact on the development of mathematics and consequences and applications of their theories are still being realized today in the 21st century. Many consequences have evolved from their theories, some resolving the impossibility of several geometric constructions that the Greeks had worked hard at for many years.

Among the first notable consequences of the nonexistence results of Abel and Galois was that pure mathematics would no longer be adequate as a reliable means for rootfinding, and the need for numerical methods became manifest. In the sections that follow we will introduce some **iterative** methods for finding a root of an equation $f(x) = 0$ that is known to exist. In each, a sequence of approximations x_n is constructed that, under appropriate hypotheses, will

---

Berlin, but the good news came too late; Abel had died from tuberculosis shortly before Crelle's letter arrived.

[3] Galois was born to a politically active family in a time of much political unrest in France. His father was mayor in a city near Paris who had committed suicide in 1829 after a local priest had forged the former's name on some libelous public documents. This loss affected Galois considerably and he became quite a political activist. His mathematics teachers from high school through university were astounded by his talent, but he was expelled from his university for publicly criticizing the director of the university for having locked the students inside to prevent them from joining some political riots. He joined a conservative National Guard that was accused of plotting to overthrow the government. His political activities caused him to get sent to prison twice, the second term for a period of six months. While in prison, he apparently fell in love with Stephanie-Felice du Motel, a prison official's daughter. Soon after he got out from prison he was challenged to a duel, the object of which had to do with Stephanie. In this duel he perished. The night before the duel, he wrote out all of his main mathematical discoveries and passed them on to a friend. It was only after this work was posthumously published that Galois's deep achievements were discovered. There is some speculation that Galois's fatal duel was set up to remove him from the political landscape.

"converge" to an actual root r. Convergence here simply means that the error $|r - x_n|$ goes to zero as n gets large. The speed of convergence will depend on the particular method being used and possibly also on certain properties of the function $f(x)$.

6.2: THE BISECTION METHOD

This method, illustrated in Figure 6.5, is very easy to understand and write a code for; it has the following basic assumptions:

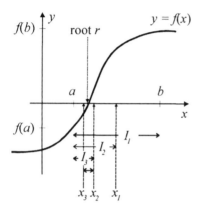

ASSUMPTIONS: $f(x)$ is continuous on $[a,b]$, $f(a), f(b)$ have opposite signs, and we are given an error tolerance = tol > 0.

Because of the assumptions, the intermediate value theorem from calculus tells us that $f(x)$ has at least one **root** (meaning a solution of the equation $f(x) = 0$) within the interval $[a,b]$. The method will iteratively construct a sequence x_n that converges to a root r and will stop when it can be guaranteed that $|r - x_n| < $ tol.

FIGURE 6.5: Illustration of the bisection method. The points x_n are the midpoints of the intervals I_n that get halved in length at each iteration.

The philosophy of the method can be paraphrased as "divide and conquer." In English, the algorithm works as follows:

We start with $x_1 = (a+b)/2$ being the midpoint of the interval $[a,b] \equiv [a_1,b_1] \equiv I_1$. We test $f(x_1)$. If it equals zero, we stop since we have found the exact root. If $f(x_1)$ is not zero, then it will have opposite signs either with $f(a)$ or with $f(b)$. In the former case we next look at the interval $[a,x_1] \equiv [a_2,b_2] \equiv I_2$ that now must contain a root of $f(x)$; in the latter case a root will be in $[x_1,b] \equiv [a_2,b_2] \equiv I_2$. The new interval I_2 has length equal to half of that of the original interval. Our next approximation is the new midpoint $x_2 = (a_2 + b_2)/2$. As before, either x_2 will be an exact root or we continue to approximate a root in I_3 that will be in either the left half or right half of I_2. Note that at each iteration, the approximation x_n lies in the interval I_{n+1}, which also contains an actual root. From this it follows that:

$$\text{error} = |x_n - r| \leq \text{length}(I_{n+1}) = \frac{\text{length}(I_1)}{2^n} = \frac{b-a}{2^n}. \tag{1}$$

We wish to write a MATLAB M-file that will perform this bisection method for us. We will call our function bisect('function', a, b, tol). This one has four input variables. The first one is an actual mathematical function (with the generic name: function) for which a root is sought. The second two variables, a and b, denote the endpoints of an interval at which the function has opposite signs, and the last variable tol denotes the maximum tolerated error. The program should cause the iterations to stop after the error gets below this tolerance (the estimate (1) will be useful here). Before attempting to write the MATLAB M-file, it is always recommended that we work some simple examples by hand. The results will later be used to check our program after we write it.

EXAMPLE 6.1: Consider the function $f(x) = x^5 - 9x^2 - x + 7$.

(a) Show that $f(x)$ has a root on the interval $[1,2]$.

(b) Use the bisection method to approximate this root with an error < 0.01.

(c) How many iterations would we need to use in the bisection method to guarantee an error < 0.00001?

SOLUTION: Part (a): Since $f(x)$ is a polynomial, it is continuous everywhere. Since $f(1) = 1 - 9 - 1 + 7 = -2 < 0$, and $f(2) = 32 - 36 - 2 + 7 = 1 > 0$, it follows from the intermediate value theorem that $f(x)$ must have a root in the interval $[1, 2]$.

Part (b): Using (1), we can determine the number of iterations required to achieve an error guaranteed to be less than the desired upper bound $0.01 = 1/100$. Since $b - a = 1$, the right side of (1) is $1/2^n$, and clearly $n = 7$ is the first value of n for which this is less than $1/100$. ($1/2^7 = 1/128$.) By the error estimate (1), this means we will need to do 7 iterations. At each step, we will need to evaluate the function $f(x)$ at the new approximation value x_n. If we computed these iterations directly on MATLAB, we would need to enter the formula only once, and make use of MATLAB's editing features. Another way to deal with functions on MATLAB would be to simply store the mathematical function as an M-file. But this latter approach is not so suitable for situations like this where the function gets used only for a particular example. We now show a way to enter a function temporarily into a MATLAB session as a so-called "inline" function:

<fun_name>=inline('<math expression>') →	Causes a mathematical function to be defined (temporarily, only for the current MATLAB session), the name will be <fun_nam>, the formula will be given by <math expression> and the input variables determined by MATLAB when it scans the expression.

<fun_name>=inline('<math expression>', 'x1', 'x2', ..., 'xn') →	Works as above but specifies input variables to be x1, x2, ..., xn in the same order.

We enter our function now as an inline function giving it the convenient and generic name "f."

```
>> f=inline('x^5-9*x^2-x+7')     →f = Inline function:   f(x) = x^5-9*x^2-x+7
```

We may now work with this function in MATLAB just as with other built-in mathematical functions or stored M-file mathematical functions. Its definition will be good for as long as we are in the MATLAB session in which we created this inline function. For example, to evaluate $f(2)$ we can now just type:

```
>> f(2)          → 1
```

For future reference, in writing programs containing mathematical functions as variables, it is better to use the following equivalent (but slightly longer) method:

```
>> feval(f,2)    →1
```

feval(<funct>, a1,a2,...,an) →	returns the value of the stored or inline function funct(x1,x2, ..., xn) of n variables at the values x1=a1, x2=a2, ... , xn=an

Let's now use MATLAB to perform the bisection method:

```
>> a1=1; b1=2; x1=(a1+b1)/2, f(x1) %    [a1,b1]=[a,b] and x1 (first
>>%  approximation) is the midpoint.  We need to test f(x1).
→x1 = 1.5000 (=first approximation), ans = -7.1563 (value of function at first approximation)

>> a2=x1; b2=b1; x2=(a2+b2)/2, f(x2) %the bisected interval [a2,b2]
>> % is always chosen to be the one where function changes sign.
→x2 =1.7500, ans =-5.8994 (n=2 approximation and value of function)

>> a3=x2; b3=b1; x3=(a3+b3)/2, f(x3)      →x3 =1.8750, ans = -3.3413

>> a4=x3; b4=b3; x4=(a4+b4)/2, f(x4)      →x4 = 1.9375, ans =-1.4198

>> a5=x4; b5=b4; x5=(a5+b5)/2, f(x5)      →x5 = 1.9688, ans = -0.2756

>> a6=x5; b6=b5; x6=(a6+b6)/2, f(x6)
→x6 =1.9844, ans =0.3453 (n=6 approximation and corresponding y-coordinate)

>>a7=a6; b7=x6; x7=(a7+b7)/2, f(x7)       →x7 =1.9766, ans =0.0307
```

The above computations certainly beg to be automated by a loop and this will be done soon in a program.

Part (c): Let's use a MATLAB loop to find out how many iterations are required to guarantee an error < 0.00001

```
>> n=1; while 1/2^(n)>=0.00001
```

```
n=n+1;
end
>> n    →n =17
```

EXERCISE FOR THE READER 6.1:   In the example above we found an approximation (x7) to a root of $f(x)$ that was accurate with an error less than 0.01. For the actual root $x = r$, we of course have $f(r) = 0$, but $f(x7) = 0.0307$. Thus the error of the y-coordinate is over three times as great as that for the x-coordinate.   Use calculus to explain this discrepancy.

EXERCISE FOR THE READER 6.2:   Consider the function
$$f(x) = \cos(x) - x .$$
(a) Show that $f(x)$ has a exactly one root on the interval $[0, \pi/2]$.
(b) Use the bisection method to approximate this root with an error < 0.01.
(c) How many iterations would we need to use in the bisection method to guarantee an error $< 10^{-12}$?

With the experience of the last example behind us, we should now be ready to write our program for the bisection method.   In it we will make use of the following built-in MATLAB functions.

sign(x) →	=(the sign of the real number x)= $\begin{cases} 1, & \text{if } x > 0 \\ 0, & \text{if } x = 0 \\ -1, & \text{if } x < 0 \end{cases}$

Recall that with built-in functions such as quad, some of the input variables were made optional with default values being used if the variables are not specified in calling the function.   In order to build such a feature into a function, the following command is useful in writing such an M-file:

nargin (inside the body of a funtion M-file)→	Gives the number of input arguments (that are specified when a function is called).

PROGRAM 6.1: An M-file for the bisection method.

```
function [root, yval] = bisect(varfun, a, b, tol)
% input variables: varfun, a, b, tol
% output variables:  root, yval
% varfun = the string representing a mathematical function (built-in,
% M-file, or inline) of one variable that is assumed to have opposite
% signs at the points x=a,  and x=b.  The program will perform the
% bisection method to approximate a root of varfun in [a,b] with an
% error < tol.  If the tol variable is omitted a default value of
% eps*max(abs(a),abs(b),1) is used.

%we first check to see if there is the needed sign change
ya=feval(varfun,a); yb=feval(varfun,b);
if sign(ya)==sign(yb)
```

```
     error('function has same sign at endpoints')
end

%we assign the default tolerance, if none is specified
if nargin < 4
   tol=eps*max([abs(a) abs(b) 1]);
end

%we now initialize the iteration
an=a; bn=b; n=0;

%finally we set up a loop to perform the bisections
while (b-a)/2^n >= tol
   xn=(an + bn)/2; yn=feval(varfun, xn); n=n+1
   if yn==0
      fprintf('numerically exact root')
      root=xn; yval=yn;
      return
   elseif sign(yn)==sign(ya)
      an=xn; ya=yn;
   else
      bn=xn; yb=yn;
   end
end

root=xn; yval=yn;
```

We will make some more comments on the program as well as the algorithm, but first we show how it would get used in a MATLAB session.

EXAMPLE 6.2: (a) Use the above `bisect` program to perform the indicated approximation of parts (b) and (c) of Example 6.1.
(b) Do the same for the approximation problem of Exercise for the Reader 6.1

SOLUTION: Part (a): We need only run these commands to get the first approximation (with tol = 0.01):[4]

```
>> f=inline('x^5-9*x^2-x+7', 'x');
>> bisect(f,1,2,.01)
→ans = 1.9766 (This is exactly what we got in Example 6.1(b))
```

By default, a function M-file with more than one output variable will display only the first one (stored into the temporary name "ans"). To display all of the output variables (so in this case also the y-coordinate), use the following syntax:

```
>> [x,y]=bisect(f,1,2,.01)
→x = 1.9844, y =0. 0307
```

To obtain the second approximation, we should use more decimals.

```
>> format long
```

---

[4] We point out that if the function f were instead stored as a function M-file, the syntax for `bisect` would change to `bisect('f',…)` or `bisect(@f,…)`.

```
>> [x,y]=bisect(f,1,2,0.00001)
```
→x =1.97579193115234, y= 9.717120432028992e-005

Part (b): There is nothing very different needed to do this second example:

```
>> g=inline('cos(x)-x')
```
→g = Inline function: g(x) = cos(x)-x

```
>> [x,y]=bisect(g,0,pi/2,0.01)
```
→x =0.74858262448819, y = -0.01592835281578

```
>> [x,y]=bisect(g,0,pi/2,10^(-12))
```
→x =0.73908513321527, y = -1.888489364887391e-013

Some additional comments about our program are now in order. Although it may have seemed a bit more difficult to follow this program than Example 6.1, we have considerably economized by overwriting an or bn at each iteration, as well as xn and yn. There is no need for the function to internally construct vectors of all of the intermediate intervals and approximations if all that we are interested in is the final approximation and perhaps also its y-coordinate. We used the error('message') flag command in the program. If it ever were to come up (i.e., only when the corresponding if-branch's condition is met), then the error 'message' inside would be printed and the function execution would immediately terminate. The general syntax is as follows:

error('message') → (inside the body of a function)	causes the message to display on the command window and the execution of the function to be immediately terminated.

Notice also that we chose the default tolerance to be $eps \cdot \max(|a|,|b|,1)$, where eps is MATLAB's unit roundoff. Recall that the unit roundoff is the maximum relative error arising from approximating a real number by its floating point representative (see Chapter 5). Although the program would still work if we had just used eps as the default tolerance, in cases where $\max(|a|,|b|)$ is much larger than 1, the additional iterations would yield the same floating point approximation as with our chosen default tolerance. In cases where $\max(|a|,|b|)$ is much smaller than one, our default tolerance will produce more accurate approximations. As with all function M-files, after having stored bisect, if we were to type help bisect in the command window, MATLAB would display all of the adjacent block of comment lines that immediately follow the function definition line. It is good practice to include comment lines (as we have) that explain various parts of a program.

EXERCISE FOR THE READER 6.3: In some numerical analysis books, the while loop in the above program bisect is rewritten as follows:

```
while (b-a)/2^n >= tol
   xn=(an + bn)/2; yn=feval(varfun, xn); n=n+1;
   if yn*ya > 0
      an=xn; ya=yn;
```

```
     else
          bn=xn; yb=yn;
     end
end
```

   The only difference with the corresponding part in our program is with the
condition in the if-branch, everything else is identical.
(a) Explain that mathematically, the condition in the if-branch above is equivalent
to the one in our program (i.e., both always have the same truth values).
(b) In mathematics there is no smallest positive number. As in Chapter 5, numbers
that are too small in MATLAB will underflow to 0. Depending on the version of
MATLAB you are using, the smallest positive (distinguishable from 0) number in
MATLAB is something like 2.2251e-308. Anything smaller than this will be
converted (underflow) to zero. (To see this enter 10^(-400)). Using these
facts, explain why the for loop in our program is better to use than the above
modification of it.
(c) Construct a continuous function $f(x)$ with a root at $x = 0$ so that if we apply
the bisection program on the interval $[-1, 3]$ with tol = 0.001, the algorithm will
work as it is supposed to, however, if we apply the (above) modified program the
output will not be within the tolerance 0.001 of $x = 0$.

   We close this section with some further comments on the bisection method. It is
the oldest of the methods for rootfinding. It is theoretically guaranteed to work as
long as the hypotheses are satisfied. Recall the assumptions are that $f(x)$ is
continuous on $[a,b]$ and that $f(a)$ and $f(b)$ are of opposite signs. In this case it
is said that the interval $[a,b]$ is a **bracket** of the function $f(x)$. The bisection
method unfortunately cannot be used to locate zeros of functions that do not
possess brackets. For example, the function $y = x^2$ has a zero only at $x = 0$ but
otherwise y is always positive so this function has no bracket. Although the
bisection method converges rather quickly, other methods that we will introduce
will more often work much faster. For a single rootfinding problem, the difference
in speed is not much of an issue, but for more complicated or advanced problems
that require numerous rootfinding "subproblems," it will be more efficient to use
other methods. A big advantage of the bisection method over other methods we
will introduce is that the error analysis is so straightforward and we are able to
determine the number of necessary iterations quite simply before anything else is
done. The **residual** of an approximation x_n to a root $x = r$ of $f(x)$ is the value
$f(x_n)$. It is always good practice to examine the residual of approximations to a
root. Theoretically the residuals should disintegrate to zero as the approximations
get better and better, so it would be a somewhat awkward situation if your
approximation to a root had a very large residual. Before beginning any
rootfinding problem, it is often most helpful to begin with a (computer-generated)
plot of the function.

EXERCISES 6.2:

1. The function $f(x) = \sin(x)$ has a root at $x = \pi$. Find a bracket for this root and use the bisection method with tol $= 10^{-12}$ to obtain an approximation of π that is accurate to 12 decimals. What is the residual?

2. The function $\ln(x) - 1$ has a root at $x = e$. Find a bracket for this root and use the bisection method with tol $= 10^{-12}$ to obtain an approximation of e that is accurate to 12 decimals. What is the residual?

3. Apply the bisection method to find a root of the equation $x^6 + 6x^2 + 2x = 20$ in the interval $[0,2]$ with tolerance 10^{-7}.

4. Apply the bisection method to find a root of the equation $x^9 + 6x^2 + 2x = 3$ in the interval $[-2,-1]$ with tolerance 10^{-7}.

5. Use the bisection method to approximate the smallest positive root of the equation $\tan(x) = x$ with error $< 10^{-10}$.

6. Use the bisection method to approximate the smallest positive root of the equation $e^{2x} = \sin(x) + 1$ with error $< 10^{-10}$.

7. (*Math Finance*) It can be shown[5] that if equal monthly deposits of *PMT* dollars are made into an annuity (interest-bearing account) that pays $100r\%$ annual interest compounded monthly, then the value $A(t)$ of the account after t years will be given by the formula

    $$A(t) = PMT \frac{(1+r/12)^{12t} - 1}{r/12}.$$

    Suppose Mr. Jones is 30 years old and can afford monthly payments of $350.00 into such an annuity. Mr. Jones would like to plan to be able to retire at age 65 with a $1million nest egg. Use the bisection method to find the minimum interest rate (= $100r$ %) Mr. Jones will need to shop for in order to reach his retirement goal.

8. (*Math Finance*) It can be shown that to pay off a 30-year house mortgage for an initial loan of PV dollars with equal monthly payments of *PMT* dollars and a fixed annual interest rate of $100r\%$ compounded monthly, the following equation must hold:

    $$PV = PMT \frac{1 - (1+r/12)^{-360}}{r/12}.$$

    (For a 15-year mortgage, change 360 to 180.) Suppose the Bradys wish to buy a house that costs $140,000. They can afford monthly payments of $1,100 to pay off the mortgage. What kind of interest rate $100r\%$ would they need to be able to afford this house with a 30-year mortgage? How about with a 15-year mortgage? If they went with a 30 year mortgage, how much interest would the Bradys need to pay throughout the course of the loan?

---

[5] See, for example, Chapter 3 and Appendix B of [BaZiBy-02] for detailed explanations and derivations of these and other math finance formulas.

9. Modify the `bisect` program in the text to create a new one, `bisectvv` (stands for: bisection algorithm, vector version), that has the same input variables as `bisect`, but the output variables will now be two vectors x and y that contain all of the successive approximations ($x = [x_1, x_2, \cdots, x_n]$) and the corresponding residuals ($y = [f(x_1), f(x_2), \cdots, f(x_n)]$). Run this program to redo Exercise 3, print only every fourth component, $x_1, y_1, x_5, y_5, x_9, y_9, \cdots$ and also the very last components, x_n, y_n.

10. Modify the 'bisect' program in the text to create a new one, `bisectte` (stands for: bisection algorithm, tell everything), that has the same input variables as `bisect`, but this one has no output variables. Instead, it will output at each of the iterations the following phrase: "Iteration n=< k >, approximation = < xn >, residual = < yn >," where the values of k, xn, and yn at each iteration will be the actual numbers. < k > should be an integer and the other two should be floating point numbers. Apply your algorithm to the function $f(x) = 5x^3 - 8x^2 + 2$ with bracket [1, 2] and tolerance = 0.002.

11. Apply the `bisect` program to $f(x) = \tan(x)$ with tol = 0.0001 to each of the following sets of intervals. In each case, is the output (final approximation) within the tolerance of a root? Carefully explain what happens in each case.
    (a) $[a, b] = [5, 7]$,   (b) $[a, b] = [4, 7]$,   (c) $[a, b] = [4, 5]$.

12. In applying the bisection method to a function $f(x)$ using a bracket $[a, b]$ on which $f(x)$ is known to have exactly one root r, is it possible that x_2 is be a better approximation to r than x_5? (This means $|x_2 - r| < |x_5 - r|$.) If no, explain why not; if yes, supply a specific counterexample.

6.3: NEWTON'S METHOD

Under most conditions, when Newton's method works, it converges very quickly, much faster indeed than the bisection method. It is at the foundation of all contemporary state-of-the-art rootfinding programs. The error analysis, however, is quite a bit more awkward than with the bisection method and this will be relegated to Section 6.5. Here we examine various situations in which Newton's method performs outstandingly, where it can fail, and in which it performs poorly.

ASSUMPTIONS: $f(x)$ is a differentiable function that has a root $x = r$ which we wish to accurately approximate. We would like the approximation to be accurate to MATLAB's machine precision of about 15 significant digits.

The idea of the method will be to repeatedly use tangent lines of the function situated at successive approximations to "shoot at" the next approximation. More precisely, the next approximation will equal the x-intercept of the tangent line to the graph of the function taken at the point on the graph corresponding to the current approximation to the root. See Figure 6.6 for an illustration of Newton's method. We begin with an initial approximation x_0, which was perhaps obtained from a plot. It is straightforward to obtain a recursion formula for the next approximation x_{n+1} in terms of the current approximation x_n. The tangent line to the graph of $y = f(x)$ at $x = x_n$ is given by the first-order Taylor polynomial centered at $x = x_n$, which has equation (see equation (3) of Chapter 2): $y = f(x_n) + f'(x_n)(x - x_n)$. The next approximation is the x-intercept of this line and is obtained by setting $y = 0$ and solving for x. Doing this gives us the recursion formula:

$$x_{n+1} = x_n - \frac{f(x_n)}{f'(x_n)}, \qquad (2)$$

where it is required that $f'(x_n) \neq 0$.

It is quite a simple task to write a MATLAB program for Newton's method, but following the usual practice, we will begin working through an example "by hand".

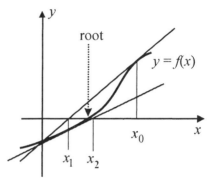

FIGURE 6.6: Illustration of Newton's method.   To go from the initial approximation (or guess) x_0 to the next approximation x_1, we simply take the x_1 to be the x-intercept of the tangent line to the graph of $y = f(x)$ at the point $(x_0, f(x_0))$. This procedure gets iterated to obtain successive approximations.

EXAMPLE 6.3: Use Newton's method to approximate $\sqrt[4]{2}$ by performing five iterations on the function $f(x) = x^4 - 2$ using initial guess $x = 1.5$. (Note that $[1, 2]$ is clearly a bracket for the desired root.)

SOLUTION: Since $f'(x) = 4x^3$, the recursion formula (2) becomes:

$$x_{n+1} = x_n - \frac{f(x_n)}{f'(x_n)} = x_n - \frac{x_n^4 - 2}{4x_n^3}.$$

Let's now get MATLAB to find the first five iterations along with the residuals and the errors. For convenient display, we store this data in a 5×3 matrix with the first column containing the approximations, the second the residuals, and the third the errors.

```
>> x(1)=1.5; %initialize, remember zero can't be an index
>> for n=1:5
x(n+1)=x(n)-(x(n)^4-2)/(4*x(n)^3);
A(n, :) = [ x(n+1)    (x(n+1)^4-2)    abs(x(n+1)-2^(1/4))];
end
```

To be able to see how well the approximation went, it is best to use a different format (from the default `format short`) when we display the matrix A.

```
>> format long e
>> A
```

We display this matrix in Table 6.1.

TABLE 6.1: The successive approximations, residuals, and errors resulting from applying Newton's method to $f(x) = x^4 - 2$ with initial approximation $x_0 = 1.5$.

n	x_n	$f(x_n)$	Error $= \lvert r - x_n \rvert$
1	1.2731481481481e+000	6.2733693232248e-001	8.3941033145427e-002
2	1.1971498203523e+000	5.3969634451728e-002	7.9427053495620e-003
3	1.1892858119092e+000	5.2946012602728e-004	7.8696906514741e-005
4	1.1892071228136e+000	5.2545275686100e-008	7.8109019252537e-009
5	1.1892071150027e+000	1.3322676295502e-015	2.2204460492503e-016

The table shows quite clearly just how fast the errors are disintegrating to zero. As we mentioned, if the conditions are right, Newton's method will converge extremely quickly. We will give some clarifications and precise limitations of this comment, but let us first write an M-file for Newton's method.

PROGRAM 6.2: An M-file for Newton's method.[6]

```
function [root, yval] = newton(varfun, dvarfun,  x0, tol, nmax)
% input variables: varfun, dvarfun, x0, tol, nmax
% output variables:  root, yval
% varfun = the string representing a mathematical function (built-in,
% M-file, or inline) and dvarfun = the string representing the
% derivative, x0 = the initial approx.  The program will perform
% Newton's method to approximate a root of varfun near x=x0 until
% either successive approximations differ by less than tol or nmax
% iterations have been completed, whichever comes first.  If the tol
% and nmax variables are omitted, default values of
% eps*max(abs(a),abs(b),1) and 30 are used.
% we now assign the default tolerance and maximum number of iterations if
% none are specified
if nargin < 4
   tol=eps*max([abs(a) abs(b) 1]); nmax=30;
end

%we now initialize the iteration
xn=x0;

%finally we set up a loop to perform the approximations
for n=1:nmax
```

---

[6] When one needs to use an apostrophe in a string argument of an `fprintf` statement, the correct syntax is to use a double apostrophe. For example, `fprintf('Newton's')` would produce an error message but `fprintf(Newton''s)` would produce → Newton's.

```
      yn=feval(varfun, xn); ypn=feval(dvarfun, xn);
      if yn == 0
         fprintf('Exact root found\r')
         root = xn; yval = 0;
         return
      end
      if ypn == 0
        error('Zero derivative encountered, Newton''s method failed,
                                         try changing x0')
      end
      xnew=xn-yn/ypn;
      if abs(xnew-xn)<tol
            fprintf('Newton''s method has converged\r')
            root = xnew; yval = feval(varfun, root);
            return
      elseif n==nmax
            fprintf('Maximum number of iterations reached\r')
            root = xnew; yval = feval(varfun, root);
            return
      end
      xn=xnew;
end
```

EXAMPLE 6.4: (a) Use the above `newton` program to find a root of the equation of Example 6.3 (again using $x_0 = 1.5$). (b) Next use the program to approximate e by finding a root of the equation $\ln(x) - 1 = 0$. Check the error of this latter approximation.

SOLUTION: Part (a): We temporarily construct some in-line functions. Take careful note of the syntax.

```
>> f = inline('x^4-2')        →f =Inline function:  f(x) = x^4-2
>> fp = inline('4*x^3')       →fp =Inline function:  fp(x) = 4*x^3
>> format long
>> newton(f, fp, 1.5)
→Newtons method has converged   →ans = 1.18920711500272

>> [x,y]=newton(f,fp,1.5) %to see also to see the y-value
→Newton's method has converged
→x =1.18920711500272, y = -2.220446049250313e-016
```

Part (b):

```
>> f=inline('log(x)-1'); fp=inline('1/x');
>> [x,y]=newton(f,fp,3)
→Newton's method has converged   →x =2.71828182845905, y = 0
>>abs(exp(1)-x)                   →ans = 4.440892098500626e-016
```

We see that the results of part (a) nicely coincide with the final results of the previous example.

EXERCISE FOR THE READER 6.4: In part (b) of Example 6.4, show using calculus that the y-coordinate corresponding to the approximation of the root e

found is about as far from zero as the x-coordinate is from the root. Explain how floating point arithmetic caused this y-coordinate to be outputted as zero, rather than something like 10^{-17}. (Refer to Chapter 5 for details about floating point arithmetic.)

We next look into some pathologies that can cause Newton's method to fail. Later, in Section 6.5, we will give some theorems that will give some guaranteed error estimates with Newton's method, provided certain hypotheses are satisfied. The first obvious problem (for which we built an error message into our program) is if at any approximation x_n we have $f'(x_n) = 0$. Unless the function at hand is highly oscillatory near the root, such a problem can often be solved simply by trying to reapply Newton's method with a different initial value x_0 (perhaps after examining a more careful plot); see Figure 6.7.

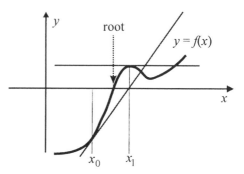

FIGURE 6.7: A zero derivative encountered in Newton's method. Here x_2 is undefined. Possible remedy: Use a different initial value x_0.

A less obvious problem that can occur in Newton's method is **cycling**. Cycling is said to occur when the sequence of x_ns gets caught in an infinite loop by continuing to run through the same set of fixed values.

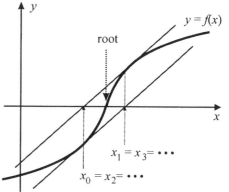

FIGURE 6.8: A cycling phenomenon encountered in Newton's method. Possible remedy: Take initial approximation closer to actual root.

We have illustrated the cycling phenomenon with a cycle having just two values. It is possible for such a "Newton cycle" to have any number of values.

EXERCISE FOR THE READER 6.5: (a) Construct explicitly a polynomial $y = p(x)$ with an initial approximation x_0 to a root such that Newton's method will cause cycling.
(b) Draw a picture of a situation where Newton's method enters into a cycle having exactly four values (rather than just two as in Figure 6.8).

Another serious problem with Newton's method is that the approximations can actually sometimes continue to move away from a root. An illustration is provided by Figure 6.8 if we move x_0 to be a bit farther to the left. This is another reason why it is always recommended to examine residuals when using Newton's method. In the next example we will use a single function to exhibit all three phenomena in Newton's method: convergence, cycling, and divergence.

EXAMPLE 6.5: Consider the function $f(x) = \arctan(x)$, which has a single root at $x = 0$. The graph is actually similar in appearance (although horizontally shifted) to that of Figure 6.8. Show that there exists a number $a > 0$ such that if we apply Newton's method to find the root $x = 0$ (a purely academic exercise since we know the root) with initial approximation $x_0 > 0$, the following will happen:

(i) If $x_0 < a$, then $x_n \to 0$ (convergence to the root, as desired).

(ii) If $x_0 = a$, then x_n will cycle back and forth between a and $-a$.

(iii) If $x_0 > a$, then $|x_n| \mapsto \infty$ (the approximations actually move farther and farther away from the root).

Next apply the bisection program to approximate this critical value $x = a$ and give some examples of each of (i) and (iii) using the Newton method program.

SOLUTION: Since $f'(x) = \dfrac{1}{1 + x^2}$, Newton's recursion formula (2) becomes:

$$x_{n+1} = x_n - \frac{f(x_n)}{f'(x_n)} = x_n - (1 + x_n^2)\arctan(x_n) \equiv g(x_n).$$

(The function g(x) is defined by the above formula.) Since $g(x)$ is an odd function (i.e., $g(-x) = -g(x)$), we see that we will enter into a cycle (with two numbers) exactly when $(x_1 =)g(x_0) = -x_0$. (Because then $x_2 = g(x_1) = g(-x_0)$ $= -(-x_0) = x_0$, and so on.) Thus, Newton cycles can be found by looking for the positive roots of $g(x) + x = 0$. Notice that $(g(x) + x)' = 1 - 2x\arctan(x)$ so that the function in parentheses increases (from its initial value of 0 at $x = 0$) until x reaches a certain positive value (the root of $1 - 2x\arctan(x) = 0$) and after this value of x it is strictly decreasing and will eventually become zero (at some value $x = a$) and after this will be negative. Again, since $g(x)$ is an odd function, we can summarize as follows: (i) for $0 < |x| < a$, $|g(x)| < |x|$, (ii) $g(\pm a) = \mp a$, and (iii) for $|x| > a$, $|g(x)| > |x|$. Once x is situated in any of these three ranges, $g(x)$ will thus be in the same range and by the noted properties of $g(x)$ and of $g(x) + x$ we can conclude the assertions of convergence, cycling, and divergence to infinity as x_0 lies in one of these ranges. We now provide some numerical data that will demonstrate each of these phenomena.

First, since $g(a) = -a$, we may approximate $x = a$ quite quickly by using the bisection method to find the positive root of the function $h(x) = g(x) + x$. We must be careful not to pick up the root $x = 0$ of this function.

```
>> h=inline('2*x-(1+x^2)*atan(x)');
>> h(0.5), h(5) %will show a bracket to the unique positive root
→ans = 0.4204, ans =-25.7084

>> format long
>> a=bisect(h,.5,5)          → a=1.39174520027073
```

To make things more clear in this example, we use a modification of the newton algorithm, called newtonsh, that works the same as newton, except the output will be a matrix of all of the successive approximations x_n and the corresponding y-values. (Modifying our newton program to get newtonsh is straightforward and is left to the reader.)

```
>> format long e
>> B=newtonsh(f,fp,1)
→Newton's method has converged
 (We display the matrix B in Table 6.2.)
```

TABLE 6.2: The result of applying Newton's method to the function $f(x) = \arctan(x)$ with $x_0 = 1 < a$ (critical value). Very fast convergence to the root $x = 0$.

n	x_n	$f(x_n)$
1	-5.707963267948966e-001	-5.186693692550166e-001
2	1.168599039989131e-001	1.163322651138959e-001
3	-1.061022117044716e-003	-1.061021718890093e-003
4	7.963096044106416e-010	7.963096044106416e-010
5	0	0
6	0	0

```
>> B=newtonsh(f,fp,a)        →Maximum number of iterations reached (see Table 6.3)
```

TABLE 6.3: The result of applying Newton's method to the function $f(x) = \arctan(x)$ with $x_0 = a$ (critical value). The approximations cycle.

n	x_n	$f(x_n)$
1	-1.391745200270735e+000	-9.477471335169905e-001
2	1.391745200270735e+000	9.477471335169905e-001
...	...	...
28	1.391745200270735e+000	9.477471335169905e-001
29	-1.391745200270735e+000	-9.477471335169905e-001
30	1.391745200270735e+000	9.477471335169905e-001

```
>> B=newtonsh(f,fp,1.5)
→??? Error using ==> newtonsh
→zero derivative encountered, Newton's method failed, try changing x0
```

We have got our own error flag. We know that the derivative $1/(1+x^2)$ of arctan(x) is never zero. What happened here is that the approximations x_n were getting so large, so quickly (in absolute value) that the derivatives underflowed to zero. To get some output, we redo the above command with a cap on the number of iterations; the results are displayed in Table 6.4.

```
>> B=newtonsh(f,fp,1.5, 0.001, 9)
→Maximum number of iterations reached      →B =
```

TABLE 6.4: The result of applying Newton's method to the function $f(x) = \arctan(x)$ with $x_0 = 1.5 > a$ (critical value). The successive approximations alternate between positive and negative values and their absolute values diverge very quickly to infinity. The corresponding y-values will, of course, alternate between tending to $\pm\pi/2$, the limits of $f(x) = \arctan(x)$ as $x \to \pm\infty$.

n	x_n	$f(x_n)$
1	-1.694079600553820e+000	-1.037546359137891e+000
2	2.321126961438388e+000	1.164002042421975e+000
3	-5.114087836777514e+000	-1.377694528702752e+000
4	3.229568391421002e+001	1.539842326908012e+000
5	-1.575316950821204e+003	-1.570161533990085e+000
6	3.894976007760884e+006	1.570796070053906e+000
7	-2.383028897355213e+013	-1.570796326794855e+000
8	8.920280161123818e+026	1.570796326794897e+000
9	-1.249904599365711e+054	-1.570796326794897e+000

In all of the examples given so far, when Newton's method converged to a root, it did so very quickly. The main reason for this is that each of the roots being approximated was a **simple root**. Geometrically this means that the graph of the differentiable function was not tangent to the x-axis at this root. Thus a root $x = r$ is a simple root of $f(x)$ provided that ($f(r) = 0$ and) $f'(r) \neq 0$. A root r that is not simple is called a **multiple root**

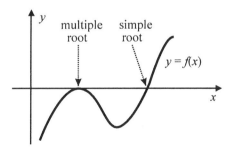

FIGURE 6.9: Illustration of the two types of roots a function can have. Newton's method performs much more effectively in approximating simple roots.

of **order** M ($M > 1$) if $f(r) = f'(r) = f''(r) = \cdots \quad f^{(M-1)}(r) = 0$ but $f^{(M)}(r) \neq 0$ (see Figure 6.9). These definitions were given for polynomials in

Exercises 3.2. Multiple roots of order 2 are sometimes called **double roots**, order-3 roots are **triple roots**, and so on. If $x = r$ is an order-M root of $f(x)$, it can be shown that $f(x) = (x-r)^M h(x)$ for some continuous function $h(x)$ (see Exercise 13).

EXAMPLE 6.6: How many iterations will it take for Newton's method to approximate the multiple root $x = 0$ of the function $f(x) = x^{21}$ using an initial approximation of $x = 1$ if we want an error < 0.01? How about if we want an error < 0.001?

SOLUTION: We omit the MATLAB commands, but summarize the results. If we first try to run $\texttt{newton}$, (or better a variant of it that displays some additional output variables) with a tolerance of 0.01, and a maximum number of iterations = 50, we will get the message that the method converged. It did so after a whopping 33 iterations and the final value of root = 0.184 (which is not within the desired error tolerance) and a (microscopically small) yval = 1.9842e-015. The reason the program gave us the convergence message is because the adjacent root approximations differed by less than the 0.01 tolerance. To get the root to be less than 0.01 we would actually need about 90 iterations! And to get to an approximation with a 0.001 tolerated error, we would need to go through 135 iterations. This is a pathetic rate of convergence; even the (usually slower) bisection method would only take 7 iterations for such a small tolerance (why?).

EXERCISES 6.3:

1. For each of the functions shown below, find Newton's recursion formula (2). Next, using the value of x_0 that is given, find each of x_1, x_2, x_3.

    (a) $f(x) = x^3 - 2x + 5$; $x_0 = -3$ (c) $f(x) = xe^{-x}$; $x_0 = 0.5$

    (b) $f(x) = e^x - 2\cos(x)$; $x_0 = 1$ (d) $f(x) = \ln(x^4) - \cos(x)$; $x_0 = 1$

2. For each of the functions shown below, find Newton's recursion formula (2). Next, using the value of x_0 that is given, find each of x_1, x_2, x_3.

    (a) $f(x) = x^3 - 15x^2 + 24$; $x_0 = -3$ (c) $f(x) = \ln(x)$; $x_0 = 0.5$

    (b) $f(x) = e^x - 2e^{-x} + 5$; $x_0 = 1$ (d) $f(x) = \sec(x) - 2e^{x^2}$; $x_0 = 1$

3. Use Newton's method to find the smallest positive root of each equation to 12 digits of accuracy. Indicate the number of iterations used.

    (a) $\tan(x) = x$

    (b) $x\cos(x) = 1$

    (c) $4x^2 = e^{x/2} - 2$

    (d) $(1+x)\ln(1+x^2) = \cos(\sqrt{x})$

    Suggestion: You may wish to modify our $\texttt{newton}$ program so as to have it display another output variable that gives the number of iterations.

4. Use Newton's method to find the smallest positive root of each equation to 12 digits of accuracy. Indicate the number of iterations used.

(a) $e^{-x} = x$ (c) $x^4 - 2x^3 + x^2 - 5x + 2 = 0$

(b) $e^x - x^8 = \ln(1 + 2^x)$ (d) $e^x = x^\pi$

5. For the functions given in each of parts (a) through (d) of Exercise 1, use the Newton's method program to find all roots with an error at most 10^{-10}.

6. For the functions given in each of parts (a) through (c) of Exercise 2, use the Newton's method program to find all roots with an error at most 10^{-10}. For part (d) find only the two smallest positive roots.

7. Use Newton's method to find all roots of the polynomial $x^4 - 5x^2 + 2$ with each being accurate to about 15 decimal places (MATLAB's precision limit). What is the multiplicity of each of these roots?

8. Use Newton's method to find all roots of the polynomial $x^6 - 4x^4 - 12x^2 + 2$ with each being accurate to about 15 decimal places (MATLAB's precision limit). What is the multiplicity of each of these roots?

9. (*Finance-Retirement Plans*) Suppose a worker puts in PW dollars at the end of each year into a 401(k) (supplemental retirement plan) annuity and does this for NW years (working years). When the worker retires, he would like to withdraw from the annuity a sum of PR dollars at the end of each year for NR years (retirement years). The annuity pays $100r\%$ annual interest compounded annually on the account balance. If the annuity is set up so that at the end of the NR years, the account balance is zero, then the following equation must hold:

$$PW[(1+r)^{NW} - 1] = PR[1 - (1+r)^{-NR}]$$

(see [BaZiBy-02]). The problem is for the worker to decide on what interest rate is needed to fund this retirement scheme. (Of course, other interesting questions arise that involve solving for different parameters, but any of the other variables can be solved for explicitly.) Use Newton's method to solve the problem for each of the following parameters:
(a) $PW = 2,000, PR = 10,000, NW = 35, NR = 25$
(b) $PW = 5,000, PR = 20,000, NW = 35, NR = 25$
(c) $PW = 5,000, PR = 80,000, NW = 35, NR = 25$
(d) $PW = 5,000, PR = 20,000, NW = 25, NR = 25$

10. For which values of the initial approximation x_0 will Newton's method converge to the root $x = 1$ of $f(x) = x^2 - 1$?

11. For which values of the initial approximation x_0 will Newton's method converge to the root $x = 0$ of $f(x) = \sin^2(x)$?

12. For (approximately) which values of the initial approximation $x_0 > 0$ will Newton's method converge to the root of
(a) $f(x) = \ln(x)$ (c) $f(x) = \sqrt[3]{x}$

(b) $f(x) = x^3$ (d) $f(x) = e^x - 1$

13. The following algorithm for calculating the square root of a number $A > 0$ actually was around for many years before Newton's method:

$$x_{n+1} = \frac{1}{2}\left(x_n + \frac{A}{x_n}\right).$$

(a) Run through five iterations of it to calculate $\sqrt{10}$ starting with $x_0 = 3$. What is the error?

(b) Show that this algorithm can be derived from Newton's method.

14. Consider each of the following two schemes for approximating π:

SCHEME 1: Apply Newton's method to $f(x) = \cos(x) + 1$ with initial approximation $x_0 = 3$.

SCHEME 2: Apply Newton's method to $f(x) = \sin(x)$ with initial approximation $x_0 = 3$.

Discuss the similarities and differences of each of these two schemes. In particular, explain how accurate a result each scheme could yield (working on MATLAB's precision). Finally use one of these two schemes to approximate π with the greatest possible accuracy (using MATLAB).

15. Prove that if $f(x)$ has a root $x = r$ of multiplicity M then we can write: $f(x) = (x - r)^M h(x)$ for some continuous function $h(x)$.

Suggestion: Try using L'Hôpital's rule.

6.4: THE SECANT METHOD

When conditions are ripe, Newton's method works very nicely and efficiently. Unlike the bisection method, however, it requires computations of the derivative in addition to the function. Geometrically, the derivative was needed to obtain the tangent line at the current approximation x_n (more precisely at $(x_n, f(x_n))$) whose x-intercept was then taken as the next approximation x_{n+1}. If instead of this tangent line, we use the secant line obtained using the current as well as the previous approximation (i.e., the line passing through the points $(x_n, f(x_n))$ and $(x_{n-1}, f(x_{n-1}))$)

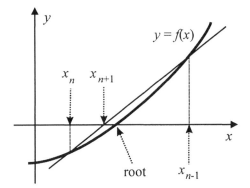

FIGURE 6.10: Illustration of the secant method. Current approximation x_n and previous approximation x_{n-1} are used to obtain a secant line through the graph of $y = f(x)$. The x-intercept of this line will be the next approximation x_{n+1}.

and take the next approximation x_{n+1} to be the x-intercept of this line, we get the so-called **secant method**; see Figure 6.10. Many of the problems that plagued Newton's method can also cause problems for the secant method. In cases where it is inconvenient or expensive to compute derivatives, the secant method is a good replacement for Newton's method. Under certain hypotheses (which were hinted at in the last section) the secant method will converge much faster than the bisection method, although not quite as fast as Newton's method. We will make these comments precise in the next section.

To derive the recursion formula for the secant method we first note that since the three points $(x_{n-1}, f(x_{n-1}))$, $(x_n, f(x_n))$, and $(x_{n+1}, 0)$ all lie on the same (secant) line, we may equate slopes:

$$\frac{0 - f(x_n)}{x_{n+1} - x_n} = \frac{f(x_n) - f(x_{n-1})}{x_n - x_{n-1}}.$$

Solving this equation for x_{n+1} yields the desired recursion formula:

$$x_{n+1} = x_n - f(x_n) \cdot \frac{x_n - x_{n-1}}{f(x_n) - f(x_{n-1})}. \tag{3}$$

Another way to obtain (3) is to replace $f'(x_n)$ in Newton's method (3) with the difference quotient approximation $(f(x_n) - f(x_{n-1}))/(x_n - x_{n-1})$.

EXAMPLE 6.7: We will recompute $\sqrt[4]{2}$ by running through five iterations of the secant method on the function $f(x) = x^4 - 2$ using initial approximations $x_0 = 1.5$ and $x_1 = 1$. Recall in the previous section we had done an analogous computation with Newton's method.

SOLUTION: We can immediately get MATLAB to find the first five iterations along with the residuals and the errors. For convenient display, we store this data in a 5×3 matrix with the first column containing the approximations, the second the residuals, and the third the errors.

```
>> x(1)=1.5; x(2)=1; %initialize, recall zero cannot be an index
>> for n=2:6
x(n+1)=x(n)-f(x(n))*(x(n)-x(n-1))/(f(x(n))-f(x(n-1)));
A(n-1, :) = [ x(n+1)    (x(n+1)^4-2)    abs(x(n+1)-2^(1/4))];
end
>> A
```

Again we display the matrix in tabular form (now in format long).

TABLE 6.5: The successive approximations, residuals and errors resulting from applying the secant method to $f(x) = x^4 - 2$ with initial approximation $x_0 = 1.5$.

n	x_n	$f(x_n)$	ERROR = $\|r - x_n\|$
2	1.12307692307692	-0.40911783200868	0.06613019192580
3	1.20829351390874	0.13152179071884	0.01908639890601
4	1.18756281243219	-0.01103858431975	0.00164430257053
5	1.18916801020327	-0.00026305171011	0.00003910479945
6	1.18920719620308	0.00000054624878	0.00000008120036

Compare the results in Table 6.5 with those of Table 6.1, which documented Newton's method for the same problem.

EXERCISE FOR THE READER 6.6:   (a) Write a MATLAB M-file called secant that has the following input variables: varfun, the string representing a mathematical function, x0 and x1, the (different) initial approximations, tol, the tolerance, nmax, the maximum number of iterations; and output variables: root, yval ,= varfun(root), and niter, the number of iterations used. The program should perform the secant method to approximate the root of varfun(x) near x0, x1 until either successive approximations differ by less than tol or nmax iterations have been completed.   If the tol and nmax variables are omitted, default values of eps*max(abs(a),abs(b),1) and 50 are used.
(b) Run your program on the rootfinding problem of Example 6.4.   Give the resulting approximation, the residual, the number of iterations used, and the actual (exact) error. Use x0 = 2 and x1 = 1.5.

   The secant method uses the current and immediately previous approximations to get a secant line and then shoots the line into the x-axis for the next approximation. A related idea would be to use the current as well as the past two approximations to construct a second-order equation (in general a parabola) and then take as the next approximation the root of this parabola that is closest to the current approximation.   This idea is the basis for Muller's method, which is further developed in the exercises of this section. We point out that Muller's method in general will converge a bit quicker than the secant method but not quite as quickly as Newton's method. These facts will be made more precise in the next section.

EXERCISES 6.4:

1.   For each of the functions shown below, find the secant method recursion formula (3). Next, using the values of x_0 and x_1 that are given, find each of x_2, x_3, x_4 .

   (a) $f(x) = x^3 - 2x + 5$; $x_0 = -3, x_1 = -2$

   (b) $f(x) = e^x - 2\cos(x)$; $x_0 = 1, x_1 = 2$

   (c) $f(x) = xe^{-x}$; $x_0 = 0.5, x_1 = 1.0$

   (d) $f(x) = \ln(x^4) - \cos(x)$; $x_0 = 1, x_1 = 1.5$

2.   For each of the functions shown below, find the secant method recursion formula (3). Next, using the values of x_0 and x_1 that are given, find each of x_2, x_3, x_4 .

   (a) $f(x) = x^3 - 15x^2 + 24$; $x_0 = -3, x_1 = -4$

   (b) $f(x) = e^x - 2e^{-x} + 5$; $x_0 = 1, x_1 = 2$

   (c) $f(x) = \ln(x)$; $x_0 = 0.5, x_1 = 2$

   (d) $f(x) = \sec(x) - 2e^{x^2}$; $x_0 = 1, x_1 = 1.5$

3. Use the secant method to find the smallest positive root of each equation to 12 decimals of accuracy. Indicate the number of iterations used.

    (a) $\tan(x) = x$

    (c) $4x^2 = e^{x/2} - 2$

    (b) $x\cos(x) = 1$

    (d) $(1+x)\ln(1+x^2) = \cos(\sqrt{x})$

    Suggestion: You may wish to modify your `secant` program so as to have it display another output variable that gives the number of iterations.

4. Use the secant method to find the smallest positive root of each equation to 12 decimals of accuracy. Indicate the number of iterations used.

    (a) $e^{-x} = x$

    (c) $x^4 - 2x^3 + x^2 - 5x + 2 = 0$

    (b) $e^x - x^8 = \ln(1 + 2^x)$

    (d) $e^x = x^\pi$

5. For which values of the initial approximation x_0 will the secant method converge to the root $x = 1$ of $f(x) = x^2 - 1$?

6. For which values of the initial approximation x_0 will the secant method converge to the root $x = 0$ of $f(x) = \sin^2(x)$?

7. For (approximately) which values of the initial approximation $x_0 > 0$ will the secant method converge to the root of

    (a) $f(x) = \ln(x)$

    (c) $f(x) = \sqrt[3]{x}$

    (b) $f(x) = x^3$

    (d) $f(x) = e^x - 1$

NOTE: The next three exercises develop Muller's method, which was briefly described at the end of this section. It will be convenient to introduce the following notation for **divided differences** for a continuous function $f(x)$ and distinct points x_1, x_2, x_3:

$$f[x_1, x_2] \equiv \frac{f(x_2) - f(x_1)}{x_2 - x_1}, \quad f[x_1, x_2, x_3] \equiv \frac{f[x_2, x_3] - f[x_1, x_2]}{x_3 - x_1}$$

8. Suppose that $f(x)$ is a continuous function and x_0, x_1, x_2 are distinct x-values. Show that the second order polynomial

    $$p(x) = f(x_2) + (x - x_2)f[x_2, x_1] + (x - x_2)(x - x_1)f[x_2, x_1, x_0]$$

    is the unique polynomial of degree at most two that passes through the three points: $(x_i, f(x_i))$ for $i = 0, 1, 2$.

    Suggestion: Since $p(x)$ has degree at most two, you need only check that $p(x_i) = f(x_i)$ for $i = 0, 1, 2$. Indeed, if $q(x)$ were another such polynomial then $D(x) \equiv p(x) - q(x)$ would be a polynomial of at most second degree with three roots and this would force $D(x) \equiv 0$ and so $p(x) \equiv q(x)$.

9. (a) Show that the polynomial $p(x)$ of Exercise 8 can be rewritten in the following form:

    $$p(x) = f(x_2) + B(x - x_2) + f[x_2, x_1, x_0](x - x_2)^2,$$

    where

    $$B = f[x_2, x_1] + (x_2 - x_1)f[x_2, x_1, x_0] = f[x_2, x_1] + f[x_2, x_0] - f[x_0, x_1].$$

    (b) Next, using the quadratic formula show that the roots of $p(x)$ are given by (first thinking of it as a polynomial in the variable $x - x_2$):

$$x - x_2 = \frac{-B \pm \sqrt{B^2 - 4f(x_2)f[x_2, x_1, x_0]}}{2f[x_2, x_1, x_0]}$$

and then show that we can rewrite this as:

$$x = x_2 - \frac{2f(x_2)}{B \pm \sqrt{B^2 - 4f(x_2)f[x_2, x_1, x_0]}} .$$

10.   Given the first three approximations for a root of a continuous function $f(x)$: x_0, x_1, x_2, Muller's method will take the next one, x_3, to be that solution in Exercise 9 that is closest to x_2 (the most current approximation). It then continues the process, replacing x_0, x_1, x_2 by x_1, x_2, x_3 to construct the next approximation, x_4.
(a) Show that the latter formula in Exercise 9 is less susceptible to floating point errors than the first one.
(b) Write an M-file, call it `muller` that will perform Muller's method to find a root. The syntax should be the same as that of the `secant` program in Exercise for the Reader 6.6, except that this one will need three initial approximations in the input rather than 2.
(c) Run your program through six iterations using the function $f(x) = x^4 - 2$ and initial approximations $x_0 = 1, x_1 = 1.5, x_2 = 1.25$ and compare the results and errors with the corresponding ones in Example 6.7 where the secant method was used.

11.   Redo Exercise 3 parts (a) through (d), this time using Muller's method as explained in Exercise 10.

12.   Redo Exercise 4 parts (a) through (d), this time using Muller's method as explained in Exercise 10.

6.5: ERROR ANALYSIS AND COMPARISON OF ROOTFINDING METHODS

We will shortly show how to accelerate Newton's method in the troublesome cases of multiple roots. This will require us to get in to some error analysis. Since some of the details of this section are a bit advanced, we recommend that readers who have not had a course in mathematical analysis simply skim over the section and pass over the technical comments.[7]

The following definition gives us a way to quantify various rates of convergence of rootfinding schemes.

DEFINITION: Suppose that a sequence $\langle x_n \rangle$ converges to a real number r. We say that the **convergence is of order** α (where $\alpha \geq 1$) provided that for some positive number A, the following inequality will eventually be true:

---

[7] Readers who wish to get more comfortable with the notions of mathematical analysis may wish to consult either of the excellent texts [Ros-96] or [Rud-64].

$$|r - x_{n+1}| \leq A|r - x_n|^\alpha \tag{4}$$

For $\alpha = 1$, we need to stipulate $A < 1$ (why?). The word "eventually" here means "for all values of n that are sufficiently large." It is convenient notation to let $e_n = |r - x_n|$ = the error of the nth approximation, which allows us to rewrite (4) in the compact form: $e_{n+1} \leq Ae_n^\alpha$. For a given sequence with a certain order of convergence, different values of A are certainly possible (indeed, if (4) holds for some number A it will hold also for any bigger number being substituted for A). It is even possible that (4) may hold for all positive numbers A (of course, smaller values of A may require larger starting values of n for validity). In case the greatest lower bound $\hat{A}$ of all such numbers A is positive (i.e., (4) will eventually hold for any number $A > \hat{A}$ but not for numbers $A < \hat{A}$), then we say that $\hat{A}$ is the **asymptotic error constant** of the sequence. In particular, this means there will always be arbitrarily large values of n, for which the error of the $(n+1)$st term is *essentially* proportional to the αth power of that of the nth term and the proportionality constant is approximately $\hat{A}$:

$$|r - x_{n+1}| \approx \hat{A}|r - x_n|^\alpha \quad \text{or} \quad e_{n+1} \approx \hat{A}e_n^\alpha. \tag{5}$$

The word "essentially" here means that we can get the ratios e_{n+1}/e_n^α as close to $\hat{A}$ as desired. In the notation of mathematical analysis, we can paraphrase the definition of $\hat{A}$ by the formula:

$$\hat{A} \equiv \limsup_{n \to \infty} e_{n+1}/e_n^\alpha$$

provided that this limsup is positive. In the remaining case that this greatest lower bound of the numbers A is zero, the asymptotic error constant is undefined (making $\hat{A} = 0$ may seem reasonable but it is not a good idea since it would make (5) fail), but we say that we have **hyper convergence of order α.**

When $\alpha = 1$, we say there is **linear convergence** and when $\alpha = 2$ we say there is **quadratic convergence**. In general, higher values of α result in speedier convergences and for a given α, smaller values of A result in faster convergences. As an example, suppose $e_n = 0.001 = 1/1000$. In case $\alpha = 1$ and $\hat{A} = 1/2$ we have (approximately and for arbitrarily large indices n) $e_{n+1} \approx 0.0005 = 1/2000$, while if $A = 1/4$, $e_{n+1} \approx 0.00025 = 1/4000$. If $\alpha = 2$ even for $A = 1$ we would have $e_{n+1} \approx (0.001)^2 = 0.000001$ and for $A = 1/4$, $e_{n+1} \approx (1/4)(0.001)^2 = 0.00000025$.

EXERCISE FOR THE READER 6.7: This exercise will give the reader a feel for the various rates of convergence.

(a) Find (if possible) the highest order of convergence of each of the following sequences that have limit equal to zero. For each find also the asymptotic error constant whenever it is defined or whether there is hyper convergence for this order

(i) $e_n = 1/(n+1)$, (ii) $e_n = 2^{-n}$, (iii) $e_n = 10^{-3^n/2^n}$, (iv) $e_n = 10^{-2^n}$, (v) $e_n = 2^{-2^n-n}$

(b) Give an example of a sequence of errors $\langle e_n \rangle$ where the convergence to zero is of order 3.

One point that we want to get across now is that quadratic convergence is extremely fast. We will show that under certain hypotheses, the approximations in Newton's method will converge quadratically to a root whereas those of the bisection method will in general converge only linearly. If we use the secant method the convergence will in general be of order $(1+\sqrt{5})/2 = 1.62...$ We now state these results more precisely in the following theorem.

THEOREM 6.1: (*Convergence Rates for Rootfinding Programs*) Suppose that one of the three methods, bisection, Newton's, or the secant method, is used to produce a sequence $\langle x_n \rangle$ that converges to a root r of a continuous function $f(x)$.

PART A: If the bisection method is used, the convergence is **essentially linear** with constant 1/2. This means that there exist positive numbers $e_n' \geq e_n = |x_n - r|$ that (eventually) satisfy $e_{n+1}' \leq (1/2)e_n'$.

PART B: If Newton's method is used and the root r is a simple root and if $f''(x)$ is continuous near the root $x = r$ then the convergence is quadratic with asymptotic error constant $|f''(r)/(2f'(r))|$, except when $f''(r)$ is zero, in which case we have hyperquadratic convergence. But if the root $x = r$ is a multiple root of order M, then the convergence is only linear with asymptotic error constant $A = (M-1)/M$.

PART C: If the secant method is used and if again $f''(x)$ is continuous near the root $x = r$ and the root is simple then the convergence will be of order $(1+\sqrt{5})/2 = 1.62...$

Furthermore, the bisection method will always converge as long as the initial approximation x_0 is taken to be within a bracket of $x = r$. Also, under the additional hypothesis that $f''(x)$ is continuous near the root $x = r$, both Newton's and the secant method will always converge to $x = r$, as long as the initial approximation(s) are sufficiently close to $x = r$.

REMARKS: This theorem has a lot of practical use. We essentially already knew what is mentioned about the bisection method. But for Newton's and the secant method, the last statement tells us that as long as we start our approximations "sufficiently close" to a root $x = r$ that we are seeking to approximate, the methods will produce sequences of approximations that converge to $x = r$. The "sufficiently close" requirement is admittedly a bit vague, but at least we can keep retrying initial approximations until we get one that will produce a "good" sequence of approximations. The theorem also tells us that once we get any sequence (from Newton's or the secant method) that converges to a root $x = r$, it will be one that converges at the stated orders. Thus, any initial approximation that produces a sequence converging to a root will produce a great sequence of approximations. A similar analysis of Muller's method will show that if it converges to a simple root, it will do so with order 1.84..., a rate quite halfway between that of the secant and Newton's methods. In general, one cannot say that the bisection method satisfies the definition of linear convergence. The reason for this is that it is possible for some x_n to be coincidentally very close to the root while x_{n+1} is much farther from it (see Exercise 14).

Sketch of Proof of Theorem 6.1: A few details of the proof are quite technical so we will reference out parts of the proof to a more advanced text in numerical analysis. But we will give enough of the proof so as to give the reader a good understanding of what is going on. The main ingredient of the proof is Taylor's theorem. We have already seen the importance of Taylor's theorem as a practical tool for approximations; this proof will demonstrate also its power as a theoretical tool. As mentioned in the remarks above, all statements pertaining to the bisection method easily follow from previous developments. The error analysis estimate (1) makes most of these comments transparent. This estimate implies that $e_n = |r - x_n|$ is at most $(b-a)/2^n$, where $b - a$ is simply the length of the bracket interval $[a,b]$. Sometimes we might get lucky since e_n could conceivably be a lot smaller, but in general we can only guarantee this upper bound for e_n. If we set $e_n' = (b-a)/2^n$, then we easily see that $e_{n+1}' = (1/2)e_n'$ and we obtain the said order of convergence in part A.

The proofs of parts B and C are more difficult. A good approach is to use Taylor's theorem. Let us first deal with the case of a simple root and first with part B (Newton's method).

Since x_n converges to $x = r$, Taylor's theorem allows us to write:

$$f(r) = f(x_n) + f'(x_n)(r - x_n) + \tfrac{1}{2}f''(c_n)(r - x_n)^2 ,$$

where c_n is a number between $x = r$ and $x = x_n$ (as long as n is large enough for f'' to be continuous between the $x = r$ and $x = x_n$).

The hypotheses imply that $f'(x)$ is nonzero for x near $x = r$. (Reason: $f'(r) \neq 0$ because $x = r$ is a simple root. Since $f'(x)$ (and $f''(x)$) are continuous near $x = r$ we also have $f'(x) \neq 0$ for x close enough to $x = r$.) Thus we can divide both sides of the previous equation by $f'(x_n)$ and since $f(r) = 0$ (remember $x = r$ is a root of $f(x)$), this leads us to:

$$0 = \frac{f(x_n)}{f'(x_n)} + r - x_n + \frac{1}{2}\frac{f''(c_n)}{f'(x_n)}(r - x_n)^2 .$$

But from Newton's recursion formula (2) we see that the first term on the right of this equation is just $x_n - x_{n+1}$ and consequently

$$0 = x_n - x_{n+1} + r - x_n + \frac{1}{2}\frac{f''(c_n)}{f'(x_n)}(r - x_n)^2 .$$

We cancel the x_n s and then can rewrite the equation as

$$r - x_{n+1} = \frac{-f''(c_n)}{2f'(x_n)}(r - x_n)^2 .$$

We take absolute values in this equation to get the desired proportionality relationship of errors:

$$e_{n+1} = \left|\frac{f''(c_n)}{2f'(x_n)}\right| e_n^2 . \tag{6}$$

Since the functions f' and f'' are continuous near $x = r$ and x_n (and hence also c_n) converges to r, the statements about the asymptotic error constants or the hyperquadratic convergence now easily follow from (6).

Moving on to part C (still in the case of a simple root), a similar argument to the above leads us to the following corresponding proportionality relationship for errors in the secant method (see Section 2.3 of [Atk-89] for the details):

$$e_{n+1} \approx e_n e_{n-1}\left|\frac{f''(r)}{2f'(r)}\right| . \tag{7}$$

It will be an exact inequality if r is replaced by certain numbers near $x = r$ —as in (6). In (7) we have assumed that $f''(r) \neq 0$. The special case where this second derivative is zero will produce a faster converging sequence, so we deal only with the worse (more general) general case and the estimates we get here will certainly apply all the more to this special remaining case. From (7) we can actually deduce the precise order $\alpha > 0$ of convergence. To do this we temporarily define the proportionality ratios A_n by the equations $e_{n+1} = A_n e_n^{\alpha}$ (cf, equation (5)).

We can now write:

$$e_{n+1} = A_n e_n^\alpha = A_n (A_{n-1} e_{n-1}^\alpha)^\alpha = A_n A_{n-1}^\alpha e_{n-1}^{\alpha^2},$$

which gives us that

$$\frac{e_{n+1}}{e_n e_{n-1}} = \frac{A_n A_{n-1}^\alpha e_{n-1}^{\alpha^2}}{A_{n-1} e_{n-1}^\alpha e_{n-1}} = A_n A_{n-1}^{\alpha-1} e_{n-1}^{\alpha^2-\alpha-1}.$$

Now, as $n \to \infty$, (7) shows the left side of the above equation tends to some positive number. On the right side, however, since $e_n \to 0$, assuming that the A_n's do not get too large (see Section 2.3 of [Atk-89] for a justification of this assumption) this forces the exponent $\alpha^2 - \alpha - 1 = 0$. This equation has only one positive solution, namely $\alpha = (1 + \sqrt{5})/2$.

Next we move on to the case of a multiple root of multiplicity M. We can write $f(x) = (x-r)^M h(x)$ where $h(x)$ is a continuous function with $h(r) \neq 0$. We additionally assume that $h(x)$ is sufficiently differentiable. In particular, we will have $f'(x) = M(x-r)^{M-1}h(x) + (x-r)^M h'(x)$, and so we can rewrite Newton's recursion formula (2) as:

$$x_{n+1} = x_n - \frac{(x_n - r)h(x_n)}{Mh(x_n) + (x_n - r)h'(x_n)} = g(x_n),$$

where

$$g(x) = x - \frac{(x-r)h(x)}{Mh(x) + (x-r)h'(x)}.$$

Since

$$g'(x) = 1 - \frac{h(x)}{Mh(x) + (x-r)h'(x)} - (x-r)\left(\frac{h(x)}{Mh(x) + (x-r)h'(x)}\right)',$$

we have that

$$g'(r) = 1 - \frac{1}{M} = (M-1)/M > 0$$

(since $M \geq 2$). Taylor's theorem now gives us that

$$x_{n+1} = g(x_n) = g(r) + g'(r)(x_n - r) + g''(c_n)(x_n - r)^2/2$$
$$= r + g'(r)(x_n - r) + g''(c_n)(x_n - r)^2/2$$

(where c_n is a number between r and x_n). We can rewrite this as:

$$e_{n+1} = [(M-1)/M]e_n + g''(c_n)e_n^2/2.$$

Since $e_n \to 0$, (assuming g'' is continuous at $x = r$), we can divide both sides of this inequality by e_n to get the asserted linear convergence. For details on this latter convergence and also for proofs of the actual convergence guarantee (we have only given sketches of the proofs of the rates of convergence assuming the approximations converge to a root), we refer the interested reader to Chapter 2 of [Atk-89]. QED

From the last part of this proof, it is apparent that in the case we are using Newton's method for a multiple root of order $M > 1$, it would be a better plan to use the modified recursion formula:

$$x_{n+1} = x_n - M \frac{f(x_n)}{f'(x_n)}. \tag{8}$$

Indeed, the proof above shows that with this modification, when applied to an order M multiple root, Newton's method will again converge quadratically (Exercise 13). This formula, of course, requires knowledge about M.

EXAMPLE 6.8: Modify the Program 6.2, `newton`, into a more general one, call it `newtonmr` that is able to effectively use (8) in cases of multiple roots. Have your program run with the function $f(x) = x^{21}$ again with initial approximation $x = 1$, as was done in Example 6.6 with the ordinary `newton` program.

SOLUTION: We only indicate the changes needed to be made to `newton` to get the new program. We will need one new variable (a sixth one), call it `rootord`, that denotes the order of the root being sought after. In the first if-branch of the program (`if nargin < 4`) we also add the default value `rootord =1`. The only other change needed will be to replace the analogue of (2) in the program with that of (8). If we now run this new program, we will find the exact root $x = 0$. In fact, as you can check with (8) (or by slightly modifying the program to give as output the number of iterations), it takes only a single iteration. Recall from Example 6.6 that if we used the ordinary Newton's method for this function and initial approximation, we would need about 135 iterations to get an approximation (of zero) with error less than 0.001!

In order to effectively use this modified Newton's method for multiple roots, it is necessary to determine the order of a multiple root. One way to do this would be to compare the graphs of the function at hand near the root r in question, together with graphs of successive derivatives of the function, until it is observed that a certain order derivative no longer has a root at r; see also Exercise 15. The order of this derivative will be the order of the root. MATLAB can compute derivatives (and indefinite integrals) provided that you have the Student Version, or the package you have installed includes the Symbolic Math Toolbox; see Appendix A.

For polynomials, MATLAB has a built-in function, `roots`, that will compute all of its roots. Recall that a polynomial of degree n will have exactly n real or complex roots, if we count them according to their multiplicities. Here is the syntax of the `roots` command:

`roots([an ... a2 a1 a0])` $\rightarrow$	Computes (numerically) all of the n real and complex roots of the polynomial whose coefficients are given by the inputted vector: $$p(x) = a_n x^n + a_{n-1} x^{n-1} + \cdots + a_2 x^2 + a_1 x + a_0$$

EXAMPLE 6.9: Use MATLAB to find all of the roots of the polynomials

$$p(x) = x^8 - 3x^7 + (9/4)x^6 - 3x^5 + (5/2)x^4 + 3x^3 + (9/4)x^2 + 3x + 1,$$
$$q(x) = x^6 + 2x^5 - 6x^4 - 10x^3 + 13x^2 + 12x - 12.$$

SOLUTION: Let us first store the coefficients of each polynomial as a vector:

```
>>pv=[1 -3  9/4 -3  5/2  3  9/4  3  1]; qv= [1 2  -6  -10 13 12 -12];
>> roots(pv) %this single command will get us all of the roots of
p(x)
→  2.0000 + 0.0000i
   2.0000 - 0.0000i
   0.0000 + 1.0000i
   0.0000 - 1.0000i
  -0.0000 + 1.0000i
  -0.0000 - 1.0000i
  -0.5000 + 0.0000i
  -0.5000 - 0.0000i
```

Since some of these roots are complex, they are all listed as complex numbers. The distinct roots are $x = 2$, $i - i$, and .5, each of which are double roots. Since $(x+i)(x-i) = x^2 + 1$ these roots allow us to rewrite $p(x)$ in factored form: $p(x) = (x^2 + 1)^2 (x-2)^2 (x+0.5)^2$. The roots of $q(x)$ are similarly obtained:

```
>> roots(qv)
→  1.7321
  -2.0000
  -2.0000
  -1.7321
   1.0000
   1.0000
```

Since the roots of $q(x)$ are all real, they are written as real numbers. We see that $q(x)$ has two double roots, $x = -2$ and $x = 1$, and two simple roots that turn out to be $\pm\sqrt{3}$.

EXERCISE FOR THE READER 6.8: (*Another Approach to Multiple Roots with Newton's Method*). Suppose that $f(x)$ has multiple roots. Show that the function $f(x)/f'(x)$ has the same roots as $f(x)$, but they are all simple. Thus Newton's method could be applied to the latter function with quadratic convergence to determine each of the roots of $f(x)$. What are some problems that could crop up with this approach?

EXERCISES 6.5:

1. Find the highest order of convergence (if defined) of each of the following sequences of errors:

    (a) $e_n = 1/n^5$;   (c) $e_n = n^{-n}$

    (b) $e_n = e^{-n}$ (d) $e_n = 2^{-n^2}$

2. Find the highest order of convergence (if defined) of each of the following sequences of errors:

    (a) $e_n = 1/\ln(n)^n$;   (c) $e_n = 1/\exp(\exp(\exp(n)))$

    (b) $e_n = 1/\exp(\exp(n))$ (d) $e_n = 1/n!$

3. For each of the sequences of Exercise 1 that had a well-defined highest order of convergence, determine the asymptotic error constant or indicate if there is hyperconvergence.

4. For each of the sequences of Exercise 2 that had a well-defined highest order of convergence, determine the asymptotic error constant or indicate if there is hyperconvergence.

5. Using just Newton's method or the improvement (8) of it for multiple roots, determine all (real) roots of the polynomial

    $$x^8 + 4x^7 - 17x^6 - 84x^5 + 60x^4 + 576x^3 + 252x^2 - 1296x - 1296 .$$

    Give also the multiplicity of each root and justify these numbers.

6. Using just Newton's method or the improvement (8) of it for multiple roots, determine all (real) roots of the polynomial $x^{10} + x^9 + x^8 - 18x^6 - 18x^5 - 18x^4 + 81x^2 + 81x + 81$. Give also the multiplicity of each root and justify these numbers.

7. (*Fixed Point Iteration*) (a) Assume that $f(x)$ has a root in $[a, b]$, that $g(x) = x - f(x)$ satisfies $a \le g(x) \le b$ for all x in $[a, b]$ and that $|g'(x)| \le \lambda < 1$ for all x in $[a, b]$. Show that the following simple iteration scheme: $x_{n+1} = g(x_n)$, will produce a sequence that converges to a root of $f(x)$ in $[a, b]$.

    (b) Show that $f(x)$ has a unique root in $[a, b]$, provided that all of the hypotheses in part (a) are satisfied.

8. The following algorithm computes the square root of a number $A > 0$:

    $$x_{n+1} = \frac{x_n(x_n^2 + 3A)}{3x_n^2 + A} .$$

    (a) Show that it has order of convergence equal to 3 (assuming x_0 has been chosen sufficiently close to $\sqrt{A}$).

    (b) Perform 3 iterations of it to calculate $\sqrt{10}$ starting with $x_0 = 3$. What is the error?

    (c) Compute the asymptotic error constant.

9. Can you devise a scheme for computing cube roots of positive numbers that, like the one in Exercise 8, has order of convergence equal to 3? If you find one, test it out on $\sqrt[3]{30}$.

10. Prove: If $\beta > \alpha$ and we have a sequence that converges with order β, then the sequence will also converge with order α.

11. Is it possible to have quadratic convergence with asymptotic error constant equal to 3? Either provide an example or explain why not.

12. Prove formula (7) in the proof of Theorem 6.1, in case $f''(r) \neq 0$.

13. Give a careful explanation of how (8) gives quadratic convergence in the case of a root of order $M > 1$, provided that x_0 is sufficiently close to the root.
    Suggestion: Carefully examine the last part of the proof of Theorem 6.1.

14. (*Nonlinear Convergence of the Bisection Method*) (a) Construct a function $f(x)$ that has a root r in an interval $[a, b]$ and that satisfies the requirements of the bisection method but such that x_n does not converge linearly to r.
    (b) Is it possible to have $\limsup_{n \to \infty} e_{n+1} / e_n = \infty$ with the bisection method for a function that satisfies the conditions of part (a)?

15. (a) Explain how Newton's method could be used to detect the order of a root, and then formulate and prove a precise result.
    (b) Use the idea of part (a) to write a MATLAB M-file, newtonorddetect, having a similar syntax to the newton M-file of Program 6.2. Your program should first detect the order of the root, and then use formula (8) (modified Newton's method) to approximate the root. Run your program on several examples involving roots of order 1, 2, 3, and compare the number of iterations used with that of the ordinary Newton's method. In your comparisons, make sure to count the total number of iterations used by newtonorddetect, both in the detection process as well as in the final implementation.
    (c) Run your program of part (b) on the problem of Example 6.8.
    Note: For the last comparisons asked for in part (b), you should modify newton to output the number of iterations used, and include such an output variable in your newtonorddetect program.

Chapter 7: Matrices and Linear Systems

7.1:   MATRIX   COMPUTATIONS   AND   MANIPULATIONS   WITH MATLAB

I saw my first matrix in my first linear algebra course as an undergraduate, which came after the calculus sequence. A matrix is really just a spreadsheet of numbers, and as computers are having an increasing impact on present-day life and education, the importance of matrices is becoming paramount.   Many interesting and important problems can be solved using matrices, and the basic concepts for matrices are quite easy to introduce. Presently, matrices are making their way down into lower levels of mathematics courses and, in some instances, even elementary school curricula. Matrix operations and calculations are simple in principle but in practice they can get quite long. It is often not feasible to perform such calculations by hand except in special situations. Computers, on the other hand, are ideally suited to manipulate matrices and MATLAB has been specially designed to effectively manipulate them. In this section we introduce the basic matrix operations and show how to perform them in MATLAB. We will also present some of the numerous tricks, features, and ideas that can be used in MATLAB to store and edit matrices. In Section 7.2 we present some applications of basic matrix operations to computer graphics and animation. The very brief Section 7.3 introduces concepts related to linear systems and Section 7.4 shows ways to use MATLAB to solve linear systems.   Section 7.5 presents an algorithmic and theoretical development of Gaussian elimination and related concepts. In Section 7.6, we introduce norms with the goal of developing some error estimates for numerical solutions of linear systems. Section 7.7 introduces iterative methods for solving linear systems. When conditions are ripe, iterative methods can perform much more quickly and effectively than Gaussian elimination.

We first introduce some definitions and notations from mathematics and then translate them into MATLAB's language. A **matrix** A is a rectangular array of numbers, called *entries*. A generic form for a matrix is shown in Figure 7.1. The **rows** of a matrix are its horizontal strips of entries (labeled from the top) and the **columns** are the vertical strips of entries (labeled from the left). The entry of A that lies in row i and in column j is written as a_{ij} (if either of the indices i or j is greater than a single digit, then the notation inserts a comma: $a_{i,j}$). The matrix A is said to be of size n by m (or an $n \times m$ matrix) if it has n rows and m columns.

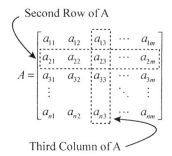

Second Row of A

Third Column of A

FIGURE 7.1: The anatomy of a matrix A having n rows and m columns. The entry that lies in the second row and the third column is written as a_{23}.

In mathematical notation, the matrix A in Figure 7.1 is sometimes written in the abbreviated form

$$A = [a_{ij}],$$

where its size either is understood from the context or is unimportant. With this notation it is easy to explain how **matrix addition/subtraction** and **scalar multiplication** work. The matrix A can be multiplied by any scalar (real number) α with the obvious definition:

$$\alpha A = [\alpha a_{ij}].$$

Matrices can be added/subtracted only when they have the same size, and the definition is the obvious one: Corresponding entries get added/subtracted, i.e.,

$$[a_{ij}] \pm [b_{ij}] = [a_{ij} \pm b_{ij}].$$

Matrix multiplication, on the other hand, is not done in the obvious way. To explain it we first recall the definition of the **dot product** of two vectors of the same size. If $a = [a_1, a_2, \cdots, a_n]$ and $b = [b_1, b_2, \cdots, b_n]$ are two vectors of the same length n (for this definition it does not matter if these vectors are row or column vectors), the dot product of these vectors is defined as follows:

$$a \cdot b = a_1 b_1 + a_2 b_2 + \cdots + a_n b_n = \sum_{k=1}^{n} a_k b_k.$$

Now, if $A = [a_{ij}]$ is an $n \times m$ matrix and $B = [b_{ij}]$ is an $m \times r$ matrix (i.e., the number of columns of A equals the number of rows of B), then the matrix product $C = AB$ is defined and it will be another matrix $C = [c_{ij}]$ of size $n \times r$. To get an entry c_{ij}, we simply take the dot product of the ith row vector of A and the jth column vector of B. Here is a simple example:

EXAMPLE 7.1: Given the matrices:

$$A = \begin{bmatrix} 1 & 0 \\ -3 & 8 \end{bmatrix}, \quad B = \begin{bmatrix} 4 & 1 \\ 3 & -6 \end{bmatrix}, \quad M = \begin{bmatrix} 4 & -9 \\ 3 & 1 \\ -2 & 5 \end{bmatrix}, \quad \text{compute the following:} \quad A - 2B,$$

AB, BA, AM, MA

SOLUTION: $A - 2B = \begin{bmatrix} 1 & 0 \\ -3 & 8 \end{bmatrix} - \begin{bmatrix} 8 & 2 \\ 6 & -12 \end{bmatrix} = \begin{bmatrix} -7 & -2 \\ -9 & 20 \end{bmatrix}$,

$$AB = \begin{bmatrix} 1 & 0 \\ -3 & 8 \end{bmatrix} \begin{bmatrix} 4 & 1 \\ 3 & -6 \end{bmatrix} = \begin{bmatrix} 4 & 1 \\ 12 & -51 \end{bmatrix}.$$

(In the product AB, the indicated second row, first column entry was computed by taking the dot product of the corresponding second row of A (shown) with the first column of B (also indicated).) Similarly,

$$BA = \begin{bmatrix} 1 & 8 \\ 21 & -48 \end{bmatrix}, \quad \text{and} \quad MA = \begin{bmatrix} 31 & -72 \\ 0 & 8 \\ -17 & 40 \end{bmatrix}.$$

The matrix product AM is not defined since the inner dimensions of the sizes of A and M are not the same (i.e., $2 \neq 3$). In particular, these examples show that matrix multiplication is, in general, not commutative; i.e., we cannot say that $AB = BA$ even when both matrix products are possible and have the same sizes.

At first glance, the definition of matrix multiplication might seem too awkward to have much use. But later in this chapter (and in subsequent chapters as well) we will see numerous applications. We digress momentarily to translate these concepts and notations into MATLAB's language. To redo the above example in MATLAB, we would simply enter the matrices (rows separated by semicolons, as shown in Chapter 1) and perform the indicated linear combination and multiplications (since default operations are matrix operations in MATLAB, dots are not used before multiplications and other operators).

MATLAB SOLUTION TO EXAMPLE 7.1:
```
>>   A=[1 0; -3 8]; B=[4 1;3 -6]; M=[4 -9;3 1;-2 5];
>>   A-2*B, A*B, B*A, A*M, M*A
→ans = -7  -2      ans =    4   1      ans =   1    8
        -9  20               12 -51              21  -48
→??? Error using ==> *
Inner matrix dimensions must agree.
→ans = 31  -72
        0   8
       -17  40
```

By hand, matrix multiplication is feasible only for matrices of very small sizes. For example, if we multiply an $n \times n$ (square) matrix by another one of the same size, each entry involves a dot product of two vectors with n components and thus will require n multiplications and $n - 1$ additions (or subtractions). Since there are n^2 entries to compute, this yields a total of $n^2(n + n - 1) = 2n^3 - n^2$ floating point operations (flops). For a 5×5 matrix multiplication, this works out to be 225 flops and for a 7×7, this works already to 637 flops, a very tedious hand calculation pushing on the fringes of insanity. As we saw in some experiments in Chapter 4, on a PC with 256 MB of RAM and a 1.6 GHz microprocessor, MATLAB can roughly perform approximately 10 million flops in a few seconds. This means that (replacing the previous bound $2n^3 - n^2$ by the more liberal but easier to deal with bound $2n^3$, setting this equal to 10 million and solving for n) MATLAB can quickly multiply two matrices of size up to about 171×171 (check that a matrix multiplication of two such large matrices works out to about 10 million flops). Actually, not all flop calculations take equal times (this is why the word "flop" has a somewhat vague definition). It turns out that because of MATLAB's specially designed features that are mainly tailored to deal effectively with matrices, MATLAB can actually quickly multiply much larger matrices. On a PC with the aforementioned specs, the following experiment took MATLAB just a few seconds to multiply a pair of 1000×1000 matrices[1].

```
>> flops(0)
>> A=rand(1000); B=rand(1000); %constructs two 1000 by 1000 random
                                 matrices
>> A*B;
>> flops      → 2.0000e + 009
```

This calculation involved 2 billion flops! Later on we will come across applications where such large-scale matrix multiplications come up naturally. Of course, if one (or both) matrices being multiplied have a lot of zero entries, then the computations needed can be greatly reduced. Such matrices (having a very low percentage of nonzero entries) are called **sparse**, and we will later see some special features that MATLAB has to deal with sparse matrices.

We next move on to introduce several of the handy ways that MATLAB offers us to enter and manipulate matrices. For illustration, we assume that the matrices A, B and M of Example 7.1 are still entered in our MATLAB session. The exponential operator ($\hat{}$) in MATLAB works by default as a matrix power. Thus if we enter

```
>> A^2 %matrix squaring
```

we get the square of the matrix A, $A^2 = AA$,

```
→ans = 1    0
       -27  64
```

whereas if we precede the operator by a dot, then, as usual, the operator changes to its entry-by-entry analog, and we get the matrix whose entries each equal the square of the corresponding entries of A.

---

[1] As in Chapter 4, we occasionally use the obsolete flop count function (from MATLAB Version 5) when convenient for an illustration.

```
>> A.^2 %element squaring
→ans = 1    0
        9   64
```

This works the same way with other operators such as multiplication. Matrix operators such as addition/subtraction, which are the same as element-by-element addition/subtraction, make the dot a nonissue.

To refer to a specific entry a_{ij} in a matrix A, we use the syntax:

A(i,j) →	MATLAB's way of referring to the ith row jth column entry a_{ij} of a matrix A,

which was introduced in Section 4.3. Thus, for example, if we wanted to change the row 1, column 2 entry of our 2×2 matrix A (from 0) to 2, we could enter:

```
>> A(1,2)=2 %without suppressing output, MATLAB shows us the whole
                    matrix
→A =  1    2
     -3    8
```

We say that a (square) matrix $D = [d_{ij}]$ is a **diagonal matrix** if all entries, except perhaps those on the main diagonal, are zero (i.e., $d_{ij} = 0$ whenever $i \neq j$). Diagonal matrices (of the same size $n \times n$) are the easiest ones to multiply; indeed, for such a multiplication only n flops are needed:

$$\begin{bmatrix} d_1 & & \\ & d_2 & \text{\Large 0} \\ & & \ddots \\ \text{\Large 0} & & d_n \end{bmatrix} \begin{bmatrix} e_1 & & \\ & e_2 & \text{\Large 0} \\ & & \ddots \\ \text{\Large 0} & & e_n \end{bmatrix} = \begin{bmatrix} d_1 e_1 & & \\ & d_2 e_2 & \text{\Large 0} \\ & & \ddots \\ \text{\Large 0} & & d_n e_n \end{bmatrix}. \tag{1}$$

The large zeros in the above notation are to indicate that all entries in the triangular regions above and below the main diagonal are zeros. There are many matrix-related problems and theorems where things boil down to considerations of diagonal matrices, or minor variations of them.

EXERCISE FOR THE READER 7.1: Prove identity (1).

In MATLAB, we can enter a diagonal matrix using the command diag as follows. To create a 5×5 diagonal matrix D with diagonal entries (in order): 1 2 –3 4 5, we could type:

```
>> diag([1 2 -3 4 5])     →ans = 1    0    0    0    0
                                  0    2    0    0    0
                                  0    0   -3    0    0
                                  0    0    0    4    0
                                  0    0    0    0    5
```

One very special diagonal matrix is $n \times n$ (square) **identity matrix** I_n or simply I (if the size is understood or unimportant). It has all the diagonal entries equaling 1 and has the property that whenever it is multiplied (on either side) by a matrix A (so that the multiplication is possible), the product is A, i.e.,

$$AI = A = IA. \tag{2}$$

Thus, the identity matrix I plays the role in matrix theory that the number 1 plays in arithmetic; i.e., it is the "(multiplicative) identity." Even easier than with the `diag` command, we can create identity matrices with the command `eye`:

```
>> I2=eye(2),  I4=eye(4)
→I2 = 1   0       →I4 = 1   0   0   0
       0   1              0   1   0   0
                          0   0   1   0
                          0   0   0   1
```

Let us check identity (2) for our stored 2×2 matrix $A = \begin{bmatrix} 1 & 2 \\ -3 & 8 \end{bmatrix}$:

```
>> I2*A,   A*I2
→ans = 1   2     →ans = 1   2
      -3   8            -3   8
```

To be able to divide one matrix A by another one B, we will actually have to multiply A by the inverse B^{-1} of the matrix B, if the latter exists and the multiplication is possible. It is helpful to think of the analogy with real numbers: To perform a division, say $5 \div 2$, we can recast this as a multiplication $5 \cdot 2^{-1}$, where the inverse of 2 (as with any nonzero real number) is the reciprocal 1/2. The only real number that does not have an inverse is zero; thus we can always divide any real number by any other real number as long as the latter is not zero. Note that the inverse a^{-1} of a real number a has the property that when the two are multiplied the result will be 1 ($a \cdot a^{-1} = 1 = a^{-1} \cdot a$). To translate this concept into matrix theory is simple; since the number 1 translates to the identity matrix, we see that for a matrix B^{-1} to be the **inverse of a matrix** B, we must have

$$BB^{-1} = I = B^{-1}B. \tag{3}$$

In this case we also say that the matrix B is **invertible** (or **nonsingular**). The only way for the equations of (3) to be possible is if B is a square matrix. There are, however, a lot of square matrices that are not invertible. One way to tell whether a square matrix B is invertible is by looking at its determinant $\det(B)$ (which was introduced in Section 4.3), as the following theorem shows:

THEOREM 7.1: (*Invertibility of Square Matrices*)
(1) A square matrix B is invertible exactly when its determinant $\det(B)$ is nonzero.
(2) In case of a 2×2 matrix $B = \begin{bmatrix} a & b \\ c & d \end{bmatrix}$ with determinant $\det(B) \equiv ad - bc \neq 0$,
the inverse is given by the formula

$$B^{-1} = \frac{1}{\det(B)} \begin{bmatrix} d & -b \\ -c & a \end{bmatrix}.$$

For a proof of this theorem we refer to any good linear algebra textbook (for example [HoKu-71]). There is an algorithm for computing the inverse of a matrix, which we will briefly discuss later, and there are more complicated formulas for the inverse of a general $n \times n$ matrix, but we will not need to go so far in this direction since MATLAB has some nice built-in functions for finding determinants and inverses. They are as follows:

`inv(A)` →	Numerically computes the inverse of a square matrix A
`det(A)` →	Computes the determinant of a square matrix A

The `inv` command must be used with caution, as the following simple examples show. From the theorem, the matrix $M = \begin{bmatrix} 2 & 3 \\ 1 & 2 \end{bmatrix}$ is easily inverted, and MATLAB confirms the result:

```
>> M=[2 3; 1 2];
>> inv(M)
```

→ ans = 2   -3
        -1   2

However, the matrix $M = \begin{bmatrix} 3 & -6 \\ 2 & -4 \end{bmatrix}$ has $\det(M) = 0$, so from the theorem we know that the inverse does not exist. If we try to get MATLAB to compute this inverse, we get the following:

```
>> M=[3 -6; 2 -4];
>> inv(M)
```
→Warning: Matrix is singular to working precision.
ans = Inf Inf
       Inf Inf

The output does not actually tell us that the matrix is not invertible, but it gives us a meaningless answer (`Inf` is MATLAB's way of writing ∞) that seems to suggest that there is no inverse. This brings us to a subtle and important point about floating point arithmetic. Since MATLAB, or any computer system, can work only with a finite number of digits, it is not really possible for MATLAB to distinguish between zero and a very small positive or negative number. Furthermore, when doing computations (e.g., in finding an inverse of a (large) matrix,) there are (a lot of) calculations that must be performed and these will introduce roundoff errors. Because of this, something that is actually zero may appear as a nonzero but small number and vice versa (especially after the "noise" of calculations has distorted it). Because of this it is in general impossible to tell if a certain matrix is invertible or not if its determinant is very small. Here is some practical advice on computing inverses. If you get MATLAB to compute an inverse of a square matrix and get a "warning" as above, you should probably reject the output.   If you then check the determinant of the matrix, chances are good that it will be very small.   Later in this chapter we will introduce the concept of condition numbers for matrices and these will provide a more reliable way to detect so-called **poorly conditioned matrices** that are problematic in linear systems.

Building and storing matrices with MATLAB can be an art. Apart from `eye` and `diag` that were already introduced, MATLAB has numerous commands for the construction of special matrices.   Two such commonly used commands are `ones` and `zeros`.

`zeros(n,m)` →	Constructs an $n \times m$ matrix whose entries each equal 0.
`ones(n,m)` →	Constructs an $n \times m$ matrix whose entries each equal 1.

Of course, `zeros(n,m)` is redundant since we can just use `0*ones(n,m)` in its place. But matrices of zeros come up often enough to justify separate mention of this command.

EXAMPLE 7.2: A **tridiagonal** matrix is one whose nonzero entries can lie either on the main diagonal or on the diagonals directly above/below the main diagonal. Consider the 60×60 tridiagonal matrix A shown below:

$$A = \begin{bmatrix} 1 & 1 & 0 & 0 & 0 & 0 & 0 & \cdots & 0 \\ -1 & 1 & 2 & 0 & 0 & 0 & 0 & \cdots & 0 \\ 0 & -1 & 1 & 3 & 0 & 0 & 0 & \cdots & 0 \\ 0 & 0 & -1 & 1 & 1 & 0 & 0 & \cdots & 0 \\ 0 & 0 & 0 & -1 & 1 & 2 & \ddots & \cdots & 0 \\ 0 & 0 & 0 & 0 & -1 & 1 & 3 & \cdots & 0 \\ \vdots & \vdots & \vdots & \vdots & \ddots & \ddots & \ddots & \ddots & \vdots \\ 0 & 0 & 0 & 0 & \cdots & 0 & -1 & 1 & \\ 0 & 0 & 0 & 0 & 0 & 0 & 0 & \cdots & 1 \end{bmatrix}.$$

It has 1's straight down the main diagonal, −1's straight down the submain diagonal, the sequence (1,2,3) repeated on the supermain diagonal, and zeros for all other entries.
(a) Store this matrix in MATLAB
(b)   Find its determinant and compute and store the inverse matrix as B, if it exists (do not display it). Multiply the determinant of A with that of B.
(c) Print the 6×6 matrix C, which is the submatrix of A whose rows are made up of the rows 30, 32, 34, 36, 38, and 40 of A and whose columns are column numbers 30, 31, 32, 33, 34, and 60 of A .

SOLUTION: Part (a): We start with the 60×60 identity matrix:

```
>> A=eye(60);
```

To put the −1's along the submain diagonal, we can use the following for loop:

```
>> for i=1:59, A(i+1,i)=-1; end
```

(Note: there are 59 entries in the submain diagonal; they are $A(2, 1)$, $A(3, 2)$, ...,
$A(60,59)$ and each one needs to be changed from 0 to −1.)   The supermain
diagonal entries can be changed using a similar for loop structure, but since they
cycle between the three values 1, 2, 3, we could add in some branching within the
for loop to accomplish this cycling. Here is one possible scheme:

```
>> count=1; %initialize counter
>> for i=1:59
if count==1, A(i,i+1)=1;
elseif count==2, A(i,i+1)=2;
else A(i,i+1)=3;
end
count=count+1; %bumps up counter by one
if count==4, count=1; end %cycles counter back to one after it passes
                          %                        3
end
```

We can do a brief check of the upper-left 6×6 submatrix to see if A has shaped
out the way we want it; we invoke the submatrix features introduced in Section
4.3.

```
>> A(1:6,1:6)    → ans =      1   1   0   0   0   0
                             -1   1   2   0   0   0
                              0  -1   1   3   0   0
                              0   0  -1   1   1   0
                              0   0   0  -1   1   2
```

This looks like what we wanted. Here is another way to construct the supermain
diagonal of A. We first construct a vector v that contains the desired supermain
diagonal entries:

```
>> vseed=[1 2 3]; v=vseed;
>> for i=1:19
v=[v vseed]; %tacks on "vseed" onto existing v
end
```

Using this vector v, we can reset the supermain diagonal entries of A as we did
the submain diagonal entries:

```
>> for i=1:59
A(i,i+1)=v(i);
end
```

Shortly, we will give another scheme for building such banded matrices, but it
would behoove the reader to understand why the above loop construction does the
job.

Part (b):
```
>> det(A)    →ans = 3.6116e + 017
>> B=inv(A); det(A)*det(B)    →ans =1.0000
```

This agrees with a special case of a theorem in linear algebra which states that for any pair of square matrices A and B of the same size, we have:

$$\det(AB) = \det(A) \cdot \det(B).$$ (4)

Since it is easy to show that $\det(I) = 1$, from (3) and (4) it follows that $\det(A) \cdot \det(A^{-1}) = 1$.

Part (c): Once again using MATLAB's array features introduced in Section 4.3, we can easily construct the desired submatrix of A as follows:

```
>> C=A(30:2:40, [30:33 59 60])
→ C =     1   3   0   0   0   0
          0  -1   1   2   0   0
          0   0   0  -1   0   0
          0   0   0   0   0   0
          0   0   0   0   0   0
          0   0   0   0   0   0
```

Tridiagonal matrices, like the one in the above example, are special cases of **banded matrices**, which are square matrices with all zero entries except on a certain set of diagonals. Large-sized tridiagonal and banded matrices are good examples of sparse matrices. They arise naturally in the very important finite difference methods that are used to numerically solve ordinary and partial differential equations. In its full syntax, the `diag` command introduced earlier allows us to put any vector on any diagonal of a matrix.

`diag(v,k)` →	For an integer k and an appropriately sized vector v, this command creates a square matrix with all zero entries, except for the kth diagonal on which will appear the vector v. k = 0 gives the main diagonal, k = 1 gives the supermain diagonal, k = −1 the submain diagonal, etc.

For example, in the construction of the matrix A in the above example, after having constructed the 60×60 identity matrix, we could have put in the −1's in the submain diagonal by entering:

```
>>A=A+ diag(-ones(1,59),-1);
```

and with the vector v constructed as in our solution, we could put the appropriate numbers on the supermain diagonal with the command:

```
>>A=A+ diag(v(1:59),1);
```

EXERCISE FOR THE READER 7.2: (*Random Integer Matrix Generator*) (a) For testing of hypotheses about matrices, it is often more convenient to work with integer-valued matrices rather than (floating point) decimal-valued matrices. Create a function M-file, called `randint(n,m,k)`, that takes as input three positive integers n, m and k, and will output an $n \times m$ matrix whose entries are integers randomly distributed in the set $\{-k, -k+1,..., -1, 0, 1, 2,..., k-1, k\}$.

(b) Use this program to test the formula (4) given in the above example by creating two different 6×6 random integer matrices A and B and computing $\det(AB) - \det(A)\cdot\det(B)$ to see if it equals zero. Use $k = 9$ so that the matrices have single-digit entries. Repeat this experiment two times (it will produce different random matrices A and B) to see if it still checks. In each case, give the values of $\det(AB)$.

(c) Keeping $k = 9$, try to repeat the same experiment (again three times) using instead 16×16 sized matrices. Explain what has (probably) happened in terms of MATLAB's default working precision being restricted to about 15 digits.

The matrix arithmetic operations enjoy the following properties (similar to those of real numbers): Here A and B are any compatible matrices and α is any number.

Commutativity of Addition: $A + B = B + A$. (5)

Associativity: $(A + B) + C = A + (B + C)$, $(AB)C = A(BC)$. (6)

Distributive Laws: $A(B + C) = AB + AC$, $(A + B)C = AC + BC$. (7)

$$\alpha(A + B) = \alpha A + \alpha B, \qquad \alpha(AB) = (\alpha A)B = A(\alpha B).$$ (8)

Each of these matrix identities is true whenever the matrices are of sizes that make all parts defined. Experiments relating to these identities as well as their proofs will be left to the exercises. We point out that matrix multiplication is not commutative; the reader is invited to do an easy counterexample to verify this fact.

We close this section with a few more methods and features for manipulating matrices in MATLAB, which we introduce through a simple MATLAB demonstration.

Let us begin with the following 3×3 matrix:

```
>> A = [1 2 3; 4 5 6; 7 8 9]      → A =      1   2   3
                                             4   5   6
                                             7   8   9
```

We can tack on [10 11 12] as a new bottom row to create a new matrix B as follows:

```
>> B=[A;10 11 12]                 → B =      1    2    3
                                             4    5    6
                                             7    8    9
                                            10   11   12
```

If instead we wanted to append this vector as a new column to get a new matrix C, we could add it on as above to the transpose A' of A, and then take transposes again (the transpose operator was introduced in Section 1.3).[2]

```
>> C=[A';  10 11 12]'              → C =        1   2   3   10
                                               4   5   6   11
                                               7   8   9   12
```

Alternatively, we could start by defining $C = A$ and then introducing the (column) vector as the fourth column of C. The following two commands would give the same result/output.

```
>> C=A;
>> C(:,4)=[10 11 12]'
```

To delete any row/column from a matrix, we can assign it to be the empty vector "[ ]". The following commands will change the matrix A into a new 2×3 matrix obtained from the former by deleting the second row.

```
>>A(2,:)=[]       → A =     1   2   3
                           7   8   9
```

EXERCISES 7.1:

1. In this exercise you will be experimentally verifying the matrix identities (5), (6), (7), and (8) using the following square "test matrices:"

$$A = \begin{bmatrix} 1 & 2 \\ 3 & 4 \end{bmatrix}, \ B = \begin{bmatrix} 1 & 2 \\ -2 & 2 \end{bmatrix}, \ C = \begin{bmatrix} 4 & -2 \\ 7 & 6 \end{bmatrix},$$

   For these matrices verify the following:
   (a) $A + B = B + A$ (matrix addition is commutative).
   (b) $(A + B) + C = A + (B + C)$ (matrix addition is associative).
   (c) $(AB)C = A(BC)$ (matrix multiplication is associative).
   (d) $A(B + C) = AB + AC$ and $(A + B)C = AC + BC$ (matrix multiplication is distributive).
   (e) $\alpha(A + B) = \alpha A + \alpha B$ and $\alpha(AB) = (\alpha A)B = A(\alpha B)$ (for any real number α) (test this last one with $\alpha = 3$).
   Note: Such experiments do not constitute proofs. A math experiment can prove only that a mathematical claim is false, however, when a lot of experiments test something to be true, this can give us more of a reason to believe it and then pursue the proof. In the next three exercises, you will be doing some more related experiments. Later exercises in this set will ask for proofs of these identities.

2. Repeat all parts of Exercise 1 using instead the following test matrices:

$$A = \begin{bmatrix} 1 & 2 & 3 \\ 4 & 5 & 6 \\ 7 & 8 & 9 \end{bmatrix}, \ B = \begin{bmatrix} 1 & 2 & 4 \\ -2 & 2 & 4 \\ -8 & -4 & 8 \end{bmatrix}, \ C = \begin{bmatrix} 3 & 4 & 7 \\ -2 & -8 & 0 \\ 7 & 3 & 12 \end{bmatrix}.$$

---

[2] Without transposes this could be done directly with a few more keystrokes as follows:
```
C=[A,  [10; 11; 12]]
```

3.   (a) By making use of the `rand` command, create three 20×20 matrices A, B, and C each of whose (400) entries are randomly selected numbers in the interval $[0, 1]$.   In this problem you are to check that the identities given in all parts of Exercise 1 continue to test true.
(b) Repeat this by creating 50×50 sized matrices.
(c) Repeat again by creating 200×200 sized matrices.
(d) Finally do it one last time using 1000×1000 sized matrices.
Suggestion: Of course, here it is not feasible to display all of these matrices and to compare all of the entries of the matrices on both sides by eye. (For part (d) this would entail 1 million comparisons!) The following `max` command will be useful here:

`max(A)` →	If A is a (row or column) vector, this returns the maximum value of the entries of A; if A is an $n \times m$ matrix, it returns the m-length vector whose jth entry equals the maximum value in the jth column of A.
`max(max(A))` →	(From the functionality of the single "max" command) this will return the maximum value of all the entries in A.
`max(max(abs(A)))` →	Since `abs(A)` is the matrix whose entries are the absolute values of the corresponding entries of A, this command will return the maximum absolute value of all the entries in A.

Thus, an easy way to check if two matrices E and F (of the same size) are equal is to check that `max(max(abs(E-F)))` equals zero.
Note: Due to roundoff errors (that should be increasing with the larger-sized matrices), the matrices will not, in general, agree to MATLAB's working precision of about 15 decimals. Your conclusions should take this into consideration.

4.   (a) Use the function `randint` that was constructed in Exercise for the Reader 7.2 to create three 3×3 random integer matrices (use $k = 9$) A, B, C on which you are to test once again each of the parts of Exercise 1.
(b) Repeat for 20×20 random integer matrices.
(c) Repeat again for 100×100 random integer matrices.
(d) If your results checked exactly in parts (b) and (c), explain why things were able to work with such a large-sized matrix in this experiment, whereas in the experiment of checking identity (4) the Exercise for the Reader 7.2, even a moderately sized 16×16 led to problems.
Suggestion: For parts (b) and (c) you need not print the matrices; refer to the suggestion of the previous exercise.

5.   (a) Build a 20×20 "checkerboard matrix" whose entries alternate from zeros and ones, with a one in the upper-left corner and such that for each entry of this matrix, the immediate upper, lower, left, and right neighbors (if they exist) are different.
(b) Write a function M-file, call it `checker(n)`, that inputs a positive integer n and outputs an $n \times n$, and run it for $n = 2, 3, 4, 5$ and 6.

6.   Making use of the `randint` M-file of Exercise for the Reader 7.2, perform the following experiments.
(a) Run through $N = 100$ trials of taking the determinant of a 3×3 matrix with random integer values spread out among –5 through 5. What was the average value of the determinants? What percent of these 100 matrices were <u>not</u> invertible?
(b) Run through the same experiments with everything the same, except this time let the random integer values be spread out among -10 through 10.
(c) Repeat part (b) except this time using $N = 500$ trials.
(d) Repeat part (c) except this time work with 6×6 matrices.
(e) Repeat part (c) except this time work with 10×10 matrices.
What general patterns have you noticed?   Without doing the experiment, what sort of results

would you expect if we were to repeat part (c) with 20×20 matrices?

Note: You need not print all of these matrices or even their determinants in your solution. Just include the relevant MATLAB code and output needed to answer the questions along with the answers.

7.   This exercise is similar to the previous one, except this time we will be working with matrices whose entries are real numbers (rather than integers) spread out uniformly randomly in an interval. To generate such an $n \times n$ matrix, for example, if we want the entries to be uniformly randomly spread out over $(-3, 3)$, we could use the command `6*rand(n) - 3*ones(n)`.

(a) Run through $N = 100$ trials of taking the determinant of a 3×3 matrix whose entries are selected uniformly randomly as real numbers in (-3, 3). What was the average value of the determinants? What percent of these 100 matrices were <u>not</u> invertible?

(b) Run through the same experiments with everything the same, except this time let the real number entries be uniformly randomly spread out among -10 through 10.

(c) Repeat part (b) except this time using $N = 500$ trials.

(d) Repeat part (c) except this time work with 6×6 matrices.

(e) Repeat part (c) except this time work with 10×10 matrices.

What general patterns have you noticed? Without doing the experiment, what sort of results would you expect if we were to repeat part (c) with 20×20 matrices?

Note: You need not print all of these matrices or even their determinants in your solution; just include the relevant MATLAB code and output needed to answer the questions along with the answers.

8.   Let $M = \begin{bmatrix} 1 & 1 \\ 0 & 1 \end{bmatrix}$, $N = \begin{bmatrix} 1 & 1 \\ 1 & 1 \end{bmatrix}$

(a) Find $M^2, M^3, M^{26}, N^2, N^3, N^{26}$.

(b) Find general formulas for M^n and N^n where n is any positive integer.

(c) Can you find a 2×2 matrix A of real numbers that satisfies $A^2 = I$ (with $A \neq I$)?

(d) Find a 3×3 matrix $A \neq I$ such that $A^3 = I$. Can you find such a matrix that is not diagonal?

Note: Part (c) may be a bit tricky. If you get stuck, use MATLAB to run some experiments on (what you think might be) possible square roots.

9.   Let M and N be the matrices in Exercise 8.

(a) Find (if they exist) the inverses M^{-1}: and N^{-1}.

(b) Find square roots of M and N, i.e., find (if possible) two matrices S and T so that:

$$S^2 = M \ (i.e., \ S = \sqrt{M}) \text{ and } R^2 = N .$$

10. (*Discovering Properties and Nonproperties of the Determinant*)   For each of the following equations, run through 100 tests with 2×2 matrices whose entries are randomly selected integers within $[-9, 9]$ (using the `randint` M-file of Exercise for the Reader 7.2). For those that test false, record a single counter-example. For those that test true, repeat the same experiment twice more, first using 3×3 and then using 6×6 matrices of the same type. In each case, write your MATLAB code so that if all 100 matrix tests pass as "true" you will have as output (only) something like: "With 100 tests involving 2×2 matrices with random integer entries from -9 to 9, the identity always worked"; while if it fails for a certain matrix, the experiment should stop at this point and output the matrix (or matrices) that give a counterexample. What are your conclusions?

(a) $\det(A') = \det(A)$

(b) $\det(2A) = 4 \det(A)$

(c) $\det(-A) = \det(A)$

(d) $\det(A + B) = \det(A) + \det(B)$

(e) If matrix B is obtained from A by replacing one row of A by a number k times the corresponding row of A, then $\det(B) = k \det(A)$.

(f) If one row of A is a constant multiple of another row of A then $\det(A) = 0$.

(g) Two of these identities are not quite correct, in general, but they can be corrected using another of these identities that is correct. Elaborate on this.

Suggestion: For part (f) automate the experiment as follows: After a random integer matrix A is built, randomly select a first row number $r1$ and a different second row number $r2$. Then select randomly an integer k in the range

$[-9, 9]$. Replace the row $r2$ with k times row $r1$. This will be a good possible way to create your test matrices. Use a similar selection process for part (e).

11. (a) Prove that matrix addition is commutative $A + B = B + A$ whenever A and B are two matrices of the same size. (This is identity (5) in the text.)
    (b) Prove that matrix addition is associative, $(A + B) + C = A + (B + C)$ whenever A, B, and C are matrices of the same size. (This is the first part of identity (6) in the text.)

12. (a) Prove that the distributive laws for matrices: $A(B + C) = AB + AC$ and $(A + B)C = AC + BC$, whenever the matrices are of appropriate sizes so that a particular identity makes sense. (These are identities (7) in the text.)
    (b) Prove that for any real number α, we have that $\alpha(A + B) = \alpha A + \alpha B$, whenever A, B and C are matrices of the same size and that $\alpha(AB) = (\alpha A)B = A(\alpha B)$ whenever A and B are matrices with AB defined. (These are identities (8) in the text.)

13. Prove that matrix multiplication is associative, $(AB)C = A(BC)$ whenever A, B, and C are matrices so that both sides are defined. (This is the second part of identity (6) in the text.)

14. (*Discovering Facts about Matrices*) As we have seen, many matrix rules closely resemble corresponding rules of arithmetic. But one must be careful since there are some exceptions. One such notable exception we have encountered is that, unlike regular multiplication, matrix multiplication is not commutative; that is, in general we cannot say that $AB = BA$. For each of the statements below about matrices, either give a counterexample, if it is false, or give a proof if it is true. In each identity, assume that the matrices involved can be any matrices for which the expressions are all defined. Also, we use 0 to denote the zero matrix (i.e., all entries are zeros).
    (a) $0A = 0$.
    (b) If $AB = 0$, then either $A = 0$ or $B = 0$.
    (c) If $A^2 = 0$, then $A = 0$.
    (d) $(AB)' = B'A'$ (recall A' denotes the transpose of A).
    (e) $(AB)^2 = A^2B^2$.
    (f) If A and B are invertible square matrices, then so is AB and $(AB)^{-1} = B^{-1}A^{-1}$.

    Suggestion: If you are uncertain of any of these, run some experiments first (as shown in some of the preceding exercises). If your experiments produce a counterexample, you have disproved the assertion. In such a case you merely record the counterexample and move on to the next one.

7.2: INTRODUCTION TO COMPUTER GRAPHICS AND ANIMATION

Computer graphics is the generation and transformation of pictures on the computer. This is a hot topic that has important applications in science and business as well as in Hollywood (computer special effects and animated films). In this section we will show how matrices can be used to perform certain types of geometric operations on "objects." The objects can be either two- or three-dimensional but most of our illustrations will be in the two-dimensional plane. For two-dimensional objects, the rough idea is as follows. We can represent a basic object in the plane as a MATLAB graphic by using the command plot(x,y), where x and y are vectors of the same length. We write x and y as row vectors, stack x on top of y, and we get a $2 \times n$ matrix A where n is the common length of x and y. We can do certain mathematical operations to this matrix to change it into a new matrix $A1$, whose rows are the corresponding vertex vectors $x1$ and $y1$. If we look at plot(x1,y1), we get a transformed version of the original geometrical object. Many interesting geometric transformations can be realized by simple matrix multiplications, but to make this all work nicely we will need to introduce a new artificial third row of the matrix A, that will simply consist of 1's. If we work instead with these so-called homogeneous coordinates, then all of the common operations of scaling (vertically or horizontally), shifting, rotating, and reflecting can be realized by matrix multiplications of these homogeneous coordinates. We can mix and repeat such transformations to get more complicated geometric transformations; and by putting a series of such plots together we can even make movies. Another interesting application is the construction of fractal sets. Fractal sets (or fractals) are beautiful geometric objects that enjoy a certain "self-similarity property," meaning that no matter how closely one magnifies and examines the object, the fine details will always look the same.

Polygons, which we recall are planar regions bounded by a finite set of line segments, are represented by their vertices. If we store the x-coordinates of these vertices and the y-coordinates of these vertices as separate vectors (say the first two rows of a matrix) preserving the order of adjacency, then MATLAB's plot command can easily plot the polygon.

EXAMPLE 7.3: We consider the following "CAT" polygon shown in Figure 7.2. Store the x- coordinates of the vertices of the CAT as the first row vector of a matrix A, and the corresponding y-coordinates as the second row of the same matrix in such a way that MATLAB will be able to reproduce the cat by plotting the second row vector of A versus the first. Afterwards, obtain the plot from MATLAB.

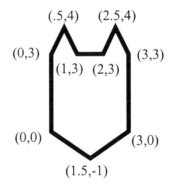

FIGURE 7.2: CAT graphic for Exampe 7.3.

SOLUTION: We can store these nine vertices in a 2×10 matrix A (the first vertex appears twice so the polygon will be closed when we plot it). We may start at any vertex we like, but we must go around the cat in order (either clockwise or counterclockwise). Here is one such matrix that begins at the vertex (0,0) and moves clockwise around the cat.

```
>> A=[0   0  .5  1  2  2.5  3  3  1.5  0; ...
      0   3   4  3  3   4   3  0  -1   0];
```

To reproduce the cat, we plot the second row of A (the y's) versus the first row (the x's):

```
>> plot(A(1,:), A(2,:))
```

   In order to get the cat to fit nicely in the viewing area (recall, MATLAB always sets the view area to just accommodate all the points being plotted), we reset the viewing range to $-2 \le x \le 5, -3 \le y \le 6$, and then use the equal setting on the axes so the cat will appear undistorted.

```
>> axis([-2 5 -3 6])
>> axis('equal')
```

The reader should check how each of the last two commands changes the cat graphic; we reproduce only the final plot in Figure 7.3(a). Figure 7.3 actually contains two cats, the original one (white) as well as a gray cat. The gray cat was obtained in the same fashion as the orginal cat, except that the plot command was replaced by the fill command, that works specifically with polygons and whose syntax is as follows:

fill(x,y,color) →	Here x and y are vectors of the x- and y-coordinates of a polygon (preserving adjacency order); color can be either one of the predefined plot colors (as in Table 1.1) in single quotes, (e.g., k would be a black fill) or an RGB-vector [r g b] (with r, g, and b each being numbers in [0,1]) to produce any color; for example, [.5 .5 .5] gives medium gray.

The elements r, g, and b in a color vector determine the amounts of red, green, and blue to use to create a color; any color can be created in this way. For example, [r g b] = [1 0 0] would be a pure-red fill; magenta is obtained with the rgb vector [1 0 1], and different tones of gray can be achieved by using equal amounts of red, green, and blue between [0 0 0] (black) and [1 1 1] (white).

For the gray cat in Figure 7.3(b), we used the command fill(A(1,:), A(2,:), [.5 .5 .5]). To get a black cat we could either set the rgb vector to [0 0 0] or replace it with k, which represents the preprogrammed color character for black (see Table 1.1).

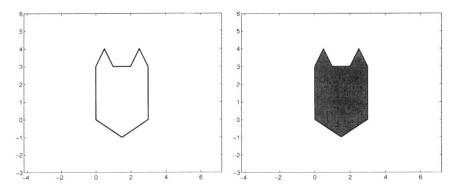

FIGURE 7.3: Two MATLAB versions of the cat polygon: (a) (left) the first white cat was obtained using the `plot` command and (b) (right) the second with the `fill` command.

EXERCISE FOR THE READER 7.3: After experimenting a bit with *rgb* color vectors, get MATLAB to produce an orange cat, a brown cat, and a purple cat. Also, try and find the rgb color vector that best matches the MATLAB built-in color cyan (from Table 1.1, the symbol for cyan is c).

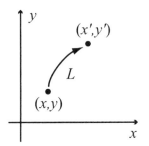

FIGURE 7.4: A linear tranformation *L* in the plane.

A **linear transformation** *L* on the plane R^2 corresponds to a 2×2 matrix *M* (Figure 7.4). It transforms any point (x, y) (represented by the column vector $\begin{bmatrix} x \\ y \end{bmatrix}$) to the point $M\begin{bmatrix} x \\ y \end{bmatrix}$ obtained by multiplying it on the left by the matrix *M* . The reason for the terminology is that a linear transformation preserves the two important linear operations for vectors in the plane: addition and scalar multiplication.   That   is,   letting $P_1 = \begin{bmatrix} x_1 \\ y_1 \end{bmatrix}$, $P_2 = \begin{bmatrix} x_2 \\ y_2 \end{bmatrix}$ be two points in the plane (represented by column vectors), and writing $L(P) = MP$, the linear transformation axioms can be expressed as follows:

$$L(P_1 + P_2) = L(P_1) + L(P_2) , \tag{9}$$

$$L(\alpha P_1) = \alpha L(P_1) . \tag{10}$$

Both of these are to be valid for any choice of vectors $P_i (i = 1, 2)$ and scalar α . Because $L(P)$ is just MP (the matrix *M* multiplied by the matrix *P*), these two identities are consequences of the general properties (7) and (8) for matrices. By the same token, if M is any $n \times n$ matrix, then the transformation $L(P) = MP$

defines a linear transformation (satisfying (9) and (10)) for the space $\mathbf{R}^n$ of n-length vectors. Of course, most of our geomtric applications will deal with the cases $n = 2$ (the plane) or $n = 3$ (3-dimensional (x, y, z) space).

Such tranformations and their generalizations are a basis for what is used in contemporary interactive graphics programs and in the construction of computer videos. If, as in the above example of the CAT, the vertices of a polygon are stored as columns of a matrix A, then, because of the way matrix multiplication works, we can transform each of the vertices at once by multiplying the matrix M of a linear transformation by A. The result will be a new matrix containing the new vertices of the transformed graphic, which can be easily plotted.

We now move on to give some important examples of transformations on the plane R^2.

(1) <u>Scalings:</u> For $a > 0$, $b > 0$ the linear transformation

$$\begin{bmatrix} x' \\ y' \end{bmatrix} = \begin{bmatrix} a & 0 \\ 0 & b \end{bmatrix} \begin{bmatrix} x \\ y \end{bmatrix} = \begin{bmatrix} ax \\ by \end{bmatrix}$$

will scale the horizontal direction with respect to $x = 0$ by a factor of a and the vertical direction with respect to $y = 0$ by a factor of b. If either factor is < 1, there is contraction (shrinkage) toward 0 in the corresponding direction, while factors > 1 give rise to an expansion (stretching) away from 0 in the corresponding direction. As an example, we use $a = 0.3$ and $b = 1$ to rescale our original CAT (Figure 7.5).

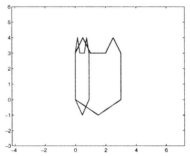

FIGURE 7.5: The scaling of the original cat using factors $a = 0.3$ for horizontal scaling and $b = 1$ (no change) for vertical scaling has produced the narrow-faced cat.

We assume we have left in the graphics window the first (white) cat of Figure 7.3(a).

```
>>M=[.3 0; 0 1]; %store scaling matrix
>>A1=M*A; %create the vertex matrix of the transformed cat;
>>hold on %leave the original cat in the window so we can compare
>>plot(A1(1,:), A1(2,:), 'r') %new cat will be in red
```

Caution: Changes in the axis ranges can also produce scale changes in MATLAB graphics.

(2) <u>Rotations:</u> For a rotation angle θ, the linear tranformation that rotates a point (x, y) an angle θ (counterclockwise) around the origin $(0,0)$ is given by the following linear tranformation:

$$\begin{bmatrix} x' \\ y' \end{bmatrix} = \begin{bmatrix} \cos\theta & -\sin\theta \\ \sin\theta & \cos\theta \end{bmatrix} \begin{bmatrix} x \\ y \end{bmatrix}.$$

(See Exercise 12 for a justification of this.) As an example, we rotate the original cat around the origin using angle $\theta = -\pi/4$ (Figure 7.6). Once again, we assume the graphics window initially contains the original cat of Figure 7.3 before we start to enter the following MATLAB commands:

```
>> M=[cos(-pi/4) -sin(-pi/4); sin(-pi/4) cos(-pi/4)];
>> A1=M*A;, hold on,plot(A1(1,:), A1(2,:), 'r')
```

(3) <u>Reflections:</u>   The linear tranformation that reflects points over the x-axis is given by

$$\begin{bmatrix} x' \\ y' \end{bmatrix} = \begin{bmatrix} -1 & 0 \\ 0 & 1 \end{bmatrix}\begin{bmatrix} x \\ y \end{bmatrix} = \begin{bmatrix} -x \\ y \end{bmatrix}.$$

Similary, to reflect points across the y-axis, the linear transformation will use the matrix $M = \begin{bmatrix} 1 & 0 \\ 0 & -1 \end{bmatrix}$. As an   example, we reflect our original CAT over the y-axis (Figure 7.7). We assume we have left in the graphics window the first cat of Figure 7.3.

```
>> M=[-1 0; 0 1];
>> A1=M*A; hold on, plot(A1(1,:), A1(2,:), 'r')
```

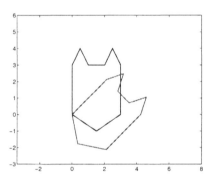

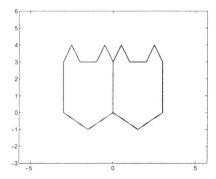

FIGURE 7.6: The rotation (red) of the original   CAT   (blue)   using   angle $\theta = -\pi/4$. The point of rotation is the origin (0,0).

FIGURE 7.7: The reflection (left) of the original CAT across the y-axis.

(4) <u>Shifts:</u> Shifts are very simple and important transformations that are not linear transformations. For a fixed (shift) vector $V_0 = (x_0, y_0) \neq (0,0)$ that we identify, when convenient, with the column vector $\begin{bmatrix} x_0 \\ y_0 \end{bmatrix}$, the **shift transformation** T_{V_0} associated with the shift vector V_0 is defined as follows:

$$(x', y') = T_{V_0}(x, y) = (x, y) + V_0 = (x + x_0, y + y_0).$$

What the shift transformation does is simply move all x-coordinates by x_0 units and move all y-coordinates by y_0 units. As an example we show the outcome of applying the shift transformation $T_{(1,1)}$ to our familiar CAT graphic. Rather than a matrix multiplication with the 2×10 CAT vertex matrix, we will need to add the corresponding 2×10 matrix, each of whose 10 columns is the shift column vector $\begin{bmatrix} 1 \\ 1 \end{bmatrix}$ that we are using (Figure 7.8).

Once again, we assume the graphics window initially contains the original (white) CAT of Figure 7.3 before we start

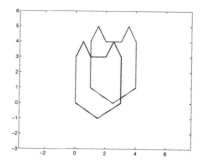

FIGURE 7.8: The shifted CAT (upper right) came from the original CAT using a shift vector (1,1). So the cat was shifted one unit to the right and one unit up.

to enter the following MATLAB commands (and that the CAT vertex matrix A is still in the workspace).

```
>>size(A)  %check size of A   →ans = 2   10
>> V=ones(2,10); A1=A+V; hold on, plot(A1(1,:),A1(2,:), 'r')
```

EXERCISE FOR THE READER 7.4: Explain why the shift transformation is never a linear transformation.

It is unfortunate that the shift transformation cannot be realized as a linear transformation, so that we cannot realize it as using a 2×2 matrix multiplication of our vertex matrix. If we could do this, then all of the important transformations mentioned thus far could be done in the same way and it would make combinations (and in particular making movies) an easier task. Fortunately there is a way around this using so-called homogeneous coordinates. We first point out a more general type of transformation than a linear transformation that includes all linear transformations as well as the shifts. We define it only on the two-dimensional space $\mathbf{R}^2$, but the definition carries over in the obvious way to the three-dimensional space $\mathbf{R}^3$ and higher-dimensional spaces as well. An **affine transformation** on $\mathbf{R}^2$ equals a linear tranformation and/or a shift (applied together). Thus, an affine transformation can be written in the form:

$$\begin{bmatrix} x' \\ y' \end{bmatrix} = M \begin{bmatrix} x \\ y \end{bmatrix} + V_0 = \begin{bmatrix} a & b \\ c & d \end{bmatrix} \begin{bmatrix} x \\ y \end{bmatrix} + \begin{bmatrix} x_0 \\ y_0 \end{bmatrix}, \tag{11}$$

The **homogeneous coordinates** of a point/vector $\begin{bmatrix} x \\ y \end{bmatrix}$ in $\mathbf{R}^2$ is the point/vector $\begin{bmatrix} x \\ y \\ 1 \end{bmatrix}$ in $\mathbf{R}^3$. Note that the third coordinate of the identified three-dimensional point is always 1 in homogeneous coordinates. Geometrically, if we identify a point (x, y) of $\mathbf{R}^2$ with the point $(x, y, 0)$ in $\mathbf{R}^3$ (i.e., we identify $\mathbf{R}^2$ with the plane $z = 0$ in $\mathbf{R}^3$) then homogeneous coordinates simply lift all of these points up one unit to the plane $z = 1$. It may seem at first glance that homogeneous coordinates are making things more complicated, but the advantage in computer graphics is given by the following result.

THEOREM 7.2: (*Homogeneous Coordinates*) Any affine transformation on $\mathbf{R}^2$ is a linear transformation if we use homogeneous coordinates. In other words, any affine transformation T on $\mathbf{R}^2$ can be expressed using homogeneous coordinates in the form:

$$\begin{bmatrix} x' \\ y' \\ 1 \end{bmatrix} = T\left(\begin{bmatrix} x \\ y \\ 1 \end{bmatrix}\right) = H \begin{bmatrix} x \\ y \\ 1 \end{bmatrix} \tag{12}$$

(matrix multiplication), where H is some 3×3 matrix.

Proof: The proof of the theorem is both simple and practical; it will show how to form the matrix H in (12) from the parameters in (11) that determine the affine transformation.

Case 1: T is a linear transformation on $\mathbf{R}^2$ with matrix $M = \begin{bmatrix} a & b \\ c & d \end{bmatrix}$, i.e.,

$T\left(\begin{bmatrix} x \\ y \end{bmatrix}\right) = M \begin{bmatrix} x \\ y \end{bmatrix} = \begin{bmatrix} a & b \\ c & d \end{bmatrix}\begin{bmatrix} x \\ y \end{bmatrix}$ (no shift). In this case, the transformation can be expressed in homogeneous coordinates as:

$$\begin{bmatrix} x' \\ y' \\ 1 \end{bmatrix} = T\left(\begin{bmatrix} x \\ y \\ 1 \end{bmatrix}\right) = \begin{bmatrix} a & b & 0 \\ c & d & 0 \\ 0 & 0 & 1 \end{bmatrix}\begin{bmatrix} x \\ y \\ 1 \end{bmatrix} = H \begin{bmatrix} x \\ y \\ 1 \end{bmatrix}. \tag{13}$$

To check this identity, we simply perform the matrix multiplication:

$$\begin{bmatrix} x' \\ y' \\ 1 \end{bmatrix} = \begin{bmatrix} a & b & 0 \\ c & d & 0 \\ 0 & 0 & 1 \end{bmatrix}\begin{bmatrix} x \\ y \\ 1 \end{bmatrix} = \begin{bmatrix} ax + by + 0 \\ cx + dy + 0 \\ 0 + 0 + 1 \end{bmatrix} \Rightarrow \begin{bmatrix} x' \\ y' \end{bmatrix} = \begin{bmatrix} ax + by \\ cx + dy \end{bmatrix} = M \begin{bmatrix} x \\ y \end{bmatrix},$$

as desired.

Case 2:   T is a shift transformation on $\mathbf{R}^2$ with shift vector $V_0 = \begin{bmatrix} x_0 \\ y_0 \end{bmatrix}$, that is,

$T\left(\begin{bmatrix} x \\ y \end{bmatrix}\right) = \begin{bmatrix} x \\ y \end{bmatrix} + \begin{bmatrix} x_0 \\ y_0 \end{bmatrix}$ (so the matrix M in (12) is the identity matrix). In this case, the transformation can be expressed in homogeneous coordinates as:

$$\begin{bmatrix} x' \\ y' \\ 1 \end{bmatrix} = T\left(\begin{bmatrix} x \\ y \\ 1 \end{bmatrix}\right) = \begin{bmatrix} 1 & 0 & x_0 \\ 0 & 1 & y_0 \\ 0 & 0 & 1 \end{bmatrix} \begin{bmatrix} x \\ y \\ 1 \end{bmatrix} = H \begin{bmatrix} x \\ y \\ 1 \end{bmatrix}. \tag{14}$$

We leave it to the reader to check, as was done in Case 1, that this homogeneous coordinate linear transformation does indeed represent the shift.

Case 3: The general case (linear transformation plus shift);

$$\begin{bmatrix} x' \\ y' \end{bmatrix} = T\left(\begin{bmatrix} x \\ y \end{bmatrix}\right)\begin{bmatrix} x \\ y \end{bmatrix} = \begin{bmatrix} a & b \\ c & d \end{bmatrix}\begin{bmatrix} x \\ y \end{bmatrix} + \begin{bmatrix} x_0 \\ y_0 \end{bmatrix},$$

can now be realized by putting together the matrices in the preceding two special cases:

$$\begin{bmatrix} x' \\ y' \\ 1 \end{bmatrix} = T\left(\begin{bmatrix} x \\ y \\ 1 \end{bmatrix}\right) = \begin{bmatrix} a & b & x_0 \\ c & d & y_0 \\ 0 & 0 & 1 \end{bmatrix}\begin{bmatrix} x \\ y \\ 1 \end{bmatrix} = H \begin{bmatrix} x \\ y \\ 1 \end{bmatrix}. \tag{15}$$

We leave it to the reader check this (using the distributive law (7)).

The basic transformations that we have so far mentioned can be combined to greatly expand the mutations that can be performed on graphics. Furthermore, by using homogeneous coordinates, the matrix of such a combination of basic transformations can be obtained by simply multiplying the matrices by the individual basic transformations that are used, in the correct order, of course. The next example illustrates this idea.

EXAMPLE 7.4: Working in homogeneous coordinates, find the transformation that will rotate the CAT about the tip of its chin by an angle of $-90°$. Express the transformation using the 3×3 matrix M for homogeneous coordinate multiplication, and then get MATLAB to create a plot of the transformed CAT along with the original.

SOLUTION: Since the rotations we have previously introduced will always rotate around the origin (0,0), the way to realize this transformation will be by combining the following three transformations (in order):
(i) First shift coordinates so that the chin gets moved to (0,0). Since the chin has coordinates $(1.5, -1)$, the shift vector should be the opposite so we will use the shift transformation

$$T_{(-1.5,1)} \sim \begin{bmatrix} 1 & 0 & -1.5 \\ 0 & 1 & 1 \\ 0 & 0 & 1 \end{bmatrix} = H_1$$

(the tilde notation is meant to indicate that the shift transformation is represented in homogeneous coordinates by the given 3×3 matrix H_1, as specified by (14)).

(ii) Next rotate (about $(0,0)$) by $\theta = -90°$. This rotation transformation R has matrix

$$\begin{bmatrix} \cos(-90°) & -\sin(-90°) \\ \sin(-90°) & \cos(-90°) \end{bmatrix} = \begin{bmatrix} 0 & 1 \\ -1 & 0 \end{bmatrix},$$

and so, by (13), in homogeneous coordinates is represented by

$$R \sim \begin{bmatrix} 0 & 1 & 0 \\ -1 & 0 & 0 \\ 0 & 0 & 1 \end{bmatrix} = H_2.$$

(iii) Finally we undo the shift that we started with in (i), using

$$T_{(1.5,-1)} \sim \begin{bmatrix} 1 & 0 & 1.5 \\ 0 & 1 & -1 \\ 0 & 0 & 1 \end{bmatrix} = H_3.$$

If we multiply each of these matrices (in order) on the left of the original homogeneous coordinates, we obtain the transformed homogeneous coordinates:

$$\begin{bmatrix} x' \\ y' \\ 1 \end{bmatrix} = H_3 H_2 H_1 \begin{bmatrix} x \\ y \\ 1 \end{bmatrix} \equiv M \begin{bmatrix} x \\ y \\ 1 \end{bmatrix},$$

that is, the matrix M of the whole transformation is given by the product $H_3 H_2 H_1$. We now turn things over to MATLAB to compute the matrix M and to plot the before and after plots of the CAT.

```
>> H1=[1 0 -1.5; 0 1 1; 0 0 1]; H2=[0 1 0; -1 0 0; 0 0 1];
>> H3=[1 0 1.5;0 1 -1; 0 0 1];
>> format rat %will give a nicer display of the matrix M
>> M=H3*H2*H1
→M =      0      1      5/2
         -1      0      1/2
          0      0      1
```

We will multiply this matrix M by the matrix AH of homogeneous coordinates corresponding to the matrix A. To form AH, we simply need to tack on a row of ones to the bottom of the matrix A. (See Figure 7.9.)

```
>> AH=A; %start with A
>> size(A) %check the size of A
→ans = 2   10
>> AH(3,:)=ones(1,10); %form the
>> %appropriately sized third row
>> %for AH
>> size(AH)  →ans = 3   10
>> hold on, AH1=M*AH;
>> plot(AH1(1,:), AH1(2,:),  'r')
```

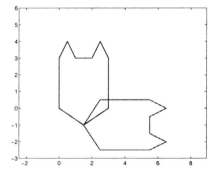

FIGURE 7.9: The red CAT was obtained from the blue cat by rotating $-90°$ about the chin. The plot was obtained using homogeneous coordinates in Example 7.3.

EXERCISE FOR THE READER 7.5:   Working in homogeneous coordinates, (a) find the transformation that will shift the CAT one unit to the right and then horizontally expand it by a factor of 2 (about $x = 0$) to make a "fat CAT". Express the transformation using the 3×3 matrix M for homogeneous coordinate multiplication, and then use MATLAB to create a plot of the transformed fat cat along with the original.
(b) Next, find four transformations each shifting the cat by one of the following shift vectors $(\pm 1, \pm 1)$ (so that all four shift vectors are used) after having rotated the CAT about the central point (1.5, 1.5) by each of the following angles: 30° for the upper-left CAT, −30° for the upper-right CAT, 45° for the lower-left cat, and −45° for the lower-right cat. Then fill in the four cats with four different (realistic cat) colors, and include the graphic.

We now show how we can put graphics transformations together to create a movie in MATLAB. This can be done in the following two basic steps:

STEPS FOR CREATING A MOVIE IN MATLAB:

Step 1: Construct a sequence of MATLAB graphics that will make up the frames of the movie. After the *j*th frame is constructed, use the command `M(:,j) = getframe;` to store the frame as the *j*th column of some (movie) matrix M.

Step 2: To play the movie, use the command `movie(M, rep, fps)`, where M is the movie matrix constructed in step 1, `rep` is a positive integer giving the number of times the movie is to be (repeatedly) played, and `fps` denotes a positive integer giving the speed, in "frames per second," at which the movie is to be played.

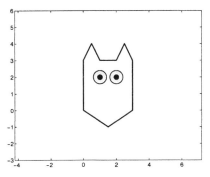

Our next example gives a very simple example of a movie. The movie star will of course be the CAT, but this time we will give it eyes (Figure 7.10). For this first example, we do not use matrix transformations, but instead we directly edit (via a loop) the code that generates the graphic. Of course, a textbook cannot play the movie, so the reader is encouraged to rework the example in front of the computer and thus replay the movie.

FIGURE 7.10:   The original CAT of Example 7.3 with eyes added, the star of our first cat movie.

EXAMPLE 7.5:   Modify the CAT graphic to have a black outline, to have two circular eyes (filled in with yellow), and with two smaller black-filled pupils at the center of the eyes. Then make a movie of the cat closing and then reopening its eyes.

SOLUTION: The strategy will be as follows: To create the new CAT with the specified eyes, we use the "hold on" command after having created the basic CAT. Then we `fill` in yellow two circles of radius 0.4 centered at (1, 2) (left eye) and at (2, 2) (right eye); after this we fill in black two smaller circles with radii 0.15 at the same centers (for the pupils). The circles will actually be polygons obtained by parametric equations. To gradually close the eyes, we use a for loop to create CATs with the same outline but whose eyes are shrinking only in the vertical direction.

This could be done with homogeneous coordinate transforms (that would shrink in the y direction each eye but maintain the centers—thus it would have to first shift the eyes down to $y = 0$, shrink and then shift back), or alternatively we could just directly modify the y parametric equations of each eye to put a shrinking scaling factor in front of the sine function to turn the eyes both directly into a shrinking (and later expanding) sequence of ellipses. We proceed with the second approach. Let us first show how to create the CAT with the indicated eyes. We begin with the original CAT (this time with black line color rather than blue), setting the `axis` options as previously, and then enter `hold on`. Assuming this has been done, we can create the eyes as follows:

```
>> t=0:.02:2*pi;   %creates time vector for parametric equations
>> x=1+.4*cos(t); y=2+.4*sin(t); %creates circle for left eye
>> fill(x,y,'y') %fills in left eye
>> fill(x+1,y, 'y') %fills in right eye
>> x=1+.15*cos(t); y=2+.15*sin(t); %creates circle for left pupil
>> fill(x,y,'k') %fills in left pupil
>> fill(x+1,y,'k') %fills in right pupil
```

To make the frames for our movie (and to "get" them), we employ a for loop that goes through the above construction of the CAT with eyes, except that a factor will be placed in front of the sine term of the y-coordinates of both eyes and pupils. This factor will start at 1, shrink to 0, and then expand back to the value of 1 again. To create such a factor, we need a function with starting value 1 that decreases to zero, then turns around and increases back to 1. One such function that we can use is $(1 + \cos x)/2$ over the interval $[0, 2\pi]$. Below we give one possible implementation of this code:

```
>>t=0:.02:2*pi; counter=1;
>>A=[0   0   .5   1   2   2.5   3   3   1.5   0;... ...
     0   3   4    3   3   4     3   0   -1    0];
>>x=1+.4*cos(t); xp=1+.15*cos(t);
>>for s=0:.2:2*pi
 factor = (cos(s)+1)/2;
 plot(A(1,:), A(2,:), 'k')
 axis([-2 5 -3 6]), axis('equal')
 y=2+.4*factor*sin(t); yp=2+.15*factor*sin(t);
 hold on
 fill(x,y,'y'), fill(x+1,y, 'y'), fill(xp,yp,'k'), fill(xp+1,yp,'k')
 M(:, counter) = getframe;
 hold off, counter=counter+1;
end
```

The movie is now ready for screening. To view it the reader might try one (or both) of the following commands.

```
>> movie(M,4,5) %slow playing movie, four repeats
>> movie(M,20,75) %much faster play of movie, with 20 repeats
```

EXERCISE FOR THE READER 7.6: (a) Create a MATLAB function M-file, called mkhom(A), that takes a $2 \times m$ matrix of vertices for a graphic (first row has x-coordinates and second row has corresponding y-coordinates) as input and outputs a corresponding $3 \times m$ matrix of homogeneous coordinates for the vertices.
(b) Create a MATLAB function M-file, called rot(Ah,x0,y0, theta) that has inputs, Ah, a matrix of homogeneous coordinates of some graphic, two real numbers,
x0, y0 that are the coordinates of

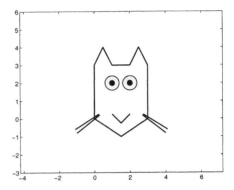

FIGURE 7.11: The more sophisticated cat star of the movie in Exercise for the Reader 7.7 (b).

the center of rotation, and theta , the angle (in radians) of rotation. The output will be the homogeneous coordinate vertex matrix gotten from Ah by rotating the graph an angle theta about the point $(x0, y0)$.

EXERCISE FOR THE READER 7.7: (a) Recreate the above movie working in homogeneous coordinate transforms on the eyes.
(b) By the same method, create a similar movie that stars a more sophisticated cat, replete with whiskers and a mouth, as shown in Figure 7.11. In this movie, the cat starts off frowning and the pupils will shift first to the left, then to the right, then back to center and finally up, down and back to center again, at which point the cat will wiggle its whiskers up and down twice and change its frown into a smile.

Fractals or **fractal sets** are complicated and interesting sets (in either the plane or three-dimensional space) that have the **self-similarity property** that if one magnifies a certain part of the fractal (any number of times) the details of the structure will look exactly the same.

The computer generation of fractals is also a hot research area and we will look at some of the different methods that are extensively used. Fractals were gradually discovered by mathematicians who were specialists in set theory or function theory, including (among others) the very famous Georg F. L. P. Cantor (1845–1918, German), Waclaw Sierpinski (1882–1969, Polish), Gaston Julia (1893–1978, French) and Giuseppe Peano (1858–1932, Italian) during the late nineteenth and early twentieth century. Initially, fractals came up as being pathological objects without any type of unifying themes. Many properties of

factals that have shown them to be so useful in an assortment of fields were discovered and popularized by the Polish/French mathematician Benoit Mandelbrot[3](Figure 7.12). The precise definition of a fractal set takes a lot of preliminaries; we refer to the references, for example, that are cited in the footnote on this page for details. Instead of this, we will jump into some examples. The main point to keep in mind is that all of the examples we give (in the text as well as in the exercises) are actually impossible to print out exactly because of the self-similarity property; the details would require a printer with infinite resolution. Despite this problem, we can use loops or recursion with MATLAB to get some decent renditions of fractals that, as far as the naked eye can tell (your printer's resolution permitting), will be accurate illustrations.
Fractal sets are usually best described by an iterative procedure that runs on forever.

EXAMPLE 7.6: (*The Sierpinski Gasket*)   To obtain this fractal set, we begin with an equilateral triangle that we illustrate in gray in Figure 7.13(a); we call this set the *zeroth generation*.   By considering the midpoints of each of the sides of this triangle, we can form four (smaller) triangles that are similar to the original. One is upside-down and the other three have the same orientation as the original.   We delete this central upside down subtriangle from the zeroth generation to form the *first generation* (Figure 7.13(b)).

FIGURE 7.12: Benoit Mandelbrot (b. 1924) Polish/ French mathematician.

---

[3] Mandelbrot was born in Poland in 1924 and his family moved to France when he was 12 years old. He was introduced to mathematics by his uncle Szolem Mandelbrojt, who was a mathematics professor at the Collège de France. From his early years, though, Mandelbrot showed a strong preference for mathematics that could be applied to other areas rahter than the pure and rather abstruse type of mathematics on which his uncle was working. Since World War II was taking place during his school years, he often was not able to attend school and as a result much of his education was done at home through self-study. He attributes to this informal education the development of his strong geometric intuition. After earning his Ph.D. in France he worked for a short time at Cal Tech and the Institute for Advanced Study (Princeton) for postdoctoral work. He then returned to France to work at the Centre National de la Recherche Scientifique. He stayed at this post for only three years since he was finding it difficult to fully explore his creativity in the formal and traditional mathematics societies that dominated France in the mid-twentieth century (the "Bourbaki School"). He returned to the United States, taking a job as a research fellow with the IBM research laboratories. He found the atmosphere extremely stimulating at IBM and was able to study what he wanted. He discovered numerous applications and properties of fractals; the expanse of applications is well demonstrated by some of the other joint appointments he has held while working at IBM. These include Professor of the Practice of Mathematics at Harvard University, Professor of Engineering at Yale, Professor of Economics at Harvard, and Professor of Physiology at the Einstein College of Medicine. Many books have been written on fractals and their applications. For a very geometric and accessible treatment (with lots of beautiful pictures of fractals) we cite [Bar-93], along with [Lau-91]; see also [PSJY-92]. More analytic (and mathematically advanced) treatments are nicely done in the books [Fal-85] and [Mat-95].

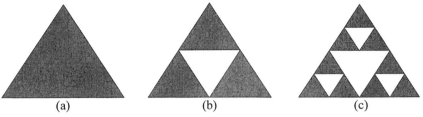

(a)   (b)   (c)

FIGURE 7.13: Generation of the Sierpinski gasket of Example 7.6: (a) the zeroth generation (equilateral triangle), (b) first generation, (c) second generation. The generations continue on forever to form the actual set.

Next, on each of the three (equilateral) triangles that make up this first generation, we again perform the same procedure of deleting the upside-down central subtriangle to obtain the generation-two set (Figure 7.13(c)). This process is to continue on forever and this is how the Sierpinski gasket set is formed. The sixth generation is shown in Figure 7.14.

FIGURE 7.14: Sixth generation of the Sierpinski gasket fractal of Example 7.6.

Notice that higher generations become indistinguishable to the naked eye, and that if we were to focus on one of the three triangles of the first generation, the Sierpinski gasket looks the same in this triangle as does the complete gasket. The same is true if we were to focus on any one of the nine triangles that make up the second generation, and so on.

EXERCISE FOR THE READER 7.8: (a) Show that the nth generation of the Sierpinski triangle is made up of 3^n equilateral triangles. Find the area of each of these nth-generation triangles, assuming that the initial sidelengths are one.
(b) Show that the area of the Sierpinski gasket is zero.
NOTE: It can be shown that the Sierpinski gasket has dimension $\log 4 / \log 3 = 1.2618...$, where the *dimension* of a set is a rigorously defined

measure of its true size. For example, any countable union of line segments or smooth arcs is of dimension one and the inside of any polygon is two-dimensional. Fractals have dimensions that are non integers. Thus a fractal in the plane will have dimension somewhere (strictly) between 1 and 2 and a fractal in three-dimensional space will have dimension somewhere strictly between 2 and 3. None of the standard sets in two and three dimensions have this property. This noninteger dimensional property is often used as a definition for fractals. The underlying theory is quite advanced; see [Fal-85] or [Mat-95] for more details on these matters.

In order to better understand the self-similarity property of fractals, we first recall from high-school geometry that two triangles are similar if they have the same angles, and consequently their corresponding sides have a fixed ratio. A **similarity transformation** (or **similitude** for short) on $\mathbf{R}^2$ is an affine transformation made up of one or more of the following special transformations: scaling (with both x- and y-factors equal), a reflection, a rotation, and/or a shift. In homogeneous coordinates, it thus follows that a similitude can be expressed in matrix form as follows:

$$
\begin{bmatrix} x' \\ y' \\ 1 \end{bmatrix} = T\left(\begin{bmatrix} x \\ y \\ 1 \end{bmatrix} \right) = \begin{bmatrix} s\cos\theta & -s\sin\theta & x_0 \\ \pm s\sin\theta & \pm s\cos\theta & y_0 \\ 0 & 0 & 1 \end{bmatrix} \begin{bmatrix} x \\ y \\ 1 \end{bmatrix} = H \begin{bmatrix} x \\ y \\ 1 \end{bmatrix}, \tag{16}
$$

where s can be any nonzero real number and the signs in the second row of H must be the same. A scaling with both x- and y-factors being equal is customarily called a **dilation**.

EXERCISE FOR THE READER 7.9: (a) Using Theorem 7.2 (and its proof), justify the correctness of (16).
(b) Show that for any two similar triangles in the plane there is a similitude that transforms one into the other.
(c) Show that if any particular feature (e.g., reflection) is removed from the definition of a similitude, then two similar triangles in the plane can be found, such that one cannot be transformed to the other by this weaker type of transformation.

The self-similarity of a fractal means, roughly, that for the whole fractal (or at least a critical piece of it), a set of similitudes $S_1, S_2, \cdots, S_K$ can be found (the number K of them will depend on the fractal) with the following property: All S_j's have the same scaling factor $s < 1$ so that F can be expressed as the union of the transformed images $F_i = S_i(F)$ and these similar (and smaller) images are **essentially disjoint** in that different ones can have only vertex points or boundary edges in common. Many important methods for the computer generation of fractals will hinge on the discovery of these similitudes $S_1, S_2, \cdots, S_K$. Finding them also has other uses in both the theory and application of fractals. These concepts will be important in Methods 1 and 2 in our solution of the following example.

EXAMPLE 7.7: Write a MATLAB function M-file that will produce graphics for the Sierpinski gasket fractal.

SOLUTION: We deliberately left the precise syntax of the M-file open since we will actually give three different approaches to this problem and produce three different M-files. The first method is a general one that will nicely take advantage of the self-similarity of the Sierpinski gasket and will use homogeneous coordinate transform methods. It was, in fact, used to produce high-quality graphic of Figure 7.14. Our second method will illustrate a different approach, called the **Monte-Carlo method**, that will involve an iteration of a random selection process to obtain points on the fractal, and will plot each of the points that get chosen. Because of the randomness of selection, enough iterations produce a reasonably representative sample of points on the fractal and the resulting plot will give a decent depiction of it. Monte-Carlo is a city on the French Riviera known for its casinos (it is the European version of Las Vegas). The method gets its name from the random (chance) selection processes it uses. Our third method works similarly to the first but the ideas used to create the M-file are motivated by the special structure of the geometry, in this case of the triangles.

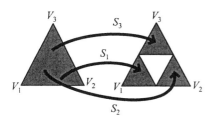

FIGURE 7.15: The three natural similitudes S_1, S_2, S_3 for the Sierpinski gasket with vertices V_1, V_2, V_3 shown on the zeroth and first generations. Since the zeroth generation is an equilateral triangle, so must be the three triangles of the first generation.

Method 1: The Sierpinski gasket has three obvious similitudes, each of which transforms it into one of the three smaller "carbon copies" of it that lie in the three triangles of the first generation (see Figure 7.15). These similitudes have very simple form, involving only a dilation (with factor 0.5) and shifts. The first transformation S_1 involves no shift. Referring to the figure, it is clear that S_2 must shift V_1 to the midpoint of the line segment V_1V_2 that is given by (as a vector) $(V_1 + V_2)/2$. The shift vector needed to do this, and hence the shift vector for S_2 is $(V_2 - V_1)/2$. (Proof: If we shift V_1 by this vector we get $V_1 + (V_2 - V_1)/2 = (V_2 + V_1)/2$). Similarly the shift vector for S_3 is $(V_3 - V_1)/2$. It follows that the corresponding matrices for these three similitudes are as given below:

$$S_1 \sim \begin{bmatrix} .5 & 0 & 0 \\ 0 & .5 & 0 \\ 0 & 0 & 1 \end{bmatrix}, S_2 \sim \begin{bmatrix} .5 & 0 & (V_2(1)-V_1(1))/2 \\ 0 & .5 & (V_2(2)-V_1(2))/2 \\ 0 & 0 & 1 \end{bmatrix}, S_3 \sim \begin{bmatrix} .5 & 0 & (V_3(1)-V_1(1))/2 \\ 0 & .5 & (V_3(2)-V_1(2))/2 \\ 0 & 0 & 1 \end{bmatrix}.$$

The program sgasket1 (V1,V2,V3,ngen) given below has four input variables: V1, V2, V3 should be the row vectors representing the vertices $(0,0)$, $(1,\sqrt{3}),(2,0)$ of a particular equilateral triangle, and ngen is the generation number of the Sierpinski gasket to be drawn. The program has no output variables, but will produce a graphic of this generation ngen of the Sierpinski gasket. The idea behind the algorithm is the following. The three triangles making up the generation-one gasket can be obtained by applying each of the three special similitudes S_1, S_2, S_3 to the single generation-zero Gasket. By the same token, each of the nine triangles that comprise the generation-two gasket can by obtained by applying one of the similitudes of S_1, S_2, S_3 to one of the generation-one triangles. In general, the triangles that make up any generation gasket can be obtained as the union of the triangles that result from applying each of the similitudes S_1, S_2, S_3 to each of the previous generation triangles. It works with the equilateral triangle having vertices $(0,0),(1,\sqrt{3}),(2,0)$. The program makes excellent use of recursion.

PROGRAM 7.1: Function M-file for producing a graphic of any generation of the Sierpinski gasket on the special equilateral triangle with vertices $(0,0),(1,\sqrt{3}),(2,0)$ (written with comments in a way to make it easily modified to work for other fractals).

```
function sgasket1(V1,V2,V3,ngen)
%input variables: V1,V2,V3 should be the vertices [0 0], [1,sqrt(3)],
%and [2,0] of a particular equilateral triangle in the plane taken as
%row vectors, ngen is the number of iterations to perform in
%Sierpinski gasket generation.
%The gasket will be drawn in medium gray color.

%first form matrices for similitudes
   S1=[.5 0 0;0 .5 0;0 0 1];
   S2=[.5 0 1; 0 .5 0;0 0 1];
   S3=[.5 0 .5; 0 .5 sqrt(3)/2;0 0 1];

if ngen == 0
   %Fill triangle
  fill([V1(1) V2(1) V3(1) V1(1)], [V1(2) V2(2) V3(2) V1(2)], [.5 .5
.5])
  hold on
else
%recursively invoke the same function on three outer subtriangles
%form homogeneous coordinate matrices for three vertices of triangle
   A=[V1; V2; V3]'; A(3,:)=[1 1 1];
   %next apply the similitudes to this matrix of coordinates
   A1=S1*A; A2=S2*A; A3=S3*A;
%finally, reapply sgasket1 to the corresponding three triangles with
%ngen bumped down by 1. Note, vertex vectors have to be made into
%row vectors using '(transpose).
   sgasket1(A1([1 2],1)', A1([1 2],2)', A1([1 2],3)', ngen-1)
   sgasket1(A2([1 2],1)', A2([1 2],2)', A2([1 2],3)', ngen-1)
   sgasket1(A3([1 2],1)', A3([1 2],2)', A3([1 2],3)', ngen-1)
end
```

To use this program to produce, for example, the generation-one graphic of Figure 7.13(b), one need only enter:

```
>> sgasket1([0 0], [1 sqrt(3)], [2 0], 1)
```

If we wanted to produce a graphic of the more interesting generation-six Sierpinski gasket of Figure 7.15, we would have only to change the last input argument from 1 to 6. Note, however, that this function left the graphics window with a `hold on`. So before doing anything else with the graphics window after having used it, we would need to first enter `hold off`. Alternatively, we could also use the following command:

| clf | → | Clears the graphics window. |

In addition to recursion, the above program makes good use of MATLAB's elaborate matrix manipulation features. It is important that the reader fully understands how each part of the program works. To this end the following exercise should be useful.

EXERCISE FOR THE READER 7.10: (a) Suppose the above program is invoked with these input variables: $V1 = [0\ 0]$, $V2 = [1\ \sqrt{3}\ ]$, $V3 = [2\ 0]$, ngen = 1. On the first run/iteration, what are the numerical values of each of the following variables:

$$A, A1, A2, A3, A1([1\ 2],2), A3([1\ 2],3)?$$

(b) Is it possible to modify the above program so that after the graphic is drawn, the screen will be left with `hold off`. If yes, show how to do it, if not, explain.
(c) In the above program, the first three input variables $V1, V2, V3$ seem a bit redundant since we are forced to input them as the vertices of a certain triangle (which gave rise to the special similitudes $S1$, $S2$, and $S3$. Is it possible to rewrite the program so that it has only one input variable ngen? If yes, show how to do it; if not, explain.

Method 2: The Monte-Carlo method also will use the special similitudes, but its philosophy is very different from that of the first method. Instead of working on a particular generation of the Sierpinki gasket fractal, it goes all out and tries to produce a decent graphic of the actual fractal. This gets done by plotting a representative set of points on the fractal, a random sample of such. Since so much gets deleted from the original triangle, a good question is What points exactly are left in the Sierpinski gasket? Certainly the vertices of any triangle of any generation will always remain. Such points will be the ones from which the Monte-Carlo method samples. Actually there are a lot more points in the fractal than these vertices, although such points are difficult to write down. See one of the books on fractals mentioned earlier for more details.

Here is an outline of how the program will work. We start off with a point we call "Float" that is a vertex of the original (generation-zero) triangle, say $V1$. We then randomly choose one of the similitudes from S_1, S_2, S_3, and apply this to "Float" to get a new point "New," that will be the corresponding vertex of the generation-one triangle associated with the similitude that was used (lower left for S_1, upper middle for S_3, and lower right for S_2). We plot "New," redefine "Float" = "New," and repeat this process, again randomly selecting one of the similitudes to apply to "Float" to get a new point "New" of the fractal that will be plotted. At the Nth iteration, "New" will be a vertex of one of the Nth-generation triangles (recall there are 3^N such triangles) that will also lie in one of the three generation-one triangles, depending on which of S_1, S_2, S_3 had been randomly chosen. Because of the randomness of choices at each iteration, the points that are plotted usually give a decent rendition of the fractal, as long as a large enough random sample is used (i.e., a large number of iterations).

PROGRAM 7.2: Function M-file for producing a Monte-Carlo approximation graphic of Sierpinski gasket, starting with the vertices *V1*, *V2*, and *V3* of any equilateral triangle (written with comments in a way that will make it easily modified to work for other fractals).

```
function [ ] = sgasket2(V1,V2,V3,niter)
%input variables: V1,V2,V3 are vertices of an equilateral triangle in
%the plane taken as row vectors, niter is the number of iterations
%used to obtain points in the fractal.  The output will be a plot of
%all of the points.  If niter is not specified, the default value
%of 5000 is used.
%if only 3 input arguments are given (nargin==3), set niter to
%default
if nargin == 3, niter = 5000; end

%Similitude matrices for Sierpinski gasket.
S1=[.5 0 0;0 .5 0;0 0 1];
S2=[.5 0 (V2(1)-V1(1))/2; 0 .5 (V2(2)-V1(2))/2;0 0 1];
S3=[.5 0 (V3(1)-V1(1))/2; 0 .5 (V3(2)-V1(2))/2;0 0 1];

%Probability vector for Sierpinski gasket has equal probabilities
%(1/3)for choosing one of the three similitudes.
P = [1/3 2/3];

%prepare graphics window for repeated plots of points
clf, axis('equal'); hold on;

%introduce "floating point" (can be any vertex) in homogeneous
%coordinates
Float=[V1(1);V1(2);1];
i = 1; %initialize iteration counter

%Begin iteration for creating new floating points and plotting each
%one that arises.
while i <= niter
   choice = rand;
   if choice < P(1);
      New = S1 * Float;
      plot (New(1), New(2));
   elseif choice < P(2);
      New = S2 * Float;
      plot (New(1), New(2));
   else        New = S3 * Float;
```

```
        plot (New(1), New(2));
    end;
    Float=New;    i = i + 1;
end
hold off
```

Unlike the last program, this one allows us to input the vertices of any equilateral triangle for the generation-zero triangle.   The following two commands will invoke the program first with the default 5000 iterations and then with 20,000 (the latter computation took several seconds).

```
>> sgasket2([0 0], [1 sqrt(3)], [2 0])
>> sgasket2([0 0], [1 sqrt(3)], [2 0], 20000)
```

The results are shown in Figure 7.16.   The following exercise should help the reader better undertstand how the above algorithm works.

EXERCISE FOR THE READER 7.11:   Suppose that we have generated the following random numbers (between zero and one): .5672, .3215, .9543, .4434, .8289, .5661 (written to 4 decimals).
(a) What would be the corresponding sequence of similitudes chosen in the above program from these random numbers?
(b) If we used the vertices [0 0], [1 sqrt(3)], [2 0]   in the above program, find the sequence of different "Float" points of the fractal that would arise if the above sequence of random numbers were to come up.
(c) What happens if the vertices entered in the program sgasket2 are those of a nonequilateral triangle?   Will the output ever look anything like a Sierpinski gasket? Explain.

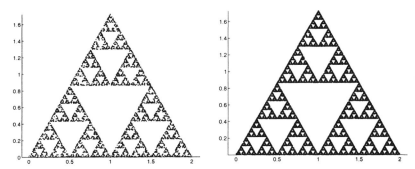

FIGURE 7.16:   Monte-Carlo renditions of the Sierpinski gasket via the program sgasket2. The left one (a) used 5000 iterations while the right one (b) used 20,000 and took noticeably more time.

Method 3:   The last program we write here will actually be the shortest and most versatile of the three. Its drawback is that, unlike the other two, which made use of the specific similitudes associated with the fractal, this program uses the special geometry of the triangle and thus will be more difficult to modify to work for other fractals. The type of geometric/mathematical ideas present in this program, however, are useful in writing other graphics programs. The program sgasket3(V1,V2,V3,ngen) takes as input three vertices V1, V2, V3 of a triangle (written as row vectors), and a positive integer ngen. It will produce a graphic of the ngen-generation Sierpinski gasket, as did the first program. It is again based on the fact that each triangle from a positive generation gasket comes in a very natural way from the triangle of the previous generation in which it lies. Instead of using similitudes and homogeneous coordinates, the program simply uses explicit formulas for the vertices of the $(N + 1)$st-generation triangles that lie within a certain Nth-generation triangle. Indeed, for any triangle from any generation of the Sierpinski gasket with vertices V_1, V_2, V_3, three subtriangles of this one form the next generation (see Figure 7.15), each has one vertex from this set, and the other two are the midpoints from this vertex to the other two. For example (again referring to Figure 7.15) the lower-right triangle will have vertices V_2, $(V_1 + V_2)/2$ = the midpoint of V_2V_1, and $(V_2 + V_3)/2$ = the midpoint of V_2V_3. This simple fact, plus recursion, is the idea behind the following program.

PROGRAM 7.3: Function M-file for producing a graphic of any generation of the Sierpinski gasket for an equilateral triangle with vertices *V1*, *V2*, and *V3*.

```
function sgasket3(V1,V2,V3,ngen)
%input variables: V1,V2,V3 are vertices of a triangle in the plane,
%written as row vectors, ngen is the generation of Sierpinski gasket
%that will be drawn in medium gray color.
if ngen == 0
%Fill triangle
  fill([V1(1) V2(1) V3(1) V1(1)],...
  [V1(2) V2(2) V3(2) V1(2)], [.5 .5 .5])
    hold on
    else
%recursively invoke the same function on three outer subtriangles
    sgasket3(V1, (V1+V2)/2, (V1+V3)/2, ngen-1)
    sgasket3(V2, (V2+V1)/2, (V2+V3)/2, ngen-1)
    sgasket3(V3, (V3+V1)/2, (V3+V2)/2, ngen-1)
end
```

EXERCISE FOR THE READER 7.12: (a) What happens if the vertices entered in the program sgasket3 are those of a nonequilateral triangle? Will the output ever look anything like a Sierpinski gasket? Explain.
(b) The program sgasket3 is more elegant than sgasket1 and it is also more versatile in that the latter program applies only to a special equilateral triangle. Furthermore, it also runs quicker since each iteration involves less computing. Justify this claim by obtaining some hard evidence by running both programs (on the standard equilateral triangle of sgasket1) and comparing tic/toc and flop counts (if available) for each program with the following values for ngen: 1, 3, 6, 8, 10.

Since programs like the one in Method 3 of the above example are usually the most difficult to generalize, we close this section with yet another exercise for the reader that will ask for such a program to draw an interesting and beautiful fractal known as the **von Koch**[4] **snowflake,** which is illustrated in Figure 7.17. The iteration scheme for this fractal is shown in Figure 7.18.

EXERCISE FOR THE READER 7.13: Create a MATLAB function, call it snow(n), that will input a positive integer n and will produce the nth generation of the so-called von Koch snowflake fractal. Note that we start off (generation 0) with an equilateral triangle with sidelength 2. To get from one generation to the next, we do the following: For each line segment on the boundary, we put up (in the middle of the segment) an equilateral triangle of 1/3 the sidelength. This construction is illustrated in Figure 7.18, which contains the first few generations of the von Koch snowflake. Run your program (and include the graphical printout) for the values: $n = 1$, $n = 2$, and $n = 6$.

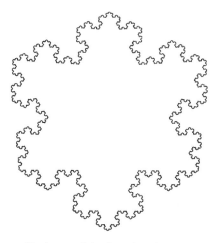

FIGURE 7.17: The von Koch snowflake fractal. This illustration was produced by the MATLAB program snow(n) of Exercise for the Reader 7.13, with an input value of 6 (generations).

Suggestions: Each generation can be obtained by plotting its set of vertices (using the plot command). You will need to set up a for loop that will be able to produce the next generation's vertices from those of a given generation. It is helpful to think in terms of vectors.

---

[4] The von Koch snowflake was introduced by Swedish mathematician Niels F. H. von Koch (1870–1924) in a 1906 paper *Une méthode géométrique élémentaire pour l'étude de certaines questions de la théorie des courbes planes*. In it he showed that the parametric equations for the curve $(x(t), y(t))$ give an example of functions that are everywhere continuous but nowhere differentiable. Nowhere differentiable, everywhere continuous functions had been first discovered in 1860 by German mathematician Karl Weierstrass (1815–1897), but the constructions known at this time all involved very complicated formulas. Von Koch's example thus gives a curve (of infinite arclength) that is continuous everywhere (no breaks), but that does not have a tangent line at any of its points. The von Koch snowflake has been used in many areas of analysis as a source of examples.

Generation $n = 0$ snowflake: Generation $n = 1$ snowflake:

   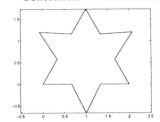

Generation $n = 2$ snowflake: Generation $n = 3$ snowflake:

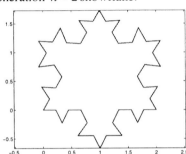   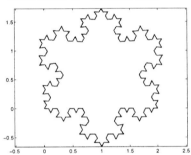

FIGURE 7.18: Some initial generations of the von Koch snowflake. Generation zero is an equilateral triangle (with sidelength 2). To get from any generation to the next, each line segment on the boundary gets replaced by four line segments each having 1/3 of the length of the original segment. The first and fourth segments are at the ends of the original segment and the middle two segments form two sides of an equilateral triangle that protrudes outward.

EXERCISES 7.2:

NOTE: In the problems of this section, the CAT refers to the graphic of Example 7.2 (Figure 7.3(a)), the "CAT with eyes" refers to the enhanced version graphic of Example 7.5 (Figure 7.10), and the "full CAT" refers to the further enhanced CAT of Exercise for the Reader 7.7(b) (Figure 7.11). When asked to print a certain transformation of any particular graphic (like the CAT) along with the original, make sure to print the original graphic in one plotting style/color along with the transformed graphic in a different plotting style/color. Also, in printing any graphic, use the `axis(equal)` setting to prevent any distortions and set the axis range to accommodate all of the graphics nicely inside the bounding box

1. Working in homogeneous coordinates, what is the transformation matrix M that will scale the CAT horizontally by a factor of 2 (to make a "fat cat") and then shift the cat vertically down a distance 2 and horizontally 1 unit to the left? Create a before and after graphic of the CAT.

2. Working in homogeneous coordinates, what is the transformation matrix M that will double the size of the horizontal and vertical dimensions of the CAT and then rotate the new CAT by an angle of 45° about the tip of its left ear (the double-sized cat's left ear, that is)? Include a before-and-after graphic of the CAT.

3. Working in homogeneous coordinates, what is the transformation matrix M that will shift the left eye and pupil of the "CAT with eyes" by 0.5 units and then expand them both by a factor of

2 (away from the centers)? Apply this transformation just to the left eye. Next, perform the analogous transformation to the CAT's right eye and then plot these new eyes along with the outline of the CAT, to get a cat with big eyes.

4. Working in homogeneous coordinates, what is the transformation matrix M that will shrink the "CAT with eyes"'s left eye and left pupil by a factor of 0.5 in the horizontal direction (toward the center of the eye) and then rotate them by an angle of 25°? Apply this transformation just to the left eye, reflect to get the right eye, and then plot these two along with the outline of the CAT, to get a cat with thinner, slanted eyes.

5. (a) Create a MATLAB function M-file, called `reflx(Ah, x0)` that has inputs, Ah, a matrix of homogeneous coordinates of some graphic and a real number x0. The output will be the homogeneous coordinate vertex matrix obtained from Ah by reflecting the graphic over the line $x = x0$. Apply this to the CAT graphic using $x0 = 2$, and give a before-and-after plot.

   (b) Create a similar function M-file `refly(Ah, y0)` for vertical reflections (about the horizontal line $y = y0$) and apply to the CAT using $y0 = 4$ to create a before and after plot.

6. (a) Create a MATLAB function M-file, called `shift(Ah,x0,y0)` that has as inputs Ah, a matrix of homogeneous coordinates of some graphic, and a pair of real numbers x0,y0. The output will be the homogeneous coordinate vertex matrix obtained from Ah by shifting the graphic using the shift vector (x0,y0). Apply this to the CAT graphic using x0 = 2 and y0 = −1 and give a before-and-after plot.

   (b) Create a MATLAB function M-file, called `scale(Ah,a,b,x0,y0)` that has inputs Ah, matrix of homogeneous coordinates of some graphic, positive numbers: a and b that represent the horizontal and vertical scaling factors, and a pair of real numbers x0, y0 that represent the coordinates about which the scaling is to be done. The output will be the homogeneous coordinate vertex matrix obtained from Ah by scaling the graphic as indicated. Apply this to the CAT graphic using $a = .25$, $b = 5$ once each with the following sets for $(x0, y0)$: (0,0), (3,0), (0,3), (2.5,4) and create a single plot containing the original CAT along with all four of these smaller, thin cats (use five different colors/plot styles).

7. Working in homogeneous coordinates, what is the transformation matrix M that will reflect an image about the line $y = x$? Create a before and after graphic of the CAT.

   Suggestion: Rotate first, reflect, and then rotate back again.

8. Working in homogeneous coordinates, what is the transformation matrix M that will shift the left eye and left pupil of the "CAT with eyes" to the left by .5 units and then expand them by a factor of 2 (away from the centers)? Apply this transformation just to the left eye, reflect to get the right eye, and then plot these two along with the outline of the "CAT with eyes," to get a cat with big eyes.

9. The **shearing** on $\mathbf{R}^2$ that shears by b in the x-direction and d in the y-direction is the linear transformation whose matrix is $\begin{bmatrix} 1 & b \\ c & 1 \end{bmatrix}$. Apply the shearing to the CAT using several different values of b when $c = 0$, then set $b = 0$ and use several different values of c, and finally apply some shearings using several sets of nonzero values for b and c.

10. (a) Show that the 2×2 matrix $\begin{bmatrix} \cos\theta & -\sin\theta \\ \sin\theta & \cos\theta \end{bmatrix}$, which represents the linear transformation for rotations by angle θ, is invertible, with inverse being the corresponding matrix for rotations by angle $-\theta$.

    (b) Does the same relationship hold true for the corresponding 3×3 homogeneous coordinate transform matrices? Justify your answer.

11.   (a) Show that the 3×3 matrix $\begin{bmatrix} 1 & 0 & x_0 \\ 0 & 1 & y_0 \\ 0 & 0 & 1 \end{bmatrix}$, which represents the shift with shift vector $\begin{bmatrix} x_0 \\ y_0 \end{bmatrix}$,

is invertible, with its inverse being the corresponding matrix for the shift using the opposite shift vector.

12.   Show that the 2×2 matrix $\begin{bmatrix} \cos\theta & -\sin\theta \\ \sin\theta & \cos\theta \end{bmatrix}$ indeed represents the linear transformation for

rotations by angle θ around the origin (0,0).

Suggestion: Let (x,y) have polar coordinates (r,α); then (x',y') has polar coordinates $(r,\alpha+\theta)$. Convert the latter polar coordinates to rectangular coordinates.

13.   (*Graphic Art: Rotating Shrinking Squares*) (a) By starting off with a square, and repeatedly shrinking it and rotating it, get MATLAB to create a graphic similar to the one shown in Figure 7.19(a).
(b) Next modify your construction to create a graph similar to the one in Figure 7.19(b) but uses alternating colors.
Note: This object is not a fractal.

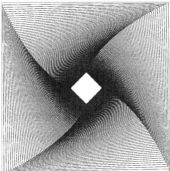 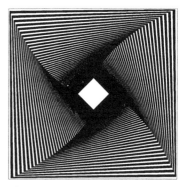

FIGURE 7.19: A rotating and shrinking square of Exercise 13: (a) (left) with no fills; (b) (right) with alternate black-and-white fills.

14.   (*Graphic Art: Cat with Eyes Mosaic*) The cat mosaic of Figure 7.20 has been created by taking the original cat, and creating new pairs of cats (left and right) for each step up.   This construction was done with a for loop using 10 iterations (so there are 10 pairs of cats above the original), and could easily have been changed to any number of iterations. Each level upward of cats got scaled to 79% of the preceding level. Also, for symmetry, the left and right cats were shifted upward and to the left and right by the same amounts, but these amounts got smaller (since the cat size did) as we moved upward.
(a) Use MATLAB to create a picture that is similar to that of Figure 7.20, but replace the "CAT with eyes" with the ordinary "CAT".
(b) Use MATLAB to create a picture that is similar to that of Figure 7.20.
(c) Use MATLAB to create a picture that is similar to that of Figure 7.20, but replace the "CAT with eyes" with the "full CAT" of Figure 7.11.
Suggestion: You should definitely use a for loop. Experiment a bit with different schemes for horizontal and vertical shifting to get your picture to look like this one.

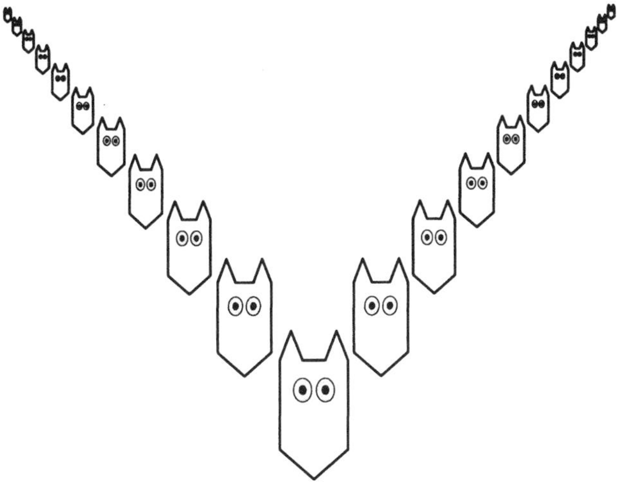

FIGURE 7.20: Cat with eyes mosaic for Exercise 14(b). The original cat (center) has been repeatedly shifted to the left and right, and up, as well as scaled by a factor of 79% each time we go up.

15. (*Movie: "Sudden Impact"*) (a) Create a movie that stars the CAT and proceeds as follows: The cat starts off at the left end of the screen. It then "runs" horizontally towards the right end of the screen. Just as its right side reaches the right side of the screen, it begins to shrink horizontally (but not vertically) until it degenerates into a vertical line segment on the right side of the screen.
    (b) Make a movie similar to the one in part (a) except that this one stars the "cat with eyes" and before it begins to run to the right, its pupils move to the right of the eyes and stay there.
    (c) Make a film similar to the one in part (b) except that this one should star the "full cat" (Figure 7.11) and upon impact with the right wall, the cat's smile changes to a frown.

16. (*Movie: "The Chase"*) (a) Create a movie that stars the "cat with eyes" and co-stars another smaller version of the same cat (scaled by factors of 0.5 in both the x- and y-directions). The movie starts off with the big cat in the upper left of the screen and the small cat to its right side (very close). Their pupils move directly toward one another to the end of the eyes, and at this point both cats begin moving at constant speed toward the right. When the smaller cat reaches the right side of the screen, it starts moving down while the big cat also starts moving down. Finally, cats stay put in the lower-right corner as their pupils move back to center.
    (b) Make the same movie except starring the "full cat" and co-starring a smaller counterpart.

17. (*Movie: "Close Encounter"*) (a) Create a movie that stars the "full CAT" (Figure 7.11) and with the following plot: The cat starts off smiling and then its eyes begin to shift all the way to the lower left. It spots a solid black rock moving horizontally directly toward its mouth level, at constant speed. As the cat spots this rock, its smile changes to a frown. It jumps upward as its pupils move back to center and just misses the rock as it brushes just past the cat's chin. The cat then begins to smile and falls back down to its original position.
    (b) Make a film similar to the one in part (a) except that it has the additional feature that the rock is rotating clockwise as it is moving horizontally.
    (c) Make a film similar to the one in part (b) except that it has the additional feature that the cat's pupils, after having spotted the rock on the left, slowly roll (along the bottom of the eyes) to the lower-right postion, exactly following the rock. Then, after the rock leaves the viewing window, have the cat's pupils move back to center postion.

18.   (*Fractal Geometry: The Cantor Square*) The Cantor square is a fractal that starts with the *unit square* in the plane: $C_0 = \{(x,y) : 0 \le x \le 1 \text{ and } 0 \le y \le 1\}$ (generation zero). To move to the next generation, we delete from this square all points such that at least one of the coordinates is inside the middle 1/3 of the original spread. Thus, to get C_1 from C_0, we delete all the points (x,y) having either $1/3 < x < 2/3$ or $1/3 < y < 2/3$. So C_1 will consist of four smaller squares each having sidelength equal to 1/3 (that of C_0) and sharing one corner vertex with C_0. Future generations are obtained in the same way. For example, to get from C_1 (first generation) to C_2 (second generation) we delete, from each of the four squares of C_1, all points (x,y) that have one of the coordinates lying in the middle 1/3 of the original range (for a certain square of C_1). What will be left is four squares for each of the squares of C_1, leaving a total of 16 squares each having sidelength equal to 1/3 that of the squares of C_1, and thus equal to 1/9. In general, letting this process continue forever, it can be shown by induction that the nth-generation Cantor square consists of 4^n squares each having sidelength $1/3^n$. The Cantor square is the set of points that remains after this process has been continued indefinitely.

(a) Identify the four similitudes S_1, S_2, S_3, S_4 associated with the Cantor square (an illustration as in Figure 7.16 would be fine) and then, working in homogeneous coordinates, find the matrices of each. Next, following the approach of Method 1 of the solution of Example 7.7, write a function M-file cantorsq1(V1,V2,V3,V4, ngen), that takes as input the vertices V1 = [0 0], V2 = [1 0], V3 = [1 1], and V4 = [0 1] of the unit square and a nonnegative integer ngen and will produce a graphic of the generation ngen Cantor square.

(b) Write a function M-file cantorsq2(V1,V2,V3,V4, niter) that takes as input the vertices V1, V2, V3 V4 of any square and a positive integer niter and will produce a Monte-Carlo generated graphic for the Cantor square as in Method 2 of the solution of Example 7.7. Run your program for the square having sidelength 1 and lower-left vertex $(-1, 2)$ using niter = 2000 and niter = 12,000.

(c) Write a function M-file cantorsq3(V1,V2,V3,V4, ngen) that takes as input the vertices V1, V2, V3 V4 of any square and a positive integer ngen and will produce a graphic for the ngen generation Cantor square as did cantorsq1 (but now the square can be any square). Run your program for the square mentioned in part (b) first with ngen = 1 then with ngen = 3. Can this program be written so that it produces a reasonable generalization of the Cantor square when the vertices are those of any rectangle?

19.   (*Fractal Geometry: The Sierpinski Carpet*) The Sierpinski carpet is the fractal that starts with the unit square $\{(x,y) : 0 \le x \le 1 \text{ and } 0 \le y \le 1\}$ with the central square of 1/3 the sidelength removed (generation zero). To get to the next generation, we punch eight smaller squares out of each of the remaining eight squares of sidelength 1/3 (generation one), as shown in Figure 7.21. Write a function M-file, scarpet2(niter), based on the Monte-Carlo method that will take only a single input variable niter and will produce a Monte-Carlo approximation of the Sierpinski carpet. You will, of course, need to find the eight similitudes associated with this fractal and get their matrices in homogeneous coordinates. Run your program with inputs niter = 1000, 2000, 5000, and 10,000.

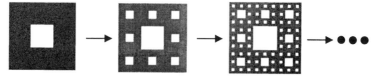

FIGURE 7.21: Illustration of generations zero (left), one (middle), and two (right) of the Sierpinski gasket fractal of Exercises 19, 20, and 21. The fractal consists of the points that remain (shaded) after this process has continued on indefinitely.

20. *(Fractal Geometry: The Sierpinski Carpet)* Read first Exercise 19 (and see Figure 7.21), and if you have not done so yet, identify the eight similitudes $S_1, S_2, \cdots, S_8$ associated with the Sierpinski carpet along with the homogeneous coordinate matrices of each. Next, following the approach of Method 1 of the solution of Example 7.7, write a function M-file scarpet1(V1,V2,V3,V4, ngen), that takes as input the vertices V1 = [0 0], V2 = [1 0], V3 = [1 1], and V4 = [0 1] of the unit square and a nonnegative integer ngen and will produce a graphic of the generation ngen Cantor square.

    Suggestions: Fill in each outer square in gray, then to get the white central square "punched out," use the hold on and then fill in the smaller square in the color white (RGB vector [1 1 1]). When MATLAB fills a polygon, by default it draws the edges in black. To suppress the edges from being drawn, use the following extra option in the fill commands: fill(xvec, yvec, rgbvec, 'EdgeColor', 'none'). Of course, another nice way to edit a graphic plot from MATLAB is to import the file into a drawing software (such as Adobe Illustrator or Corel Draw) and modify the graphic using the software.

21. *(Fractal Geometry: The Sierpinski Carpet)* (a) Write a function M-file called scarpet3(V1,V2,V3,V4, ngen) that works just like the program scarpet1 of the previous exercise, except that the vertices can be those of any square. Also, base the code not on similitudes, but rather on mathematical formulas for next-generation parameters in terms of present-generation parameters. The approach should be somewhat analogous to that of Method 3 of the solution to Example 7.7.

    (b) Is it possible to modify the sgasket1 program so that it is able to take as input the vertices of any equilateral triangle? If yes, indicate how. If no, explain why not.

22. *(Fractal Geometry: The Fern Leaf)* There are more general ways to construct fractals than those that came up in the text. One generalization of the self similarity approach given in the text allows for transformations that are not invertible (similitudes always are). In this exercise you are to create a function M-file, called fracfern(n), which will input a positive integer n and will produce a graphic for the fern fractal pictured in Figure 7.22, using the Monte-Carlo method. For this fractal the four transformations to use are (given by their homogeneous coordinate matrices)

$$S1 = \begin{bmatrix} 0 & 0 & 0 \\ 0 & .16 & 0 \\ 0 & 0 & 1 \end{bmatrix}, \quad S2 = \begin{bmatrix} .85 & .04 & 0 \\ -.04 & .85 & 1.6 \\ 0 & 0 & 0 \end{bmatrix},$$

$$S3 = \begin{bmatrix} .2 & -.26 & 0 \\ .23 & .22 & 1.6 \\ 0 & 0 & 1 \end{bmatrix}, \quad S4 = \begin{bmatrix} -.15 & .28 & 0 \\ .26 & .24 & .44 \\ 0 & 0 & 1 \end{bmatrix},$$

FIGURE 7.22: The fern leaf fractal.

and the associated probability vector is [.01 .86 .93] (i.e., in the Monte-Carlo process, 1% of the time we choose $S1$, 85% of the time we choose $S2$, 7% of the time we choose $S3$, and the remaining 7% of the time we choose $S4$).

    Suggestion: Simply modify the program sgasket2 accordingly.

23. *(Fractal Geometry: The Gosper Island)* (a) Write a function M-file gosper(n) that will input a positive integer n and will produce a graphic of the nth generation of the **Gosper island** fractal, which is defined as follows: Generation zero is a regular hexagon (with, say, unit side lengths). To get from this to generation one, we replace each of the six sides on the boundary of generation zero with three new segments as shown in Figure 7.23. The first few generations of the Gosper island are shown in Figure 7.24.

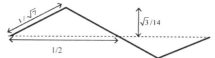

FIGURE 7.23: Iteration scheme for the definition of the Gosper island fractal of Exercise 23. The dotted segment represents a segment of a certain generation of the Gosper island, and the three solid segments represent the corresponding part of the next generation.

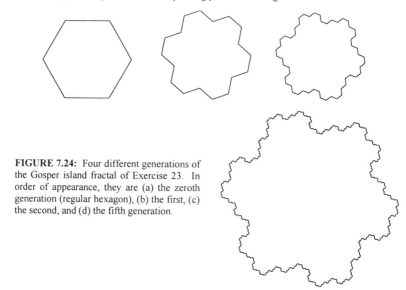

FIGURE 7.24: Four different generations of the Gosper island fractal of Exercise 23. In order of appearance, they are (a) the zeroth generation (regular hexagon), (b) the first, (c) the second, and (d) the fifth generation.

(b) (*Tessellations of the Plane*) It is well known that the only regular polygons that can tessellate (or tile) the plane are the equilateral triangle, the square, and the regular hexagon (honeybees have figured this out). It is an interesting fact that any generation of the Gosper island can also be used to tessellate the plane, as shown in Figure 7.25. Get MATLAB to reproduce each of tessellations that are shown in Figure 7.25.

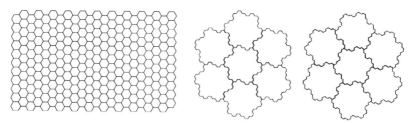

FIGURE 7.25: Tessellations with generations of Gosper islands. The top one (with regular hexagons) is the familiar honeycomb structure.

7.3: NOTATIONS AND CONCEPTS OF LINEAR SYSTEMS

The general linear system in n variables $x_1, x_2, \cdots, x_n$ and n equations can be written as

$$\begin{cases} a_{11}x_1 + a_{12}x_2 + \cdots + a_{1n}x_n = b_1 \\ a_{21}x_1 + a_{22}x_2 + \cdots + a_{2n}x_n = b_2 \\ \quad\cdots \\ a_{n1}x_1 + a_{n2}x_2 + \cdots + a_{nn}x_n = b_n \end{cases} . \tag{17}$$

Here, the a_{ij}, and b_j represent given data, and the variables $x_1, x_2, \cdots, x_n$ are the unknowns whose solution is sought. In light of how matrix multiplication is defined, these n equations can be expressed as a single matrix equation:

$$Ax = b , \tag{18}$$

where

$$A = \begin{bmatrix} a_{11} & a_{12} & a_{13} & \cdots & a_{1n} \\ a_{21} & a_{22} & a_{23} & \cdots & a_{2n} \\ \vdots & & & \ddots & \vdots \\ a_{n1} & a_{n2} & a_{n3} & \cdots & a_{nn} \end{bmatrix}, \quad x = \begin{bmatrix} x_1 \\ x_2 \\ \vdots \\ x_n \end{bmatrix}, \quad \text{and} \quad b = \begin{bmatrix} b_1 \\ b_2 \\ \vdots \\ b_n \end{bmatrix}. \tag{19}$$

It is also possible to consider more general linear systems that can contain more or fewer variables than equations, but such systems represent *ill-posed* problems in the sense that they typically do not have unique solutions. Most linear systems that come up in applications will be *well posed,* meaning that there will exist a unique solution, and thus we will be focusing most of our attention on solving well-posed linear systems. The above linear system is well posed if and only if the coefficient matrix A is invertible, in which case the solution is easily obtained, by left multiplying the matrix equation by A^{-1}: $Ax = b \implies A^{-1}Ax = A^{-1}b \implies$

$$x = A^{-1}b . \tag{20}$$

Despite its algebraic simplicity, however, this method of solution, namely computing and then left multiplying by A^{-1}, is, in general, an inefficient way of solving the system. The best general methods for solving linear systems are based on the Gaussian elimination algorithm. Such an algorithm is actually tacitly used by the MATLAB command for matrix division (left divide):

x=A\b →	Solves the matrix equation Ax = b by an elaborate Gaussian elimination procedure.

If the coefficient matrix A is invertible but "close" to being singular and/or if the size of A is large, the system (17) can be difficult to solve numerically. We will make this notion more precise later in this chapter. In case where there are two or three variables, the concepts can be illustrated effectively using geometry.

<u>CASE: $n = 2$</u> For convenience of notation, we drop the subscripts and rewrite system (17) as:

$$\begin{cases} ax + by = e \\ cx + dy = f \end{cases} . \tag{21}$$

The system (21) represents a pair of lines in the plane and the solutions, if any, are the points of intersection. Three possible situations are illustrated in Figure 7.26. We recall (Theorem 7.1) that the coefficient matrix A is nonsingular exactly when det(A) is nonzero. The singular case thus has the lines being parallel (Figure 7.26(c)). Nearly parallel lines (Figure 7.26(b)) are problematic since they are difficult to distinguish numerically from parallel lines. The determinant alone is not a reliable indicator of near singularity of a system (see Exercise for the Reader 7.14), but the condition number introduced later in this chapter will be.

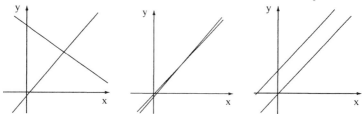

FIGURE 7.26: Three possibilities for the system (21) $\begin{cases} ax + by = e \\ cx + dy = f \end{cases}$: (a) well-conditioned, (b) ill-conditioned (nearly parallel lines), and (c) singular (parallel lines).

EXERCISE FOR THE READER 7.14: Show that for any pair of nearly parallel lines in the plane, it is possible to represent this system by a matrix equation $Ax = b$, which uses a coefficient matrix A with $\det(A) = 1$.

CASE: $n = 3$ A linear equation $ax + by + cz = d$ represents a plane in three-dimensional space $\mathbf{R}^3$. Typically, two planes in $\mathbf{R}^3$ will intersect in a line and a typical intersection of a line with a third plane (and hence the typical intersection of three planes) will be a point. This is the case if the system is nonsingular. There are several ways for such a three-dimensional system to be singular or nearly so. Apart from two of the planes being parallel (making a solution impossible), another way to have a singular system is for one of the planes to be parallel to the line of intersection of the other two. Some of these possibilities are illustrated in Figure 7.27.

For higher-order systems, the geometry is similar although not so easy to visualize since the world we live in has only three (visual) dimensions. For example, in four dimensions, a linear equation $ax + by + cz + dw = e$ will, in general, be a three-dimensional hyperplane in the four-dimensional space $\mathbf{R}^4$. The intersection of two such hyperplanes will typically be a two-dimensional plane in $\mathbf{R}^4$. If we intersect with one more such hyperplane we will (in nonsingular cases) be left with a line in $\mathbf{R}^4$, and finally if we intersect this line with a fourth such hyperplane we will be left with a point, the unique solution, as long as the system is nonsingular.

The variety of singular systems gets extremely complicated for large values of n, as is partially previewed in Figure 7.27 for the case $n = 3$. This makes the determination of near singularity of a linear system a complicated issue, which is why we will need analytical (rather than geometric) ways to detect this.

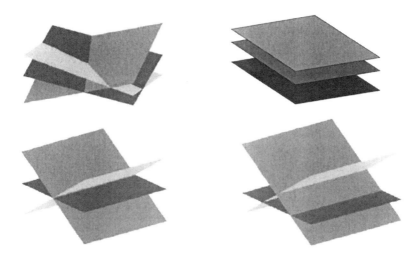

FIGURE 7.27: Four geometric possibilities for a three-dimensional system (18) $Ax = b$. (a) (upper left) represents a typical nonsingular system, three planes intersect at one common point; (b) (upper right) parallel planes, no solution, a singular system; (c) (lower left) three planes sharing a line, infinitely many solutions, a singular system; (d) (lower right) three different parallel lines arise from intersections of pairs of the three planes, no solution, singular system.[5]

7.4: SOLVING GENERAL LINEAR SYSTEMS WITH MATLAB

The best all-around algorithms for solving linear systems (17) and (18) are based on Gaussian elimination with partial pivoting. MATLAB's default linear system solver is based on this algorithm. In the next section we describe this algorithm; here we will show how to use MATLAB to solve such systems. In the following example we demonstrate three different ways to solve a (nonsingular) linear system in MATLAB by solving the following interpolation problem, and compare flop counts.

EXAMPLE 7.8: *(Polynomial Interpolation)* Find the equation of the polynomial $p(x) = ax^3 + bx^2 + cx + d$ of degree at most 3 that passes through the data points $(-5, 4)$, $(-3, 34)$, $(-1, 16)$, and $(1, 2)$.

---

[5] Note: These graphics were created on MATLAB.

SOLUTION: In general, a set of n data points (with different x-coordinates) can always be interpolated with a polynomial of degree at most $n-1$. Writing out the interpolation equations produces the following four-dimensional linear system:

$$p(-5) = 4 \Rightarrow a(-5)^3 + b(-5)^2 + c(-5) + d = 4$$
$$p(-3) = 34 \Rightarrow a(-3)^3 + b(-3)^2 + c(-3) + d = 34$$
$$p(-1) = 16 \Rightarrow a(-1)^3 + b(-1)^2 + c(-1) + d = 16$$
$$p(1) = 2 \quad \Rightarrow a \cdot 1^3 + b \cdot 1^2 + c \cdot 1 + d = -2.$$

In matrix (18) form this system becomes:

$$\begin{bmatrix} -125 & 25 & -5 & 1 \\ -27 & 9 & -3 & 1 \\ -1 & 1 & -1 & 1 \\ 1 & 1 & 1 & 1 \end{bmatrix} \begin{bmatrix} a \\ b \\ c \\ d \end{bmatrix} = \begin{bmatrix} 4 \\ 34 \\ 16 \\ -2 \end{bmatrix}.$$

We now solve this matrix equation $Ax = b$ in three different ways with MATLAB, and do a flop count[6] for each.

<u>Method 1:</u>   Left divide by A.   This is the Gaussian elimination method mentioned in the previous section.   It is the recommended method.

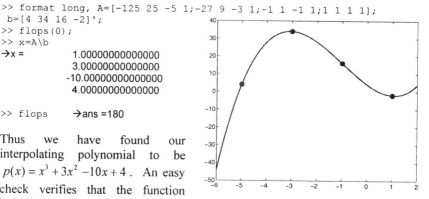

```
>> format long, A=[-125 25 -5 1;-27 9 -3 1;-1 1 -1 1;1 1 1 1];
   b=[4 34 16 -2]';
>> flops(0);
>> x=A\b
→x =          1.00000000000000
              3.00000000000000
            -10.00000000000000
              4.00000000000000

>> flops      →ans =180
```

Thus we have found our interpolating polynomial to be $p(x) = x^3 + 3x^2 - 10x + 4$. An easy check verifies that the function indeed interpolates the given data. Its graph along with the data points are shown in Figure 7.28.

FIGURE 7.28: Graph of the interpolating cubic polynomial $p(x) = x^3 + 3x^2 - 10x + 4$ for the four data points that were given in Example 7.8.[7]

---

[6] Flop counts will help us to compare efficiencies of algorithms. Later versions of MATLAB (Version 6 and later) no longer support the flop count feature. We will occasionally tap into the older Version 5 of MATLAB (using flops) to compare flop counts, for purely illustrative purposes.

[7] The MATLAB plot in the figure was created by first plotting the function, as usual, applying hold on, and then plotting each of the points as red dots using the following syntax: (e.g., for the data point (-1, 16)) plot(-1,16,'ro'). The "EDIT" menu on the graphics window can then be used to enlarge the size of the red dots after they are selected.

<u>Method 2:</u> We compute the inverse of A and multiply this inverse on the left of b.

```
>> flops(0); x=inv(A)*b
→x =              1.00000000000000
                  3.00000000000000
                -10.00000000000000
                  4.00000000000000
>> flops      →ans = 262
```

We arrived at the same (correct) answer, but with more work. The amount of extra work needed to compute the inverse rather than just solve the system gets worse with larger-sized matrices; moreover, this method is also more prone to errors. We will give more evidence and will substantiate these claims later in this section.

<u>Method 3:</u> This method is more general than those of the first two since it will work also to solve singular systems that need not have square coefficient matrices. MATLAB has the following useful command:

rref(Ab) →	Puts an augmented $n \times (m+1)$ matrix $[A \mid b]$ for the system $Ax = b$ into reduced row echelon form.

The reduced row echelon form of the augmented matrix is a form from which the general solution of the linear system can easily be obtained. In general, a linear system can have (i) no solution, (ii) exactly one solution (nonsingular case), or (iii) infinitely many solutions. We will say more about how to interpret the reduced row echelon form in singular cases later in this section, but in case of a nonsingular (square) coefficient matrix A, the reduced row echelon form of the augmented matrix $[A \mid b]$ will be $[I \mid x]$, where x is the solution vector.

Assuming A and b are still in the workspace, we construct from them the augmented matrix and then use rref

```
>> Ab=A; Ab(:,5)=b;
>> flops(0); rref(Ab)
→ ans =       1   0   0   0   1
              0   1   0   0   3
              0   0   1   0  -10
              0   0   0   1   4
>> flops      →ans = 634
```

Again we obtained the same answer, but with a lot more work (more than triple the flops that were needed in Method 1). Although rref is also based in Gaussian elimination, putting things into reduced row echelon form (as is usually taught in linear algebra courses) is a computational overkill of what is needed to solve a nonsingular system. This will be made apparent in the next section.

EXERCISE FOR THE READER 7.15: (a) Find the coefficients a, b, c, and d of the polynomial $p(x) = ax^3 + bx^2 + cx + d$ of degree at most 3 that passes through these data points $(-2, 4), (1, 3), (2, 5)$, and $(5, -22)$.

(b) Find the equation of the polynomial $p(x) = ax^8 + bx^7 + cx^6 + dx^5 + ex^4 + fx^3 + gx^2 + hx + k$ of degree at most 8 that passes through the following data points: $(-3, -14.5)$, $(-2, -12)$, $(-1, 15.5)$, $(0,2)$, $(1, -22.5)$, $(2, -112)$, $(3, -224.5)$, $(4, 318)$, $(5, 3729.5)$.   Solve the system using each of the three methods shown in Example 7.8 and compare the solutions, computation times (using tic/toc), and flop counts (if available).

   In the preceding example, things worked great and the answer MATLAB gave us was the exact one. But the matrix was small and not badly conditioned. In general, solving a large linear system involves a large sequence of arithmetic operations and the floating point errors can add up, sometimes (for poorly conditioned matrices) intolerably fast. The next example will demonstrate such a pathology, and should convince the reader of our recommendation to opt for Method 1 (left divide).

EXAMPLE 7.9:   (*The Hilbert Matrix*) A classical example of a matrix that comes up in applications and is very poorly conditioned is

$$H_n = \begin{bmatrix} 1 & 1/2 & 1/3 & \cdots & 1/n \\ 1/2 & 1/3 & 1/4 & \cdots & 1/(n+1) \\ \vdots & & & \ddots & \vdots \\ 1/n & 1/(n+1) & 1/(n+2) & \cdots & 1/(2n-1) \end{bmatrix},$$

the **Hilbert matrix** H_n **of order** n, which is defined above. This matrix can easily be entered into a MATLAB session using a for loop, but MATLAB even has a separate command hilb(n) to create it. In this example we will solve, using each of the three methods from the last example, the equation $Hx = b$, where b is the vector Hx, and where $x = (1\ 1\ 1\ \cdots\ 1)'$ with $n = 12$.

We now proceed with each of the three methods of the solution of the last example to produce three corresponding "solutions":   x_meth1, x_meth2, and x_meth3 to the linear system $Hx = b$. Since we know the exact solution to be $x = (1\ 1\ 1\ \cdots\ 1)'$, we will be able to compare both accuracy and speeds of the three methods.

```
>> x=ones(12,1); H=hilb(12); b=H*x;
>> flops(0); x_meth1=H\b10; max(abs(x-x_meth1))      → 0.2385
>> %because each component of the exact solution is 1, this can be
>> % thought of as the maximum relative error in any component.
>> flops      → 995
>> flops(0);
>> x_meth2=inv(H)*b; max(abs(x-x_meth2))→ 0.6976
>> flops      → 1997
>> flops(0);, R=rref([H b]); flops      →ans = 6484
```

If we view row reduced matrix R produced in Method 3 (or just the tenth row by entering R(12,:)), we see that the last row is entirely made up of zeros. Thus, Method 3 leads us to the false conclusion that the linear system is singular.

FIGURE 7.29: David Hilbert[8] (1862–1943), German mathematician.

The results of Method 3 are catastrophic, since the reduced row echelon form produced would imply that there are infinitely many solutions! (We will explain this conclusion shortly.) Methods 1 and 2 both had (unacceptably) large relative errors of about 24% and 70%, respectively. The increasing flop counts in the three methods also shows us that left divide (Method 1) gives us more for less work. Of course, the Hilbert matrix is quite a pathological one, but such matrices do come up often enough in applications that we must always remember to consider floating point/roundoff error when we solve linear systems. We point out that many problems that arise in applied mathematics involve very well-conditioned linear systems that can be effectively solved (using MATLAB) for dimensions of order 1000! This is the case for many linear systems that arise in the numerical solution of partial differential equations. In Section 7.6, we will examine more closely the concept

---

[8]The Hilbert matrix is but a small morsel of the vast set of mathematical contributions produced in the illustrious career of David Hilbert (Figure 7.29). Hilbert is among the very top echelon of the greatest German mathematicians, and this group contains a lot of competition. Hilbert's ability to transcend the various subfields of mathematics, to delve deeply into difficult problems, and to discover fascinating interrelations was unparalleled. Three years after earning his Ph.D., he submitted a monumental paper containing a whole new and elegant treatment of the so-called *Basis Theorem,* which had been proved by Paul Albert Gordan (1837–1912) in a much more specialized setting using inelegant computational methods. Hilbert submitted his manuscript for publication to the premier German mathematical journal *Mathematische Annalen,* and the paper was sent to Gordan to (anonymously) referee by the editor Felix Christian Klein (1849–1925), also a famous German mathematician and personal friend of Gordan's. It seemed that Gordan, the world-renowned expert in the field of Hilbert's paper, was unable to follow Hilbert's reasoning and rejected the paper for publication on the basis of its incoherence. In response, Hilbert wrote to Klein, *"... I am not prepared to alter or delete anything, and regarding this paper, I say with all modesty, that this is my last word so long as no definitive and irrefutable objection against my reasoning is raised."* Hilbert's paper finally appeared in this journal in its original form, and he continued to produce groundbreaking papers and influential books in an assortment of fields spanning all areas of mathematics. He even has a very important branch of analysis named in his honor (Hilbert space theory). In the 1900 International Conference of Mathematics in Paris, Hilbert posed 23 unsolved problems to the mathematical world. These "Hilbert problems" have influenced much mathematical research activity throughout the twentieth century. Several of these problems have been solved, and each such solution marked a major mathematical event. The remaining open problems should continue to influence mathematical thoughts well into the present millennium. In 1895, Hilbert was appointed to a "chair" at the University of Göttingen and although he was tempted with offers from other great universities, he remained at this position for his entire career. Hilbert retired in his birthplace city of Königsberg, which (since he had left it) had become part of Russia after WWII (with the name of the city changed to "Kaliningrad"). Hilbert was made an honorary citizen of this city and in his acceptance speech he gave this now famous quote: *"Wir müssen wissen, wir werden wissen (We must know, we shall know)."*

of the condition number of a matrix and its effect error bounds for numerical solutions of corresponding linear systems. Just because a matrix is poorly conditioned does not mean that all linear systems involving it will be difficult to solve numerically. The next exercise for the reader gives such an example with the Hilbert matrices.

EXERCISE FOR THE READER 7.16: In this exercise, we will be considering larger analogs of the system studied in Example 7.9. For a positive integer n, we let $c(n)$ be the least common multiple of the integers 1, 2, 3, ..., n, and we define

$b_n = H_n(c(n)e_1')$, where H_n is the nth order Hilbert matrix and e_1 is the vector $(1,0,0, ..., 0)$ (having n components). We chose $c(n)$ to be as small as possible so that the vector b_n will have all integer components. Note, in Example 7.9, we used $c(10) = 2520$.

(a) For $n = 20$, solve this system using Method 1, time it with tic/toc, and (if available) do a flop count and find the percentage of the largest error in any of the 20 components of x_meth1 to that of the largest component of the exact solution x $(= c(20))$; repeat using Method 2.

(b) Repeat part (a) for $n = 30$.

(c) In parts (a) and (b), you should have found that x_meth1 equaled the exact solution (so the corresponding relative error percentages were zero), but the relative error percentages for x_meth2 grew from 0.00496% in case $n = 10$ (Example 7.9), to about 500% in part (a) and to about 5000% in part (b). Thus the "noise" from the errors has transcended, by far, the values of the exact solution. Continue solving this system using Method 1, for $n = 40$, $n = 50$, and so on, until you start getting errors from the exact solution or the computations start to take too much time, whichever comes first.

Suggestion: MATLAB has a built-in function lcm(a,b) that will find the least common multiple of two positive integers. You can use this, along with a for loop to get MATLAB to easily compute $c(n)$, for any value of n. In each part, you may wish to use max(abs(x-x_meth1)) to detect any errors.

Note: Exercise 31 of Section 7.6 will analyze why things have gone so well with these linear systems.

We close this section by briefly explaining how to interpret the reduced row echelon form to obtain the general solution of a linear system with a singular coefficient matrix that need not be square. Suppose we have a linear system with n equations and m unknowns $x_1, x_2, \cdots, x_m$:

$$\begin{cases} a_{11}x_1 + a_{12}x_2 + \cdots + a_{1m}x_m = b_1 \\ a_{21}x_1 + a_{22}x_2 + \cdots + a_{2m}x_m = b_2 \\ \quad \cdots \\ a_{n1}x_1 + a_{n2}x_2 + \cdots + a_{nm}x_m = b_m \end{cases} . \tag{22}$$

We can write this equation in matrix form $Ax = b$, as before; but in general the coefficient matrix A need not be square. We form the **augmented matrix** of the system by tacking on the vector b as an extra column on the right of A :

$$[A \mid b] = \begin{bmatrix} a_{11} & a_{12} & a_{13} & \cdots & a_{1m} & \mid b_1 \\ a_{21} & a_{22} & a_{23} & \cdots & a_{2m} & \mid b_2 \\ \vdots & & & \ddots & \vdots & \mid \vdots \\ a_{n1} & a_{n2} & a_{n3} & \cdots & a_{nm} & \mid b_n \end{bmatrix}.$$

The augmented matrix is said to be in **reduced row echelon form** if the following four conditions are met. Each condition pertains only to the left of the partition line (i.e., the a_{ij}'s):

1. Rows of all zero entries, if any, must be grouped together at the bottom.
2. If a row is not all zeros, the leftmost nonzero entry must equal 1 (such an entry will be called a **leading one** for the row).
3. All entries above and below (in the same column as) a leading one must be zero.
4. If there are more than one leading ones, they must move to the right as we move down to lower rows.

Given an augmented matrix A, the command `rref(Ab)` will output an augmented matrix of the same size (but MATLAB will not show the partition line) that is in reduced row echelon form and that represents an **equivalent** linear system to Ab, meaning that both systems will have the same solution. It is easy to get the solution of any linear system, singular or not, if it is in reduced row echelon form. Since most of our work will be with nonsingular systems (for which `rref` should not be used), we will not say more about how to construct the reduced row echelon form. The algorithm is based on Gaussian elimination, which will be explained in the next section. For more details on the reduced row echelon form, we refer to any textbook on basic linear algebra; see, e.g., [Kol-99], or [Ant-00]. We will only show, in the next example, how to obtain the general solution from the reduced row echelon form.

EXAMPLE 7.10: (a) Which of the following augmented matrices are in reduced row echelon form?

$$M_1 = \begin{bmatrix} 1 & 2 & 0 & \mid -2 \\ 0 & 0 & 1 & \mid 3 \\ 0 & 0 & 0 & \mid 0 \end{bmatrix}, \ M_2 = \begin{bmatrix} 1 & 2 & 0 & 1 & \mid 1 \\ 0 & 1 & 0 & 2 & \mid -8 \\ 0 & 0 & 1 & 3 & \mid 4 \end{bmatrix}, \ M_3 = \begin{bmatrix} 1 & 0 & 0 & 1 & \mid 1 \\ 0 & 1 & 0 & 2 & \mid -8 \\ 0 & 0 & 0 & 0 & \mid 4 \end{bmatrix}.$$

(b) For those that are in reduced row echelon form, find the general solution of the corresponding linear system that the matrix represents.

SOLUTION: Part (a): M_1 and M_3 are in reduced row echelon form; M_2 is not. The reader who is inexperienced in this area of linear algebra should carefully verify these claims for each matrix by running through all four of the conditions (i) through (iv).

Part (b): If we put in the variables and equal signs in the three-equation linear system represented by M_1, and then solve for the variables that have leading ones in their places (here x_1 and x_3), we obtain:

$$\begin{cases} x_1 + 2x_2 & = -2 \Rightarrow x_1 = -2 - 2x_2 \\ x_3 = 3 \\ 0 = 0 \end{cases},$$

Thus there are infinitely many solutions: Letting $x_2 = t$, where t is any real number, the general solution can be expressed as

$$\begin{cases} x_1 = -2 - 2t \\ x_2 = t \\ x_3 = 3 \end{cases}, \quad t = \text{any real number}.$$

If we do the same with the augmented matrix M_3, we get $0 = 4$ for the last equation of the system. Since this is impossible, the system has no solution.

Here is a brief summary of how to solve a linear system with MATLAB. For a square matrix A, you should always try x = A\b (left divide). For singular matrices (in particular, nonsquare matrices) use rref on the augmented matrix. There will be either no solution or infinitely many solutions. No solution will always be seen in the reduced row echelon form matrix by a row of zeros before the partition line and a nonzero entry after it (as in the augmented matrix M_3 in the above example). In all other (singular) cases there will be infinitely many solutions; columns without leading ones will correspond to variables that are to be assigned arbitrary real numbers (s, t, u, ...); columns with leading ones correspond to variables that should be solved for in terms of variables of the first type.

EXERCISE FOR THE READER 7.17: Parts (a) and (b): Repeat the instructions of both parts (a) and (b) of the preceding example for the following augmented matrices:

$$M_1 = \begin{bmatrix} 1 & 0 & | & 3 \\ 0 & 1 & | & 2 \\ 0 & 0 & | & 0 \end{bmatrix}, \quad M_2 = \begin{bmatrix} 1 & -2 & 0 & 3 & | & -2 \\ 0 & 0 & 1 & -5 & | & 1 \end{bmatrix}, \quad M_3 = \begin{bmatrix} 1 & 0 & 0 & | & 1 \\ 0 & 0 & 1 & | & 3 \\ 0 & 1 & 0 & | & 4 \end{bmatrix}.$$

Part (c): Using the MATLAB command rref, find the general solutions of the following linear systems:

(i) $\begin{cases} x_1 & + & 3x_2 & + & 2x_3 & & & = & 3 \\ 2x_1 & + & 6x_2 & + & 2x_3 & - & 8x_4 & = & 4 \end{cases}$

(ii) $\begin{cases} x_1 & - & 2x_2 & + & x_3 & + & x_4 & + & 2x_5 & = & 2 \\ -2x_1 & + & 4x_2 & + & 2x_3 & + & 2x_4 & - & 2x_5 & = & 0 \\ 3x_1 & - & 6x_2 & + & x_3 & + & x_4 & + & 5x_5 & = & 4 \\ -x_1 & + & 2x_2 & + & 3x_3 & + & x_4 & + & x_5 & = & 3 \end{cases}$

EXERCISES 7.4:

1. Use MATLAB to solve the linear system $Ax = b$, with the following choices for A and b. Afterward, check to see that your solution x satisfies $Ax = b$.

(a) $A = \begin{bmatrix} 9 & -5 & 2 \\ 0 & 7 & 5 \\ -1 & -9 & 6 \end{bmatrix}$, $b = \begin{bmatrix} 67 \\ 13 \\ 27 \end{bmatrix}$ (b) $A = \begin{bmatrix} -1 & 1 & 3 & 5 \\ 3 & -4 & -1 & 5 \\ 5 & -1 & 4 & -5 \\ -2 & 3 & -5 & -4 \end{bmatrix}$, $b = \begin{bmatrix} 9 \\ 29 \\ -22 \\ -5 \end{bmatrix}$

(c) $A = \begin{bmatrix} -3 & -3 \\ 1 & -3 \end{bmatrix}$, $b = \begin{bmatrix} 2.928 \\ 3.944 \end{bmatrix}$ (d) $A = \begin{bmatrix} -12 & -20 & 10 \\ -2 & 18 & -1 \\ -3 & 14 & 1 \end{bmatrix}$, $b = \begin{bmatrix} -112.9 \\ 71.21 \\ 45.83 \end{bmatrix}$

2. (*Polynomial Interpolation*) (a) Find the equation of the polynomial of degree at most 2 (parabola) that passes through the data points $(-1, 21)$, $(1, -3)$, $(5, 69)$; then plot the function along with the data points.
(b) Find the equation of the polynomial of degree at most 3 that passes through the data points: $(-4, -58.8)$, $(2, 9.6)$, $(8, 596.4)$, $(1, 4.2)$; then plot the function along with the data points.
(c) Find the equation of the polynomial of degree at most 6 that passes through the data points: $(-2, 42)$, $(-1, -29)$, $(-0.5, -16.875)$, $(0, -6)$, $(1, 3)$, $(1.5, 18.375)$, $(2, -110)$; then plot the function along with the data points.
(d) Find the equation of the polynomial of degree at most 5 that passes through the data points: $(1, 2)$, $(2, 4)$, $(3, 8)$, $(4, 16)$, $(5, 32)$, $(6, 64)$; then plot the function along with the data points.

3. Find the general solution of each of the following linear systems:

(a) $\begin{cases} 3x_1 & + & 3x_2 & + & 2x_3 & + & 5x_4 & = & 12 \\ 2x_1 & + & 6x_2 & + & 2x_3 & - & 8x_4 & = & 4 \end{cases}$

(b) $\begin{cases} x_1 & + & x_2 & + & 3x_3 & - & x_4 & - & 5x_5 & = & 2 \\ x_1 & - & 3x_2 & + & 2x_3 & - & 2x_4 & + & x_5 & = & 2 \\ 2x_1 & - & 7x_2 & + & 3x_3 & - & x_4 & - & 4x_5 & = & 4 \\ -3x_1 & - & 2x_2 & - & 4x_3 & + & 4x_4 & - & x_5 & = & 6 \end{cases}$

(c) $\begin{cases} 2x_1 & + & 2x_2 & - & 3x_3 & + & x_4 & = & 8 \\ 4x_1 & & & - & 3x_3 & - & 2x_4 & = & 0 \\ -3x_1 & + & 4x_2 & + & x_3 & + & x_4 & = & -1 \end{cases}$

(d) $\begin{cases} 3x_1 & + & 6x_2 & + & x_3 & - & 2x_4 & + & 12x_5 & & & = & 22 \\ -8x_1 & - & 16x_2 & & & + & 16x_4 & - & 32x_5 & + & x_6 & = & -60 \\ x_1 & + & 2x_2 & & & - & 2x_4 & + & 4x_5 & & & = & 8 \end{cases}$

4. (*Polynomial Interpolation*) In polynomial interpolation problems as in Example 7.8, the coefficient matrix that arises is the so-called **Vandermonde** matrix corresponding to the vector $v = [x_0 \; x_1 \; x_2 \; \cdots \; x_n]$ that is the $(n+1) \times (n+1)$ matrix defined by

$$V = \begin{bmatrix} x_0{}^n & x_0{}^{n-1} & \cdots & x_0 & 1 \\ x_1{}^n & x_1{}^{n-1} & \cdots & x_1 & 1 \\ \vdots & \vdots & & & \ddots \\ x_n{}^n & x_n{}^{n-1} & \cdots & x_n & 1 \end{bmatrix}.$$

MATLAB (of course) has a function vander(v), that will create the Vandermonde matrix corresponding to the inputted vector v. Redo, parts (a) through (d) of Exercise 2, this time using vander(v).

(e) Write your own version myvander(v), that does what MATLAB's vander(v) does. Check the efficiency of your M-file with that of MATLAB's vander(v) by typing (after you have created and debugged your program) >>type vander to display the code of MATLAB's vander(v).

5. (*Polynomial Interpolation*) (a) Write a MATLAB function M-file called pv = polyinterp(x,y) that will input two vectors x and y of the same length (call this length "$n+1$" for now) that correspond to the x- and y-coordinates of $n+1$ data points on which we wish to interpolate with a polynomial of degree at most n. The output will be a vector $pv = [a_n \ a_{n-1} \ \cdots a_2 \ a_1 \ a_0]$ that contains the coefficients of the interpolating polynomial $p(x) = a_n x^n + a_{n-1} x^{n-1} + \cdots + a_1 x + a_0$.

(b) Use this program to redo part (b) of Exercise 2.

(c) Use this program to redo part (c) of Exercise 2.

(d) Use this program with input vectors $x = \begin{bmatrix} 0 & \pi/2 & \pi & 3\pi/2 & 2\pi & 5\pi/2 & 3\pi & 7\pi/2 & 4\pi \end{bmatrix}$, and $y = [1 \ 0 \ -1 \ 0 \ 1 \ 0 \ -1 \ 0 \ 1]$. Plot the resulting polynomial along with the data points. Include also the plot (in a different style/color) of a trig function that interpolates this data.

6. (*Polynomial Interpolation: Asking a Bit More*) Often in applications, rather than just needing a polynomial (or some other nice interpolating curve) that passes through a set of data points, we also need the interpolating curve to satisfy some additional smoothness requirements. For example, consider the design of a railroad transfer segment shown in Figure 7.30. The curved portion of "interpolating" railroad track needs to do more than just connect the

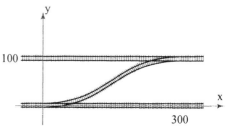

FIGURE 7.30: Exercise 6 asks to find a polynomial function modeling the junction between the two parallel sects of rails.

two parallel tracks, it must do so "smoothly" lest the trains using it would derail. Thus, if we seek a function $y = p(x)$ that models this interpolating track, we see from the figure (and the reference to the xy-coordinate system drawn in) that we would like the center curve of this interpolating track to satisfy the following conditions on the interval $0 \le x \le 300$ feet:

$$p(0) = 0, \ p(300) = 100 \text{ feet}, \ p'(0) = p'(300) = 0.$$

(The last two conditions geometrically will require the graph of $y = p(x)$ to have a horizontal tangent line at the endpoints $x = 0$ and $x = 300$, and thus connect smoothly with the existing tracks.) If we would like to use a polynomial for this interpolation, since we have four requirements, we should be working with a polynomial of degree at most 3 (that has four parameters): $p(x) = ax^3 + bx^2 + cx + d$.

(a) Set up a linear system for this interpolating polynomial and get MATLAB to solve it.

(b) Next, get MATLAB to graph the rail network (just the rails) including the two sets of parallel tracks as well as the interpolating rails. Leave a 6-foot vertical distance between each set of adjacent rails.

7. (*Polynomial Interpolation: Asking a Bit More*) Three parallel railroad tracks need to be connected by a pair of curved junction segments, as shown in Figure 7.31.

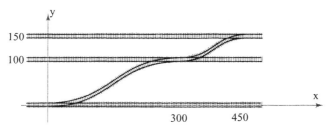

FIGURE 7.31: In Exercise 7, this set of three parallel rails is required to be joined by two sets of smooth junction rails.

(a) If we wish to use a single polynomial function to model (the center curves of) both pairs of junction rails shown in Figure 7.31, what degree polynomials should we use in our model? Set up a linear system to determine the coefficients of this polynomial, then get MATLAB to solve it and determine the polynomial.

(b) Next, get MATLAB to graph the rail network (just the rails) including the three sets of parallel tracks as well as the interpolating rails gotten from the polynomial function you found in part (a). Leave a 6-foot vertical distance between each set of adjacent rails.

(c) Do a separate polynomial interpolation for each of the two junction rails and thus find two different polynomials that model each of the two junctions. Set up the two linear systems, solve them using MATLAB, and then write down the two polynomials.

(d) Next, get MATLAB to graph the rail network (just the rails) including the three sets of parallel tracks as well as the interpolating rails gotten from the polynomial functions you found in part (c). Leave a 6-foot vertical distance between each set of adjacent rails. How does this picture compare with the one in part (b)?

NOTE: In general it is more efficient to do the piecewise polynomial interpolation that was done in part (d) rather than the single polynomial interpolation in part (b). The advantages become more apparent when there are a lot of data points. This approach is an example of what is called spline interpolation.

8. (*City Planning: Traffic Logistics*) The Honolulu street map of Figure 7.32 shows the rush-hour numbers of vehicles per hour that enter or leave the network of four one-way streets. The variables x_1, x_2, x_3, x_4 represent the traffic flows on the segments shown. For smooth traffic flow, we would like to have equilibrium at each of the four intersections; i.e., the number of incoming cars (per hour) should equal the number of outgoing cars. For example, at the intersection of Beretania and Piikoi (lower right), we should have $x_1 + 800 = x_2 + 2000$,

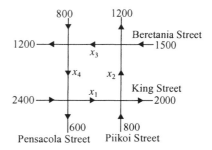

FIGURE 7.32: Rush-hour traffic on some Honolulu streets.

or (after rearranging) $x_1 - x_2 = 1200$.

(a) Obtain a linear system for the smooth traffic flow in the above network by looking at the flows at each of the four intersections.

(b) How many solutions are there? If there are solutions, which, if any, give rise to feasible traffic flow numbers.

(c) Is it possible for one of the four segments in the network to be closed off for construction (a perennial occurrence in Honolulu) so that the network will still be able to support smooth traffic flow? Explain.

9. (*City Planning: Traffic Logistics*) Figure 7.33 shows a busy network of one-way roads in the center of a city. The rush-hour inflows and outflows of vehicles per hour for the network are given as well as a listing of the nine variables that represent hourly flows of vehicles along the one-way segments in the network.

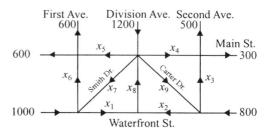

FIGURE 7.33: Rush-hour traffic in a busy city center.

(a) Following the directions of the preceding exercise, use the equilibria at each intersection to obtain a system of linear equations in the nine variables that will govern smooth traffic flow in this network.

(b) Use MATLAB's `rref` to solve this system. There are going to be infinitely many solutions, but of course not all are going to be feasible answers to the original problem. For example, we cannot have a negative traffic flow on any given street segment, and also the x_j's should be integers. Thus the solutions consist of vectors with eight components where each component is a nonnegative integer.

(c) Considering all of the feasible solutions that were contemplated in part (b), what is the maximum that x_6 can be (in a solution)? What is the minimum?

(d) Repeat part (c) for x_8.

(e) If the city wants to have a parade and close up one of the segments (corresponding to some x_j in the figure) of the town center, is it possible to do this without disrupting the main traffic flow?

(f) If you answered yes to part (e), go further and answer this question. The mayor would like to set up a "Kick the Fat" 5K run through some of the central streets in the city. How many segments (corresponding to some x_j in the figure) could the city cordon off without disrupting the main flow of traffic? Answer the same question if we require, in addition, that the streets which are cordoned off are connected together.

10. (*Economics: Input–Output Analysis*) In any large economy, major industries that are producing essential goods and services need products from other industries in order to meet their own production demands. Such demands need to be accounted for by the other industries in addition to the main consumer demands. The model we will give here is due to Russian ecomomist Wassily Leontief (1906–1999).[9] To present the main ideas, we deal only with three dependent industries: (i) electricity, (ii) steel, and (iii) water. In a certain economy, let us assume that the outside demands for each of these three industries are (annually) $d_1 =$ \$140 million for electricity, $d_2 =$ \$46 million for steel, and $d_3 =$ \$92 million for water. For each dollar that the electricity industry produces each year, assume that it will cost \$0.02 in electricity, \$0.08 in

---

[9] In the 1930s and early 1940s, Leontief did an extensive analysis of the input and output of 500 sectors of the US economy. The calculations were tremendous, and Leontief made use of the first large-scale computer (in 1943) as a necessary tool. He won the Nobel Prize in Economics in 1973 for this research. Educated in Leningrad, Russia (now again St. Petersburg, as it was before 1917), and in Berlin, Leontif subsequently moved to the United States to become a professor of economics at Harvard. His family was quite full of intellectuals: His father was also a professor of economics, his wife (Estelle Marks) was a poet, and his daughter is a professor of art history at the University of California at Berkeley.

steel, and $0.16 in water. Also for each dollar of steel that the steel industry produces each year, assume that it will cost $0.05 in electricity, $0.08 in steel, and $0.10 in water. Assume the corresponding data for producing $1 of water to be $0.12, $0.07, and $0.03. From this data, we can form the so-called **technology matrix**:

$$
\begin{array}{ccc}
 & \text{E} & \text{S} & \text{W}
\end{array}
$$

$$
M \;=\; \begin{array}{l}
\text{Electricity demand (per dollar)} \\
\text{Steel demand (per dollar)} \\
\text{Water demand (per dollar)}
\end{array}
\begin{bmatrix}
.02 & .05 & .12 \\
.08 & .08 & .07 \\
.16 & .10 & .03
\end{bmatrix}
$$

(a) Let x_1 = The amount (in dollars) of electricity produced by the electricity industry,

x_2 = The amount of steel produced by the steel industry,

x_3 = The amount of water produced by the water industry, and let

$X = \begin{bmatrix} x_1 \\ x_2 \\ x_3 \end{bmatrix}$. The matrix X is called the **output matrix**. Show/explain why the matrix MX

(called the **internal demand matrix**) gives the total internal costs of electricity, steel, and water that it will collectively cost the three industries to produce the outputs given in X.

(b) For the economy to function, the output of these three major industries must meet both the external demand and the internal demand. The external demand is given by the following *external demand matrix*:

$$
D = \begin{bmatrix} d_1 \\ d_2 \\ d_3 \end{bmatrix} = \begin{bmatrix} 140,000,000 \\ 46,000,000 \\ 92,000,000 \end{bmatrix}.
$$

(The data was given above.) Thus since the total output X of the industries must meet both the internal MX and external D demands, the matrix X must solve the matrix equation:

$$
X = MX + D \Rightarrow (I - M)X = D .
$$

It can always be shown that the matrix $I - M$ is nonsingular and thus there is always going to be a unique solution of this problem. Find the solution of this particular input/output problem.

(c) In a particular year, there is a construction boom and the demands go up to these values:

$$
d_1 = \$160 \text{ mil.}, d_2 = \$87 \text{ mil.}, d_3 = \$104 \text{ mil}
$$

How will the outputs need to change for that year?

(d) In a recessionary year, the external demands drop to the following numbers:

$$
d_1 = \$90 \text{ mil.}, d_2 = \$18 \text{ mil.}, d_3 = \$65 \text{ mil}
$$

Find the corresponding outputs.

11. (*Economics: Input-Output Analysis*) Suppose, in the economic model of the previous exercise, two additional industries are added: (iv) oil and (v) plastics. In addition to the assumptions of the previous exercise, assume further that for each dollar of electricity produced it will cost $0.18 in oil and $0.03 in plastics, for each dollar of steel produced it will cost $0.07 in oil and $0.01 in plastics, for each dollar of water produced it will cost $0.02 in plastics (but no oil), for each dollar of oil produced it will cost $0.06 in electricity, $0.02 in steel, $0.05 in water, and $0.01 in plastics (but no oil), and finally for each dollar in plastics produced, it will cost $0.02 in electricity, $0.01 in steel, $0.02 in water, $0.22 in oil, and $0.12 in plastics.

(a) Write down the technology matrix.

(b) Assuming the original external demands for the first three industries given in the preceding exercise and external demands of $d_4 = \$188$ mil. for oil and $d_5 = \$35$ mil. for plastics, solve the Leontief model for the resulting output matrix.

(c) Resolve the model using the data in part (c) of Exercise 7, along with $d_4 = \$209$ mil., $d_5 = \$60$ mil.

(d) Resolve the model using the data in part (c) of Exercise 7, along with $d_4 = \$149$ mil., $d_5 = \$16$ mil.

NOTE: (*Combinatorics: Power Sums*) It is often necessary to find the sum of fixed powers of the first several positive integers. Formulas for the sums are well known but it is difficult to remember them all. Here are the first four such *power sums*:

$$\sum_{k=1}^{n} k = 1 + 2 + 3 + \cdots + n = \frac{1}{2}n(n+1) \tag{23}$$

$$\sum_{k=1}^{n} k^2 = 1 + 4 + 9 + \cdots + n^2 = \frac{1}{6}n(n+1)(2n+1) \tag{24}$$

$$\sum_{k=1}^{n} k^3 = 1 + 8 + 27 + \cdots + n^3 = \frac{1}{4}n^2(n+1)^2 \tag{25}$$

$$\sum_{k=1}^{n} k^4 = 1 + 16 + 81 + \cdots + n^4 = \frac{1}{30}n(n+1)(2n+1)(3n^2 + 3n - 1) \tag{26}$$

Each of these formulas, and more general ones, can be proved by mathematical induction, but deriving them is more involved. It is a general fact that for any positive integer p, the power sum $\sum_{k=1}^{n} k^p$ can always be expressed as a polynomial $f(n)$ that has degree $p+1$ and has rational coefficients (fractions). See Section 3.54 (p. 199ff) of [Ros-00] for details. The next two exercises will show a way to use linear systems not only to verify, but to derive such formulas.

12. (*Combinatorics: Power Sums*) (a) Use the fact (from the general fact mentioned in the preceding note) that $\sum_{k=1}^{n} k$ can be expressed as $f(n)$ where $f(n)$ is a polynomial of degree 2: $f(n) = an^2 + bn + c$, to set up a linear system for a, b, c using $f(1) = 1$, $f(2) = 3$, $f(3) = 6$, and use MATLAB (in "format rat") to solve for the coefficients and then verify identity (23).
(b) In a similar fashion, verify identity (24).
(c) In a similar fashion, verify identity (25).
(d) In a similar fashion, verify identity (26).
Note: If you have the Student Version of MATLAB (or have access to the Symbolic Toolbox), the command factor can be used to develop your formulas in factored form—see Appendix A; otherwise leave them in standard polynomial form.

13. (*Combinatorics: Power Sums*) (a) By mimicking the approach of the previous exercise, use MATLAB to get a formula for the power sum $\sum_{k=1}^{n} k^5$. Check your formula for the values $n = 5$, and $n = 100$ (using MATLAB, of course). (b) Repeat for the sum $\sum_{k=1}^{n} k^6$.
Note: If you have the Student Version of MATLAB (or have access to the Symbolic Toolbox), the command factor can be used to develop your formulas in factored form—see Appendix A; otherwise leave them in standard polynomial form.

14. (*Combinatorics: Alternating Power Sums*) For positive integers p and n, the *alternating power sum*:

$$\sum_{k=1}^{n} (-1)^{n-k} k^p = n^p - (n-1)^p + (n-2)^p - \cdots + (-1)^{n-2}2^p + (-1)^{n-1},$$

like the power sum, can be expressed as $f(n)$ where $f(n)$ is polynomials of degree $n+1$ having rational coefficients. (For details, see [Ros-00], Section 3.17, page 152ff). As in

Exercise 5, set up linear systems for the coefficients of these polynomial, use MATLAB to solve them, and develop formulas for the alternating power sums for the following values of p. (a) $p = 1$, (b) $p = 2$, (c) $p = 3$, (d) $p = 4$. For each formula you derive, get MATLAB to check it (against the actual power sum) for the following values of n: 10, 250, and 500.

15. *(Linear Algebra: Cramer's Rule)* There is an attractive and explicit formula for solving a nonsingular system $Ax = b$ which expresses the solution of each component of the vector x entirely in terms of determinants. This formula, known as Cramer's rule,[10] is often given in linear algebra courses (because it has nice theoretical applications), but it is a very expensive one to use. Cramer's rule states that the solution $x = [x_1 \ x_2 \ \cdots \ x_n]'$ of the (nonsingular) system $Ax = b$, is given by the following formulas:

$$x_1 = \frac{\det(A_1)}{\det(A)}, \quad x_2 = \frac{\det(A_2)}{\det(A)}, \quad \cdots, \quad x_n = \frac{\det(A_n)}{\det(A)},$$

where the $n \times n$ matrix A_i is formed by replacing the ith column of the coefficient matrix A, by the column vector b.

(a) Use Cramer's rule to solve the linear system of Examples 7.7 and 7.8, and compare performance time and (if you have Version 5) flop counts and accuracies with Methods 1, 2, and 3 that were used in the text.

(b) Write a function M-file called x = cramer(A,b) that will input a square nonsingular matrix A, a column vector b of the same dimension, and will output the column vector solution of the linear system $Ax = b$ obtained using Cramer's rule. Apply this program to resolve the two systems of part (a).

Note: Of course, you should set up your calculations and programs so that you only ask MATLAB to compute $\det(A)$ once, each time you use Cramer's rule.

7.5: GAUSSIAN ELIMINATION, PIVOTING, AND LU FACTORIZATION

Here is a brief outline of this section. Our goal is a general method for solving the linear system (17) with n variables and n equations, and we will be working with the corresponding augmented matrix $[A \mid b]$ from the resulting matrix equation (18). We will first observe that if a linear system has a triangular matrix A (i.e., all entries below the main diagonal are zero, or all entries above it are zero), then the linear system is very easy to solve. We next introduce the three elementary

---

[10] Gabriel Cramer (1704–1752) was a Swiss mathematician who is credited for introducing his namesake rule in his famous book, *Introduction à l'analyse des lignes courbes algébraique.* Cramer entered his career as a mathematician with impressive speed, earning his Ph.D. at age 18 and being awarded a joint chaired professorship of mathematics at the Académie de Clavin in Geneva. With his shared appointment, he would take turns with his colleague Caldrini teaching for 2 to 3 years. Through this arrangement he was able to do much traveling and made contacts with famous mathematicians throughout Europe; all of the famous mathematicians that met him were most impressed. For example, the prominent Swiss mathematician Johann Bernoulli (1667–1748), insisted that Cramer and only Cramer be allowed to edit the former's collected works. Throughout his career, Cramer was always very productive. He put great energy into his teachings, his researches, and his correspondences with other mathematicians whom he had met. Despite the lack of practicality of Cramer's rule, Cramer actually did a lot of work in applying mathematics to practical areas such as national defense and structural design. Cramer was always in good health until an accidental fall off a carriage.   His doctor recommended for him to go to a resort city in the south of France for recuperation, but he passed away on that journey.

row operations that, when performed on an augmented matrix, will lead to another augmented matrix which represents an equivalent linear system that has a triangular coefficient matrix and thus can be easily solved.   We then explain the main algorithm, Gaussian elimination with partial pivoting, that will transform any augmented matrix (representing a nonsingular system) into an upper triangular system that is easily solved. This is less work than the usual "Gaussian elimination" taught in linear algebra classes, since the latter brings the augmented matrix all the way into reduced row echelon form. The partial pivoting aspect is mathematically redundant, but is numerically very important since it will help to cut back on floating point arithmetic errors.

FIGURE 7.34: Carl Friedrich Gauss (1777 - 1855), German Mathematician.

Gaussian elimination produces a useful factorization of the coefficient matrix, called the *LU* factorization, that will considerably cut down the amount of work needed to solve other systems having the same coefficient matrix.   We will postpone the error analysis of this procedure to the next section.   The main algorithms of this section were invented by the German mathematician Carl F. Gauss[11].

A square matrix $A = [a_{ij}]$ is called **upper triangular** if all entries below the main diagonal equal zero (i.e., $a_{ij} = 0$ whenever $i > j$).   Thus an upper triangular matrix has the following form:

$$\begin{bmatrix} a_{11} & a_{12} & a_{13} & \cdots & a_{1n} \\ & a_{22} & a_{23} & \ldots & a_{2n} \\ & & a_{33} & \cdots & \vdots \\ & \mathbf{0} & & \ddots & \vdots \\ & & & & a_{nn} \end{bmatrix}.$$

---

[11] Carl F. Gauss is acknowledged by many mathematical scholars as the greatest mathematician who ever lived. His potential was discovered early. His first mathematical discovery was the power sum identity (23), and he made this discovery while in the second grade! His teacher, looking to keep young Carl Friedrich occupied for a while asked him to perform the addition of the first 100 integers: $S = 1 + 2 + \ldots + 100$. Two minutes later, Gauss gave the teacher the answer. He did it by rewriting the sum in the reverse order $S = 100 + 99 + \ldots + 1$, adding vertically to the original to get $2S = 101 + 101 + \ldots + 101 = 100 \cdot 101$, so $S = 50 \cdot 101 = 5050$. This idea, of course, yields a general proof of the identity (23).   He was noticed by the Duke of Brunswick, who supported Gauss's education and intellectual activity for many years. Gauss's work touched on numerous fields of mathematics, physics, and other sciences. His contributions are too numerous to attempt to do them justice in this short footnote. It is said that very routinely when another mathematician would visit him in his office to present him with a recent mathematical discovery, after hearing about the theorems, Gauss would reach into his file cabinet and pull out some of his own works, which would invariably transcend that of his guest. For many years, until 2001 when the currency in Germany, as well as that of other European countries, changed to the Euro, Germany had honored Gauss by putting his image on the very common 10 Deutsche Mark banknote (value about $5); see Figure 7.34.

Similarly, a square matrix is **lower triangular** if all entries above the main diagonal are zeros. A matrix is triangular if it is of either of these two forms. Many matrix calculations are easy for triangular matrices. The next proposition shows that determinants for triangular matrices are extremely simple to compute.

PROPOSITION 7.3: If $A = [a_{ij}]$ is a triangular matrix, then the determinant of A is the product of the diagonal entries, i.e., $\det(A) = a_{11} a_{22} \cdots a_{nn}$.

The proof can be easily done by mathematical induction and cofactor expansion on the first column or row; we leave it as Exercise 14(a).

From the proposition, it follows that for a triangular matrix to be nonsingular it is equivalent that each of the diagonal entries must be nonzero. For a system $Ax = b$ with an upper triangular coefficient matrix, we can easily solve the system by starting with the last equation, solving for x_n, then using this in the second-to-last equation and solving for x_{n-1}, and continuing to work our way up. Let us make this more explicit. An upper triangular system has the form:

$$\begin{cases} a_{11}x_1 + a_{12}x_2 + \cdots \quad + a_{1n}x_n = b_1 \\ \qquad\quad a_{22}x_2 + \cdots \quad + a_{2n}x_n = b_2 \\ \qquad\qquad\qquad \ddots \qquad\qquad \vdots \\ \qquad\qquad\qquad a_{n-1,n-1}x_{n-1} + a_{n-1,n}x_n = b_{n-1} \\ \qquad\qquad\qquad\qquad\qquad\quad a_{nn}x_n = b_n \end{cases} \qquad (27)$$

Assuming A is nonsingular, we have that each diagonal entry a_{ii} is nonzero. Thus, we can start off by solving the last equation of (27):

$$x_n = b_n / a_{nn} .$$

Knowing now the value of x_n, we can then substitute this into the second-to-last equation and solve for x_{n-1}:

$$x_{n-1} = (b_{n-1} - a_{n-1,n}x_n) / a_{n-1,n-1} .$$

Now that we know both x_n and x_{n-1}, we can substitute these into the third-to-last equation and then, similarly, solve for the only remaining unknown in this equation:

$$x_{n-2} = (b_{n-2} - a_{n-2,n-1}x_{n-1} - a_{n-2,n}x_n) / a_{n-2,n-2} .$$

If we continue this process, in general, after having solved for $x_{j+1}, x_{j+2}, \cdots x_n$, we can get x_j by the formula:

$$x_j = \left(b_j - \sum_{k=j+1}^{n} a_{jk} x_k \right) / a_{jj}. \tag{28}$$

This algorithm is called **back substitution**, and is a fast and easy method of solving any upper triangular (nonsingular) linear system. Would it not be nice if all linear systems were so easy to solve? Transforming arbitrary (nonsingular) linear systems into upper triangular form will be the goal of the Gaussian elimination algorithm. For now we record for future reference a simple M-file for the back substitution algorithm.

EXAMPLE 7.11: (a) Create an M-file x=backsubst(U,b) that inputs a nonsingular upper triangular matrix U, and a column vector b of the same dimension and the output will be a column vector x which is the numerical solution of the linear system $Ux = b$ obtained from the back substitution algorithm. (b) Use this algorithm to solve the system $Ux = b$ with

$$U = \begin{bmatrix} 1 & 2 & 3 & 4 \\ 0 & 2 & 3 & 4 \\ 0 & 0 & 3 & 4 \\ 0 & 0 & 0 & 4 \end{bmatrix}, \quad b = \begin{bmatrix} 4 \\ 3 \\ 2 \\ 1 \end{bmatrix}.$$

SOLUTION: Part (a): The M-file is easily written using equation (28).

PROGRAM 7.4: Function M-file solving an upper triangular linear system $Ux = b$.

```
function x=backsubst(U,b)
%Solves the upper triangular system Ux=b by back substitution
%Inputs:   U = upper triangular matrix,   b = column vector of same
dimension
%Output:   x = column vector (solution)
[n m]=size(U);
x(n)=b(n)/U(n,n);
for j=n-1:-1:1
    x(j)=(b(j)-U(j,j+1:n)*x(j+1:n)')/U(j,j);
end
x=x';
```

Notice that MATLAB's matrix multiplication allowed us to replace the sum in (28) with the matrix product shown. Indeed U(j,j+1:n) is a row vector with the same number of entries as the column vector x(j+1:n), so their matrix product will be a real number (1×1 matrix) equaling the sum in (28). The transpose on the x-vector was necessary here to make it a column vector.

Part (b):
```
>> U=[1 2 3 4;0 2 3 4;0 0 3 4;0 0 0 4];   b=[4 3 2 1]';
>> format rat, backsubst(U,b)
→ ans =            1
                   1/2
                   1/3
                   1/4
```

The reader can check that U\b will give the same result.

A lower triangular system $Lx = b$ can be solved with an analogous algorithm called **forward substitution**. Here we start with the first equation to get x_1, then plug this result into the second equation and solve for x_2, and so on.

EXERCISE FOR THE READER 7.18: (a) Write a function M-file called x = fwdsubst (L,b), that will input a lower triangular matrix L, a column vector b of the same dimension, and will output the column vector x solution of the system $Lx = b$ solved by forward substitution. (b) Use this program to solve the system $Lx = b$, where $L = U'$, and U and b are as in Example 7.11.

We now introduce the three **elementary row operations (EROs)** that can be performed on augmented matrices.

(i)   Multiply a row by a <u>nonzero</u> constant.
(ii)   Switch two rows.
(iii) Add a multiple of one row to a different row.

EXAMPLE 7.12:   (a)   Consider the following augmented matrix:
$Ab = \begin{bmatrix} 1 & 2 & -3 & 5 \\ 2 & 6 & 1 & -1 \\ 0 & 4 & 7 & 8 \end{bmatrix}$. Perform ERO (iii) on this matrix by adding the multiple
−2 times row 1 to row 2.
(b) Perform this same ERO on I_3, the 3×3 identity matrix, to obtain a matrix M, and multiply this matrix M on the left of Ab. What do you get?

SOLUTION: Part (a): −2 times row 1 of Ab is −2[1 2 −3 5] = [−2 −4 6 −10]. Adding this row vector to row 2 of Ab, produces the new matrix:
$$\begin{bmatrix} 1 & 2 & -3 & 5 \\ 0 & 2 & 7 & -11 \\ 0 & 4 & 7 & 8 \end{bmatrix}.$$

Part (b): Performing this same ERO on I_3 produces the matrix $M = \begin{bmatrix} 1 & 0 & 0 \\ -2 & 1 & 0 \\ 0 & 0 & 1 \end{bmatrix}$,

which when multiplied by Ab gives

$$M(Ab) = \begin{bmatrix} 1 & 0 & 0 \\ -2 & 1 & 0 \\ 0 & 0 & 1 \end{bmatrix} \cdot \begin{bmatrix} 1 & 2 & -3 & 5 \\ 2 & 6 & 1 & -1 \\ 0 & 4 & 7 & 8 \end{bmatrix} = \begin{bmatrix} 1 & 2 & -3 & 5 \\ 0 & 2 & 7 & -11 \\ 0 & 4 & 7 & 8 \end{bmatrix},$$

and we are left with the same matrix that we obtained in part (a). This is no coincidence, as the following theorem shows.

THEOREM 7.4: (*Elementary Matrices*) Let A be any $n \times m$ matrix and $I = I_n$ denote the identity matrix. If B is the matrix obtained from A by performing any particular elementary row operation, and M is the matrix obtained from I by performing this same elementary row operation, then $B = MA$. Also, the matrix M is invertible, and its inverse is the matrix that results from I by performing the inverse elementary row operation on it (i.e., the elementary row operation that will transform M back into I).

Such a matrix M is called an **elementary matrix**. This result is not hard to prove; we refer the reader to any good linear algebra textbook, such as those mentioned in the last section.

It is easy to see that any of these EROs, when applied to the augmented matrix of a linear system, will not alter the solution of the system. Indeed, the first ERO corresponds to simply multiplying the first equation by a nonzero constant (the equation still represents the same line, plane, hyperplane, etc.). The second ERO merely changes the order in which the equations are written; this has no effect on the (joint) solution of the system. To see why the third ERO does not alter solutions of the system is a bit more involved, but not difficult. Indeed, suppose for definiteness that a multiple (say, 2) of the first row is added to the second. This corresponds to a new system where all the equations are the same, except for the second, which equals the old second equation plus twice the first equation. Certainly if we have all of $x_1, x_2, \cdots, x_n$ satisfying the original system, then they will satisfy the new system. Conversely, if all of the equations of the new system are solved by $x_1, x_2, \cdots, x_n$, then this already gives all but the second equation of the original system. But the second equation of the old system is gotten by subtracting twice the first equation of the new system from the second equation (of the new system) and so must also hold.

Each of the EROs is easily programmed into an M-file. We do one of them and leave the other two as exercises.

PROGRAM 7.5: Function M-file for elementary row operation (ii): switching two rows.

```
function B=rowswitch(A,i,j)
% Inputs: a matrix A, and row indices i and j
% Outputs:  the matrix gotten from A by interchanging row i and row j
[m,n]=size(A);
if i<1|i>m|j<1|j>m
   error('Invalid Index')
end
B=A;
if i==j
   return
end
B(i,:)=A(j,:);
B(j,:)=A(i,:);
```

It may seem redundant to have included that if-branch for detecting invalid indices, since it would seem that no one in their right mind would use the program with, say, an index equaling 10 with an 8×8 matrix. This program, however, might get used to build more elaborate programs and in such a program it may not always be crystal clear whether some variable expression is a valid index.

EXERCISE FOR THE READER 7.19: Write similar function M-files B=rowmult(A,i,c) for ERO (i) and B=rowcomb(A,i,j,c) for ERO (iii). The first program will produce the matrix B resulting from A by multiplying the *i*th row of the latter by c; the second program should replace the *i*th row of A by c times the *j*th row plus the *i*th row.

We next illustrate the Gaussian elimination algorithm, the partial pivoting feature as well as the *LU* decomposition, by means of a simple example. This will give the reader a feel for the main concepts of this section. Afterward we will develop the general algorithms and comment on some consequences of floating point arithmetic. Remember, the goal of Gaussian elimination is to transform, using only EROs, a (nonsingular) system into an equivalent one that is upper triangular; the latter can then be solved by back substitution.

EXAMPLE 7.13: We will solve the following linear system $Ax = b$ using Gaussian elimination (without partial pivoting).

$$\begin{cases} x_1 & + & 3x_2 & - & x_3 & = & 2 \\ 2x_1 & + & 5x_2 & - & 2x_3 & = & 3 \\ 3x_1 & + & 6x_2 & + & 9x_3 & = & 39 \end{cases}$$

SOLUTION: For convenience, we will work instead with the corresponding augmented matrix Ab for this system $Ax = b$:

$$Ab = \begin{bmatrix} 1 & 3 & -1 & | & 2 \\ 2 & 5 & -2 & | & 3 \\ 3 & 6 & 9 & | & 39 \end{bmatrix}.$$

For notational convenience, we denote the entries of this or any future augmented matrix as a_{ij}. In computer codes, what is usually done is that at each step the new matrix overwrites the old one to save on memory allocation (not to mention having to invent new variables for unnecessary older matrices). Gaussian elimination starts with the first column, clears out (makes zeros) everything below the main diagonal entry, then proceeds to the second column, and so on.

We begin by zeroing out the entry $a_{21} = 2$: Ab=rowcomb(Ab,1,2,-2).

$$Ab \rightarrow M_1(Ab) = \begin{bmatrix} 1 & 0 & 0 \\ -2 & 1 & 0 \\ 0 & 0 & 1 \end{bmatrix} \begin{bmatrix} 1 & 3 & -1 & | & 2 \\ 2 & 5 & -2 & | & 3 \\ 3 & 6 & 9 & | & 39 \end{bmatrix} = \begin{bmatrix} 1 & 3 & -1 & | & 2 \\ 0 & -1 & 0 & | & -1 \\ 3 & 6 & 9 & | & 39 \end{bmatrix},$$

where M_1 is the corresponding elementary matrix as in Theorem 7.4. In the same fashion, we next zero out the next (and last) entry $a_{31} = 3$ of the first column: Ab=rowcomb(Ab,1,3,-3).

$$Ab \to M_2(Ab) = \begin{bmatrix} 1 & 0 & 0 \\ 0 & 1 & 0 \\ -3 & 0 & 1 \end{bmatrix} \begin{bmatrix} 1 & 3 & -1 & | & 2 \\ 0 & -1 & 0 & | & -1 \\ 3 & 6 & 9 & | & 39 \end{bmatrix} = \begin{bmatrix} 1 & 3 & -1 & | & 2 \\ 0 & -1 & 0 & | & -1 \\ 0 & -3 & 12 & | & 33 \end{bmatrix}.$$

We move on to the second column. Here only one entry needs clearing, namely $a_{32} = -3$. We will always use the row with the corresponding diagonal entry to clear out entries below it. Thus, to use the second row to clear out the entry $a_{32} = -3$ in the third row, we should multiply the second row by -3 since $-3 \cdot a_{22} = -3 \cdot (-1) = 3$ added to $a_{32} = -3$ would give zero: Ab=rowcomb(Ab,2,3,-3).

$$Ab \to M_3(Ab) = \begin{bmatrix} 1 & 0 & 0 \\ 0 & 1 & 0 \\ 0 & -3 & 1 \end{bmatrix} \begin{bmatrix} 1 & 3 & -1 & | & 2 \\ 0 & -1 & 0 & | & -1 \\ 0 & -3 & 12 & | & 33 \end{bmatrix} = \begin{bmatrix} 1 & 3 & -1 & | & 2 \\ 0 & -1 & 0 & | & -1 \\ 0 & 0 & 12 & | & 36 \end{bmatrix}.$$

We now have an augmented matrix representing an equivalent upper triangular system. This system (and hence our original system) can now be solved by the back substitution algorithm:

```
>> U=Ab(:,1:3); b=Ab(:,4);
>> x=backsubst(U,b)
→ x =            2
                 1
                 3
```

NOTE: One could have gone further and similarly cleared out the above diagonal entries and then used the first ERO to scale the diagonal entries to each equal one. This is how one gets to the reduced row echelon form.

To obtain the resulting LU decomposition, we form the product of all the elementary matrices that were used in the above Gaussian elimination: $M = M_3 M_2 M_1$. From what was done, we have that $MA = U$, and hence

$$A = M^{-1}(MA) = M^{-1}U = (M_3 M_2 M_1)^{-1}U = M_1^{-1}M_2^{-1}M_3^{-1}U \equiv LU,$$

where we have defined $L = M_1^{-1}M_2^{-1}M_3^{-1}$. We have used, in the second-to-last equality, the fact that the inverse of a product of invertible matrices is the product of the inverses in the reverse order (see Exercise 7). From Theorem 7.4, each of the inverses M_1^{-1}, M_2^{-1}, and M_3^{-1} is also an elementary matrix corresponding to the inverse elementary row operation of the corresponding original elementary matrix; and furthermore Theorem 7.4 tells us how to multiply such matrices to obtain

$$L = M_1^{-1} M_2^{-1} M_3^{-1} = \begin{bmatrix} 1 & 0 & 0 \\ 2 & 1 & 0 \\ 0 & 0 & 1 \end{bmatrix} \cdot \begin{bmatrix} 1 & 0 & 0 \\ 0 & 1 & 0 \\ 3 & 0 & 1 \end{bmatrix} \cdot \begin{bmatrix} 1 & 0 & 0 \\ 0 & 1 & 0 \\ 0 & 3 & 1 \end{bmatrix} = \begin{bmatrix} 1 & 0 & 0 \\ 2 & 1 & 0 \\ 3 & -3 & 1 \end{bmatrix}.$$

We now have a factorization of the coefficient matrix A as a product LU of a lower triangular and an upper triangular matrix:

$$A = Ab = \begin{bmatrix} 1 & 3 & -1 \\ 2 & 5 & -2 \\ 3 & 6 & 9 \end{bmatrix} = \begin{bmatrix} 1 & 0 & 0 \\ 2 & 1 & 0 \\ 3 & 3 & 1 \end{bmatrix} \cdot \begin{bmatrix} 1 & 3 & -1 \\ 0 & -1 & 0 \\ 0 & 0 & 12 \end{bmatrix} = LU .$$

This factorization, which easily came from the Gaussian elimination algorithm, is a preliminary form of what is known as the LU factorization of A. Once such a factorization is known, any other (nonsingular) system $Ax = c$, having the same coefficient matrix A, can be easily solved in two steps. To see this, rewrite the system as $LUx = c$. First solve $Ly = c$ by forward substitution (works since L is lower triangular), then solve $Ux = y$ by back substitution (works since U is upper triangular). Then x will be the desired solution (Proof: $Ax = (LU)x = L(Ux) = Ly = c$).

We make some observations. Notice that we only used one of the three EROs to perform Gaussian elimination in the above example. Part of the reason for this is that none of the diagonal entries encountered was zero. If this had happened we would have needed to use the `rowswitch` ERO in order to have nonzero diagonal entries. (This is always possible if the matrix A is nonsingular.) In Gaussian elimination, the diagonal entries that are used to clear out the entries below (using `rowcomb`) are known as **pivots**. The **partial pivoting** feature, which is often implemented in Gaussian elimination, goes a bit further to assure (by switching the row with the pivot with a lower row, if necessary) that the pivot is as large as possible in absolute value. In exact arithmetic, this partial pivoting has no effect whatsoever, but in floating point arithmetic, it can most certainly cut back on errors. The reason for this is that if a pivot turned out to be nonzero, but very small, then its row would need to be multiplied by very large numbers to clear out moderately sized numbers below the pivot. This may cause other numbers in the pivot's row to get multiplied into very large numbers that, when mixed with much smaller numbers, can lead to floating point errors. We will soon give an example to demonstrate this phenomenon.

EXAMPLE 7.14: Solve the linear system $Ax = b$ of Example 7.13 using Gaussian elimination with partial pivoting.

The first step would be to switch rows 1 and 3 (to make $|a_{11}|$ as large as possible): `Ab=rowswitch(Ab,1,3)`.

$$Ab \rightarrow P_1(Ab) = \begin{bmatrix} 0 & 0 & 1 \\ 0 & 1 & 0 \\ 1 & 0 & 0 \end{bmatrix} \begin{bmatrix} 1 & 3 & -1 & | & 2 \\ 2 & 5 & -2 & | & 3 \\ 3 & 6 & 9 & | & 39 \end{bmatrix} = \begin{bmatrix} 3 & 6 & 9 & | & 39 \\ 2 & 5 & -2 & | & 3 \\ 1 & 3 & -1 & | & 2 \end{bmatrix}.$$

(We will denote elementary matrices resulting from the rowswitch ERO by P_i's.) Next we pivot on the $a_{11} = 3$ entry to clear out the entries below it. To clear out $a_{21} = 2$, we will do Ab=rowcomb(Ab,1,2,-2/3) (i.e., to clear out $a_{21} = 2$, we multiply row 1 by $-a_{21}/a_{11} = -2/3$ and add this to row 2). Similarly, to clear out $a_{31} = 1$, we will do >>Ab=rowcomb(Ab,1,3,-1/3). Combining both of these elementary matrices into a single matrix M_1, we may now write the result of these two EROs as follows:

$$Ab \rightarrow M_1(Ab) = \begin{bmatrix} 1 & 0 & 0 \\ -2/3 & 1 & 0 \\ -1/3 & 0 & 1 \end{bmatrix} \begin{bmatrix} 3 & 6 & 9 & | & 39 \\ 2 & 5 & -2 & | & 3 \\ 1 & 3 & -1 & | & 2 \end{bmatrix} = \begin{bmatrix} 3 & 6 & 9 & | & 39 \\ 0 & 1 & -8 & | & -23 \\ 0 & 1 & -4 & | & -11 \end{bmatrix}.$$

The pivot $a_{22} = 1$ is already as large as possible so we need not switch rows and can clear out the entry $a_{32} = 1$ by doing Ab=rowcomb(Ab,2,3,-1):

$$Ab \rightarrow M_2(Ab) = \begin{bmatrix} 1 & 0 & 0 \\ 0 & 1 & 0 \\ 0 & -1 & 1 \end{bmatrix} \begin{bmatrix} 3 & 6 & 9 & | & 39 \\ 0 & 1 & -8 & | & -23 \\ 0 & 1 & -4 & | & -11 \end{bmatrix} = \begin{bmatrix} 3 & 6 & 9 & | & 39 \\ 0 & 1 & -8 & | & -23 \\ 0 & 0 & 4 & | & 12 \end{bmatrix},$$

and with this the elimination is complete.

Solving this (equivalent) upper triangular system will again yield the above solution. We note that this produces a slightly different factorization: From $M_2 M_1 PA = U$, where U is the left part of the final augmented matrix, we proceed as in the previous example to get $PA = (M_2^{-1} M_1^{-1})U \equiv LU$, i.e.,

$$PA = \begin{bmatrix} 0 & 0 & 1 \\ 0 & 1 & 0 \\ 1 & 0 & 0 \end{bmatrix} \begin{bmatrix} 1 & 3 & -1 \\ 2 & 5 & -2 \\ 3 & 6 & 9 \end{bmatrix} = \begin{bmatrix} 1 & 0 & 0 \\ 2/3 & 1 & 0 \\ 1/3 & 1 & 1 \end{bmatrix} \cdot \begin{bmatrix} 3 & 6 & 9 \\ 0 & 1 & -8 \\ 0 & 0 & 4 \end{bmatrix} = LU.$$

This is the LU factorization of the matrix A. We now explain the general algorithm of **Gaussian elimination with partial pivoting** and the LU factorization. In terms of EROs and the back substitution, the resulting algorithm is quite compact.

Algorithm for Gaussian Elimination with Partial Pivoting: Given a linear system $Ax = b$ with A an $n \times n$ nonsingular matrix, this algorithm will solve for the solution vector x. The algorithm works on the $n \times (n+1)$ augmented matrix $[A \mid b]$ which we denote by Ab, but whose entries we still denote by a_{ij}.

For $k = 1$ to $n-1$
interchange rows (if necessary) to assure that $|a_{kk}| = \max_{k \le i \le n} |a_{ik}|$

---

if $|a_{kk}| = 0$, exit program with message " A is singular".

for $i = k + 1$ to n

$m_{ik} = a_{ik} / a_{kk}$

$A = \text{rowcomb}(A, k, i, -m_{ik})$

end i

end k

if $|a_{nn}| = 0$, exit program with message " A is singular".

Apply the back substitution algorithm on the final system (that is now upper triangular) to get solution to the system.

Without the interchanging rows step (unless to avoid a zero pivot), this is **Gaussian elimination** without partial pivoting. From now on, we follow the standard convention of referring to Gaussian elimination with partial pivoting simply as "Gaussian elimination," since it has become the standard algorithm for solving linear systems.

The algorithm can be recast into a matrix factorization algorithm for A. Indeed, at the kth iteration we will, in general, have an elementary matrix P_k corresponding to a row switch or permutation, followed by a matrix M_k that consists of the product of each of the elementary matrices corresponding to the "rowcomb" ERO used to clear out entries below a_{kk}. Letting U denote the upper triangular matrix left at the end of the algorithm, we thus have:

$$M_{n-1}P_{n-1} \cdots M_2 P_2 M_1 P_1 A = U .$$

The **LU factorization** (or the **LU decomposition**) of A, in general, has the form (see Section 4.4 of [GoVL-83]):

$$PA = LU , \tag{29}$$

where

$$P = P_{n-1}P_{n-2} \cdots P_2 P_1 \quad \text{and} \quad L = P(M_{n-1}P_{n-1} \cdots M_1 P_1)^{-1} , \tag{30}$$

and L is lower triangular.[12] Also, by Theorem 7.4, the matrix PA corresponds to sequentially switching the rows of the matrix A, first corresponding to P_1, next by P_2, and so on. Thus the LU factorization A, once known, leads to a quick and practical way to solve any linear system $Ax = b$. First, permute the order of the equations as dictated by the permutation matrix P (do this on the augmented matrix so that b's entries get permuted as well), relabel the system as $Ax = b$, and rewrite it as $LUx = b$. First solve $Ly = c$ by forward substitution (works since L

---

[12] The permutation matrix P in (29) cannot, in general, be dispensed with; see Exercise 6.

is lower triangular), then solve $Ux = y$ by back substitution (works since U is upper triangular). Then x will be the desired solution (Proof:

$$PA = LU \Rightarrow Ax = P^{-1}LUx = P^{-1}L(Ux) = P^{-1}L(y) = P^{-1}Pb = b \ .)$$

This approach is useful if it is needed to solve a lot of linear systems with the same coefficient matrix A. For such situations, we mention that MATLAB has a built-in function lu to compute the LU factorization of a nonsingular matrix A. The syntax is as follows:

`[L, U, P]=lu(A)` →	For a square singular matrix A, this command will output the lower triangular matrix L, the upper triangular matrix U, and the permutation matrix P of the LU factorization (29) and (30) of the matrix A.

For example, applying this command to the matrix A of Example 7.14 gives:

```
>>A=[1 3 -1; 2 5 -2; 3 6 9]; format rat, [L, U, P]=lu(A)
```

```
→L =        1      0      0
           2/3     1      0
           1/3     1      1

 U =        3      6      9
            0      1     -8
            0      0      4

 P =        0      0      1
            0      1      0
            1      0      0
```

We now wish to translate the above algorithm for Gaussian elimination into a MATLAB program. Before we do this, we make one remark about MATLAB's built-in function max, which we have encountered previously in its default format (the first syntax below):

`max(v)` →	For a vector v, this command will give the maximum of its components.
`[max, index]=max(v)` →	With an optional second output variable (that must be declared), `max(v)` will also give the first index at which this maximum value occurs.

Here, v can be either a row or a column vector. A simple example will illustrate this functionality.

```
>> v=[1 -3  5  -7  9  -11];
>> max(v)                    →ans = 9
>> [max, index] = max(v)     →max = 9, index = 5
>> [max, index] = max(abs(v)) →max = 11, index = 6
```

Another useful tool for programming M-files is the error command:

`error('message')` →	If this command is encountered with an execution of any M-file, the M-file stops running immediately and displays the message.

PROGRAM 7.6: Function M-file for Gaussian elimination (with partial pivoting) to solve the linear system $Ax = b$, where A is a square nonsingular matrix. This program calls on the previous programs backsubst, rowswitch and rowcomb.

```
function x=gausselim(A,b)
%Inputs:   Square matrix A, and column vector b of same dimension
%Output:   Column vector solution x of linear system Ax = b obtained
%by Gaussian elimination with partial pivoting, provided coefficient
%matrix A is nonsingular.
[n,n]=size(A);
Ab=[A';b']';  %form augmented matrix for system

for k=1:n
    [biggest, occured] = max(abs(Ab(k:n,k)));
    if biggest == 0
    error('the coefficient matrix is numerically singular')
    end
    m=k+occured-1;
    Ab=rowswitch(Ab,k, m);
    for j=k+1:n
        Ab=rowcomb(Ab,k,j,-Ab(j,k)/Ab(k,k));
    end
end
% BACK SUBSTITUTION
x=backsubst(Ab(:,1:n),Ab(:,n+1));
```

EXERCISE FOR THE READER 7.20: Use the program gausselim to resolve the Hilbert systems of Example 7.9 and Exercise for the Reader 7.16, and compare to the results of the left divide method that proved most successful in those examples. Apart from additional error messages (regarding condition numbers), how do the results of the above algorithm compare with those of MATLAB's default system solver?

We next give a simple example that will demonstrate the advantages of partial pivoting.

EXAMPLE 7.15: Consider the following linear system: $\begin{bmatrix} 10^{-3} & 1 \\ 1 & 2 \end{bmatrix}\begin{bmatrix} x_1 \\ x_2 \end{bmatrix} = \begin{bmatrix} 1 \\ 3 \end{bmatrix}$,

whose exact solution (starts) to look like $\begin{bmatrix} x_1 \\ x_2 \end{bmatrix} = \begin{bmatrix} 1.0020... \\ .9989... \end{bmatrix}$.

(a) Using floating point arithmetic with three significant digits and chopped arithmetic, solve the system using Gaussian elimination with partial pivoting.
(b) Repeat part (a) in the same arithmetic, except without partial pivoting.

SOLUTION: For each part we show the chain of augmented matrices. Recall, after each individual computation, answers are chopped to three significant digits before any subsequent computations.
Part (a): (with partial pivoting)

$$\begin{bmatrix} .001 & 1 & | & 1 \\ 1 & 2 & | & 3 \end{bmatrix} \xrightarrow{\text{rowswitch(A,1,2)}} \begin{bmatrix} 1 & 2 & | & 3 \\ .001 & 1 & | & 1 \end{bmatrix} \xrightarrow{\text{rowcomb(A,1,2,-.001)}} \begin{bmatrix} 1 & 2 & | & 3 \\ 0 & .998 & | & .997 \end{bmatrix}.$$

Now we use back substitution: $x_2 = .997 / .998 = .998$, $x_1 = (3 - 2x_2)/1$ $= (3 - 1.99)/1 = 1.01$. Our computed answer is correct to three decimals in x_2, but has a relative error of about 0.0798% in x_1.

Part (b): (without partial pivoting)

$$\begin{bmatrix} .001 & 1 & | & 1 \\ 1 & 2 & | & 3 \end{bmatrix} \xrightarrow{\text{rowcomb(A,1,2,-1000)}} \begin{bmatrix} 1 & 1 & | & 1 \\ 0 & -998 & | & -997 \end{bmatrix}.$$

Back substitution now gives $x_2 = -997 / -998 = .998$, $x_1 = (1 - 1 \cdot x_2)/1$ $= (1 - .998)/1 = .002$. The relative error here is unacceptably large, exceeding 100% in the second component!

EXERCISE FOR THE READER 7.21: Rework the above example using rounded arithmetic rather than chopped, and keeping all else the same.

We close this section with a bit of information on flop counts for the Gaussian elimination algorithm. Note that the partial pivoting adds no flops since permuting rows involves no arithmetic. We will assume the worst-case scenario, in that none of the entries that come up are zeros. In counting flops, we refer to the algorithm above, rather than the MATLAB program. For $k = 1$, we will need to perform $n - 1$ divisions to compute the multipliers m_{i1}, and for each of these multipliers, we will need to do a rowcomb, which will involve n multiplications and n additions/subtractions. (Note: Since the first column entry will be zero and need not be computed, we are only counting columns 2 through $n + 1$ of the augmented matrix.) Thus, associated with the pivot a_{11}, we will have to do $n - 1$ divisions, $(n-1)n$ multiplications, and $(n-1)n$ additions/subtractions. Grouping the divisions and multiplications together, we see that at the $k = 1$ (first) iteration, we will need $(n-1) + (n-1)n = (n-1)(n+1)$ multiplications/divisions and $(n-1)n$ additions/subtractions. In the same fashion, the calculations associated with the pivot a_{22} will involve $n - 2$ divisions plus $(n-2)(n-1)$ multiplications which is $(n-2)n$ multiplications/divisions and $(n-2)(n-1)$ additions/ subtractions. Continuing in this fashion, when we get to the pivot a_{kk}, we will need to do $(n-k)(n-k+2)$ multiplications/divisions and $(n-k)(n-k+1)$ additions/subtractions. Summing from $k = 1$ to $n - 1$ gives the following:

Total multiplications/divisions $\equiv M(n) = \sum_{k=1}^{n-1} (n-k)(n-k+2)$,

Total additions/subtractions $\equiv A(n) = \sum_{k=1}^{n-1} (n-k)(n-k+1)$.

Combining these two sums and regrouping gives:

Grand total flops ≡

$$F(n) = \sum_{k=1}^{n-1}(n-k)(2(n-k)+3) = 2\sum_{k=1}^{n-1}(n-k)^2 + 3\sum_{k=1}^{n-1}(n-k).$$

If we reindex the last two sums, by substituting $j = n - k$, then as k runs from 1 through $n - 1$, so will j (except in the reverse order), so that

$$F(n) = 2\sum_{j=1}^{n-1} j^2 + 3\sum_{j=1}^{n-1} j .$$

We now invoke the power sum identities (23) and (24) to evaluate the above two sums (replace n with $n - 1$ in the identities) and thus rewrite the flop count $F(n)$ as:

$$F(n) = \frac{1}{3}(n-1)n(2n-1) + \frac{3}{2}(n-1)n = \frac{2}{3}n^3 + \text{lower power terms} .$$

The "lower power terms" in the above flop counts can be explicitly computed (simply multiply out the polynomial on the left), but it is the highest order term that grows the fastest and thus is most important for roughly estimating flop counts. The flop count does not include the back substitution algorithm; but a similar analysis shows flop count for the back substitution to be just n^2 (see the Exercise for the Reader 7.22), and we thus can summarize with the following result.

PROPOSITION 7.5: (*Flop Counts for Gaussian Elimination*) In general, the number of flops needed to perform Gaussian elimination to solve a nonsingular system $Ax = b$ with an $n \times n$ coefficient matrix A is

$$\frac{2}{3}n^3 + \text{lower order terms} .$$

EXERCISE FOR THE READER 7.22: Show that for the back substitution algorithm, the number of multiplications/divisions will be $(n^2 + n)/2$, and the number of additions/subtractions will be $(n^2 - n)/2$. Hence, the grand total flops required will be n^2.

By using the natural algorithm for computing the inverse of a matrix, a similar analysis can be used to show the flop count for finding the inverse of a matrix of a nonsingular $n \times n$ matrix to be

$$(8/3)n^3 + \text{lower power terms} ,$$

or, in other words, essentially four times that for a single Gaussian elimination (see Exercise 16). Actually, it is possible to modify the algorithm (of Exercise 16) to a more complicated one that can bring this flop count down to

$2n^3$ + lower power terms ; but this is still going to be a more expensive and error-prone method than Gaussian elimination, so we reiterate: For solving a single general linear system, Gaussian elimination is the best all-around method.

The next example will give some hard evidence of the rather surprising fact that the computer time required (on MATLAB) to perform an addition/subtraction is about the same as that required to perform a multiplication/division.

EXAMPLE 7.16: In this example, we perform a short experiment to record the time and flops required to add 100 pairs of random floating point numbers. We then do the related experiment involving the same number of divisions.

```
>> A=rand(100); B=rand(100);,
>> tic, for i=1:100, C=A+B; end, toc
→ Elapsed time is 5.778000 seconds.

>> tic, for i=1:100, C=A./B;, end, toc
→ Elapsed time is 5.925000 seconds.
```

The times are roughly of the same magnitude and, indeed, the flop counts are identical and close to the actual number of mathematical operations performed. The reason for the discrepancy in the latter is that, as previously mentioned, a flop is "approximately" equal to one arithmetic operation on the computer; and this is the most useful way to think about a flop.

EXERCISES 7.5:

NOTE: As mentioned in the text, we take "Gaussian elimination" to mean Gaussian elimination with partial pivoting.

1. Solve each of the following linear systems $Ax = b$ using three-digit chopped arithmetic and Gaussian elimination (i) without partial pivoting and then (ii) with partial pivoting. Finally redo the problems (iii) using MATLAB's left divide operator, and then (iv) using exact arithmetic (any method).

$$\text{(a)} \quad A = \begin{bmatrix} 2 & -9 \\ 1 & 7.5 \end{bmatrix}, \ b = \begin{bmatrix} 2 \\ -3 \end{bmatrix} \qquad\qquad \text{(b)} \quad A = \begin{bmatrix} .99 & .98 \\ 101 & 100 \end{bmatrix}, \ b = \begin{bmatrix} 1 \\ -1 \end{bmatrix}$$

$$\text{(c)} \quad A = \begin{bmatrix} 2 & 3 & -1 \\ 4 & 2 & -1 \\ -8 & 2 & 0 \end{bmatrix}, \ b = \begin{bmatrix} 0 \\ 21 \\ -4 \end{bmatrix}$$

2. Parts (a) through (c): Repeat all parts of Exercise 1 using two-digit rounded arithmetic.

3. For each square matrix specified, find the LU factorization of the matrix (using Gaussian elimination). Do it first using (i) three-digit chopped arithmetic, then using (ii) exact arithmetic; and finally (iii) compare these with the results using MATLAB's built-in function lu.
    (a)   The matrix A in Exercise 1, part (a).
    (b)   The matrix A in Exercise 1, part (b).
    (c)   The matrix A in Exercise 1, part (c).

4. Parts (a) through (c): Repeat all parts of Exercise 3 using two-digit rounded arithmetic in (i).

5. Consider the following linear system involving the 3×3 Hilbert matrix H_3 as the coefficient matrix:

$$\begin{cases} x_1 + \dfrac{1}{2}x_2 + \dfrac{1}{3}x_3 = 2 \\ \dfrac{1}{2}x_1 + \dfrac{1}{3}x_2 + \dfrac{1}{4}x_3 = 0 \,, \\ \dfrac{1}{3}x_1 + \dfrac{1}{4}x_2 + \dfrac{1}{5}x_3 = -1 \end{cases}$$

 (a) Solve the system using two-digit chopped arithmetic and Gaussian elimination without partial pivoting.
 (b) Solve the system using two-digit chopped arithmetic and Gaussian elimination.
 (c) Solve the system using exact arithmetic (any method).
 (d) Find the LU decomposition of the coefficient matrix H_3 by using 2-digit chopped arithmetic and Gaussian elimination.
 (e) Find the exact LU decomposition of H_3.

6. (a) Find the LU factorization of the matrix $A = \begin{bmatrix} 0 & 1 \\ 1 & 0 \end{bmatrix}$. (b) Is it possible to find a lower triangular matrix L and an upper triangular matrix U (not necessarily those in part (a)) such that $A = LU$? Explain why or why not.

7. Suppose that $M_1, M_2, \cdots, M_k$ are invertible matrices of the same size. Prove that their product is invertible with $(M_1 \cdot M_2 \cdots M_k)^{-1} = M_k^{-1} \cdots M_2^{-1} \cdot M_1^{-1}$. In words, "The inverse of the product is the reverse order product of the inverses."

8. (*Storage and Computational Savings in Solving Tridiagonal Systems*)   Just as with any (nonsingular) matrix, we can apply Gaussian elimination to solve tridiagonal systems:

$$\begin{bmatrix} d_1 & a_1 & & & & & \\ b_2 & d_2 & a_2 & & & \text{\Large 0} & \\ & b_3 & d_3 & a_3 & & & \\ & & b_4 & d_4 & a_4 & & \\ & & & & \ddots & & \\ & \text{\Large 0} & & & b_{n-1} & d_{n-1} & a_{n-1} \\ & & & & & b_n & d_n \end{bmatrix} \begin{bmatrix} x_1 \\ x_2 \\ x_3 \\ x_4 \\ \vdots \\ x_{n-1} \\ x_n \end{bmatrix} = \begin{bmatrix} r_1 \\ r_2 \\ r_3 \\ r_4 \\ \vdots \\ r_{n-1} \\ r_n \end{bmatrix}. \qquad (31)$$

Here, d's stand for diagonal entries, b's for below-diagonal entries, a's for above diagonal entries, and r's for right-side entries. We can greatly cut down on storage and unnecessary mathematical operations with zero by making use of the special sparse form of the tridiagonal matrix. The main observation is that at each step of the Gaussian elimination process, we will always be left with a banded matrix with perhaps one additional band above the a's diagonal. (Think about it, and convince yourself. The only way a switch can be done in selecting a pivot is with the row immediately below the diagonal pivot entry.) Thus, we may wish to organize a special algorithm that deals only with the tridiagonal entries of the coefficient matrix.
(a) Show that the Gaussian elimination algorithm, with unnecessary operations involving zeros being omitted, will require no more than $8(n-1)$ flops (multiplications, divisions, additions, subtractions), and the corresponding back substitution will require no more than $5n$ flops. Thus the total number of flops for solving such a system can be reduced to less than $13n$.
(b) Write a program, x = tridiaggauss(d, b,a, r) that inputs the diagonal vector d (of length n) and the above and below diagonal vectors a and b (of length $n-1$) of a nonsingular tridiagonal matrix, the column vector r and will solve the tridiagonal system (31) using the Gaussian elimination algorithm but which overwrites only the four relevant diagonal vectors

(described above; you need to create an $n-2$ length vector for the extra diagonal) and the vector r rather than on the whole matrix. The output should be the solution column vector x.

(c) Test out your algorithm on the system (31) with $n=2$, $n=100$, $n=500$, and $n=1000$ using the following data in the matrices $d_i = 4$, $a_i = 1$, $b_i = 1$, $r = [1 \ -1 \ 1 \ -1 \ \cdots]$ and compare results and flop counts with MATLAB's left divide. You should see that your algorithm is much more efficient.

NOTE: The upper bound $13n$ on flops indicated in part (a) is somewhat liberal; a more careful analysis will show that the coefficient 13 can actually be made a bit smaller (How small can you make it?) But even so, the savings on flops (not to mention storage) are incredible. If we compare $13n$ with the bound $2n^3/3$, for large values of n, we will see that this modified method will allow us to solve extremely large tridiagonal systems that previously would have been out of the question. (For example, when $n = 10,000$, this modified method would require storage of $2n + 2(n-1) = 39,998$ entries and less than $13n = 130,000$ flops (this would take a few seconds on MATLAB even on a weak computer); whereas the ordinary Gaussian elimination would require the storage of $n^2 + n = 100,010,000$ entries and approximately $2n^3/3 = 6.66... \times 10^{11}$ flops, an unmanageable task!

9. *(The Thomas Method, an Even Faster Way to Solve Tridiagonal Systems)* By making a few extra assumptions that are usually satisfied in most tridiagonal systems that arise in applications, it is possible to slightly modify the usual Gaussian elimination algorithm to solve the triadiagonal system (31) in just $8n-7$ flops (compared to the upper bound $13n$ of the last problem). The algorithm, known as the **Thomas method**,[13] differs from the usual Gaussian elimation by scaling the diagonal entry to equal 1 at each pivot, and by not doing any row changes (i.e., we forgo partial pivoting, or assume the matrix is of a form that makes it unnecessary; see Exercise 10). This will mean that we will have to keep track only of the above diagonal entries (the a's vector) and the right side vector r. The Thomas method algorithm thus proceeds as follows:

Step 1: (Results from `rowmult(A, 1, 1/d_1)`): $a_1 = a_1/d_1$, $r_1 = r_1/d_1$.

(We could also add $d_1 = 1$, but since the diagonal entries will always be scaled to equal one, we do not need to explicitly record this change.)

Steps $k = 2$ through $n - 1$: (Results from `rowcomb(A, k - 1,k, -b_k)` and then `rowscale(A, k, 1/(d_k - b_k a_{k-1})))`:

$$a_k = a_k /(d_k - b_k a_{k-1}), \quad r_k = (r_k - b_k r_{k-1})/(d_k - b_k a_{k-1}).$$

Step n: (Results from same procedure as in steps 2 through $n - 1$, but there is no a_n):

$$r_n = (r_n - b_n r_{n-1})/(d_n - b_n a_{n-1}).$$

This variation of Gaussian elimination has transformed the tridiagonal system into an upper triangular system with the following special form:

$$\begin{bmatrix} 1 & a_1 & & & \\ & 1 & a_2 & \mathbf{0} & \\ & & \ddots & \ddots & \\ \mathbf{0} & & & 1 & a_{n-1} \\ & & & & 1 \end{bmatrix} \begin{bmatrix} x_1 \\ x_2 \\ \vdots \\ x_{n-1} \\ x_n \end{bmatrix} = \begin{bmatrix} r_1 \\ r_2 \\ \vdots \\ r_{n-1} \\ r_n \end{bmatrix},$$

---

[13] The method is named after the renowned physicist Llewellyn H. Thomas; but it was actually discovered independently by several different individuals working in mathematics and related disciplines. W.F. Ames writes in his book [Ame-77] (p. 52): "The method we describe was discovered independently by many and has been called the Thomas algorithm. Its general description first appeared in widely distributed published form in an article by Bruce et al. [BPRR-53]."

for which the back substitution algorithm takes on the particularly simple form: $x_n = r_n$; then for

$k = n-1, n-2, ..., 2, 1$: $x_k = r_k - a_k x_{k+1}$.

(a) Write a MATLAB M-file, x=thomas(d, b, a, r) that performs the Thomas method as described above to solve the tridiagonal system (31). The inputs should be the diagonal vector d (of length n) and the above and below diagonal vectors a and b (of length $n-1$) of a nonsingular tridiagonal matrix, and the column vector r. The output should be the computed solution, as a column vector x. Write your program so that it only overwrites the vectors a and r.

(b) Test out your program on the systems of part (c) of Exercise 8, and compare results and flop counts with those for MATLAB's left divide solver. If you have done part (c) of Exercise 8, compare also with the results from the program of the previous exercise.

(c) Do a flop count on the Thomas method to show that the total number of flops needed is $8n-7$.

NOTE: Looking over the Thomas method, we see that it assumes that $d_1 \neq 0$, and $d_k \neq b_k a_{k-1}$ (for $k = 2$ through n). One might think that to play it safe, it may be better to just use the slightly more expensive modification of Gaussian elimination described in the previous exercise, rather than risk running into problems with the Thomas method. For at most all applications, it turns out that the requirements for the Thomas method indeed are satisified. Such triadiagonal systems come up naturally in many applications, in particular in finite difference schemes for solving differential equations. One safe approach would be to simply build in a deferral to the previous algorithm in cases where the Thomas algorithm runs into a snag.

10. We say that a square matrix $A = [a_{ij}]$ is **strictly diagonally dominant (by columns)** if for each index k, $1 \leq k \leq n$, the following condition is met:

$$|a_{kk}| > \sum_{\substack{j=1 \\ j \neq k}}^{n} |a_{kj}|.$$ (32)

This condition merely states that each diagonal entry is larger, in absolute value, than the sum of the absolute values of all of the other entries in its column.

(a) Explain why when Gaussian elimination is applied to solve a linear system $Ax = b$ whose coefficient matrix is strictly diagonally dominant by columns, then no row changes will be required.

(b) Explain why the LU factorization of a diagonally dominant by columns matrix A will not have any permutation matrix.

(c) Explain why the requirements for the Thomas method (Exercise 9) will always be met if the coefficient matrix is strictly diagonally dominant by columns.

(d) Which, if any, of the above facts will continue to remain true if the strict diagonal dominance condition (32) is weakened to the following?

$$|a_{kk}| > \sum_{j=k+1}^{n} |a_{kj}|.$$

(That is, we are now only assuming that each diagonal entry is larger, in absolute value, than the sum of the absolute values of the entries that lie in the same column but below it.)

11. Discuss what conditions on the industries must hold in order for the technology matrix M of the Leontief input/output model of Exercise 10 from Section 7.4 to be diagonally dominant by columns (see the preceding exercise).

12. (*Determinants Revisited: Effects of Elementary Row/Column Operations on Determinants*) Prove the following facts about determinants, some of which were previewed in Exercise 10 of Section 7.1.

(a) If the matrix B is obtained from the square matrix A by multiplying one of the latter's rows by a number c (and leaving all other rows the same, i.e., B=rowmult(A,i,c)), then $\det(B) = c\det(A)$.

(b) If the matrix B is obtained from the square matrix A by adding a multiple of the ith row of A

to the jth row ($i \neq j$) (i.e., B = rowcomb(A,i,j,c)), then det(B) = det(A).

(c)   If the matrix B results from the matrix A by switching two rows of the latter (i.e., B=rowswitch(A,i,j)), then det(B) = −det(A).

(d)   If two rows of a square matrix are the same, then det(A) = 0.

(e)   If B is the transpose of A then det(B) = det(A).

NOTE: In light of the result of part (e), each of the statements in the other parts regarding the effect of a row operation on a determinant has a valid counterpart for the effect of the corresponding column operation on the determinant.

Suggestions: You should make use of identity (20) det(AB) = det(A)det(B), as well as Proposition 7.3 and Theorem 7.4.   The results of (a), (b), and (c) can then be proved by calculating determinants of certain elementary matrices.   The only difficult thing is for part (c) to show that the determinant of a permutation matrix gotten from the identity matrix by switching two rows equals −1.   One way this can be done is by an appropriate (but not at all obvious) matrix factorization.   Here is one way to do it for the (only) 2×2 permutation matrix:

$$\begin{bmatrix} 0 & 1 \\ 1 & 0 \end{bmatrix} = \begin{bmatrix} 1 & -1 \\ 0 & 1 \end{bmatrix}\begin{bmatrix} 1 & 0 \\ 1 & 1 \end{bmatrix}\begin{bmatrix} 1 & 0 \\ 0 & -1 \end{bmatrix}\begin{bmatrix} 1 & 1 \\ 0 & 1 \end{bmatrix}.$$

(Check this!) All of the matrix factors on the right are triangular so the determinants of each are easily computed by multiplying diagonal entries (Proposition 7.3), so using (20), we get

$$\det\left(\begin{bmatrix} 0 & 1 \\ 1 & 0 \end{bmatrix}\right) = 1 \cdot 1 \cdot (-1) \cdot 1 = -1 .$$

In general, this argument can be made to work for any permutation matrix (obtained by switching two rows of the identity matrix), by carefully generalizing the factorization. For example, here is how the factorization would generalize for a certain 3×3 permutation matrix:

$$\begin{bmatrix} 0 & 0 & 1 \\ 0 & 1 & 0 \\ 1 & 0 & 0 \end{bmatrix} = \begin{bmatrix} 1 & 0 & -1 \\ 0 & 1 & 0 \\ 0 & 0 & 1 \end{bmatrix}\begin{bmatrix} 1 & 0 & 0 \\ 0 & 1 & 0 \\ 1 & 0 & 1 \end{bmatrix}\begin{bmatrix} 1 & 0 & 0 \\ 0 & 1 & 0 \\ 0 & 0 & -1 \end{bmatrix}\begin{bmatrix} 1 & 0 & 1 \\ 0 & 1 & 0 \\ 0 & 0 & 1 \end{bmatrix}.$$

Part (d) can be proved easily from part (c); for part (e) use mathematical induction and cofactor expansion.

13.   (*Determinants Revisited: A Better Way to Compute Them*)   The Gaussian elimination algorithm provides us with an efficient way to compute determinants.   Previously, the only method we gave to compute them was by cofactor expansion, which we introduced in Chapter 4. But we saw that this was an extremely expensive way to compute determinants. A new idea is to use the Gaussian elimination algorithm to transform a square matrix A into an upper triangular matrix. From the previous exercise, each time a rowcomb is done, there will be no effect on the determinant, but each time a rowswitch is done, the determinant is negated. By Proposition 7.3, the determinant of the diagonal matrix is just the product of the diagonal entries. Of course, in the Gaussian elimination algorithm, the column vector b can be removed (if all we are interested in is the determinant). Also, if a singularity is detected, the algorithm should exit and assign det(A) = 0.

(a) Create a function M-file, called y=gaussdet(A), that inputs a square matrix M and outputs the determinant using this algorithm.

(b) Test your program out by computing the determinants of matrices with random integer entries from −9 to 9 of sizes 3×3, 8×8, 20×20, and 80×80 (you need not print the last two matrices) that you can construct using the M-file randint of Exercise for the Reader 7.2. Compare the results, computing times and (if you have Version 5) flop counts with those for MATLAB's built-in det function applied to the same matrices.

(c) Go through an analysis similar to that done at the end of the section to prove a result similar to that of Proposition 7.5 that will give an estimate of the total flop counts for this algorithm, with the highest order term being accurate.

(d) Obtain a similar flop count for the cofactor expansion method and compare with the answer you got in (c). (The highest order term will involve factorials rather than powers.)

(e) Use your answer in (c) to obtain a flop count for the amount of flops needed to apply

Cramer's rule to solve a nonsingular linear system $Ax = b$ with A being an $n \times n$ nonsingular matrix.

14. (a) Prove Proposition 7.3.
    (b) Prove an analogous formula for the determinant of a square matrix that is upper-left triangular in the sense that all entries above the off-main diagonal are zeros. More precisely, prove that any matrix of the following form,

$$A = \begin{bmatrix} a_{11} & a_{12} & a_{13} & \cdots & a_{1n} \\ a_{21} & a_{22} & a_{23} & & \\ \vdots & & \ddots & & \\ a_{n-1,1} & a_{n-1,2} & & & \text{\Large 0} \\ a_{n1} & & & & \end{bmatrix},$$

has determinant given by $\det(A) = (-1)^k a_{1n} \cdot a_{2,n-1} \cdot a_{3,n-2} \cdots a_{n-1,2} \cdot a_{n1}$, where $n = 2k + i$ $(i = 0,1)$.

Suggestion: Proceed by induction on n, where A is an $n \times n$ matrix. Use cofactor expansion along an appropriate row (or column).

15. (a) Write a function M-file, call it `[L, U, P]= mylu(A)`, that will compute the LU factorization of an inputted nonsingular matrix A.
    (b) Apply this function to each of the three coefficient matrices in Exercise 1 as well as the Hilbert matrix H_3, and compare the results (and flop counts) to those with MATLAB's built-in function `lu`. From these comparisons, does your program seem to be as efficient as MATLAB's?

16. (a) Write a function M-file, call it `B=myinv(A)`, that will compute the inverse of an inputted nonsingular matrix A, and otherwise will output the error message: "Matrix detected as numerically singular." Your algorithm should be based on the following fact (which follows from the way that matrix multiplication works). To find an inverse of an $n \times n$ nonsingular matrix A, it is sufficient to solve the following n linear equations:

$$Ax^1 = \begin{bmatrix} 1 \\ 0 \\ 0 \\ \vdots \\ 0 \end{bmatrix}, \; Ax^2 = \begin{bmatrix} 0 \\ 1 \\ 0 \\ \vdots \\ 0 \end{bmatrix}, \; Ax^3 = \begin{bmatrix} 0 \\ 0 \\ 1 \\ \vdots \\ 0 \end{bmatrix}, \; \cdots Ax^n = \begin{bmatrix} 0 \\ 0 \\ 0 \\ \vdots \\ 1 \end{bmatrix},$$

where the column vectors on the right sides of these equations are precisely the columns of the $n \times n$ identity matrix. It then would follow that $A\left[x^1 \mid x^2 \mid x^3 \mid \cdots \mid x^n \right] = I$, so that the desired inverse of A is the matrix $A^{-1} = \left[x^1 \mid x^2 \mid x^3 \mid \cdots \mid x^n \right]$. Your algorithm should be based on the LU decomposition, so it gets computed once, rather than doing a complete Gaussian elimination for each of the n equations.
    (b) Apply this function to each of the three coefficient matrices in Exercise 1 as well as the Hilbert matrix H_4, and compare the results (and flop counts) to those with MATLAB's built-in function `inv`. From these comparisons, does your program seem to be as efficient as MATLAB's?
    (c) Do a flop count, similar to the one done for Proposition 7.5 for this algorithm.
    Note: For part (a), feel free to use MATLAB's built-in function `lu`; see the comments in the text about how to use the LU factorization to solve linear systems.

7.6:   VECTOR AND MATRIX NORMS, ERROR ANALYSIS, AND EIGENDATA

In the last section we introduced the Gaussian elimination (with partial pivoting) algorithm for solving a nonsingular linear system

$$Ax = b, \tag{33}$$

where $A = [a_{ij}]$ is an $n \times n$ coefficient matrix, b is an $n \times 1$ column vector, and x is the $n \times 1$ column vector of variables whose solution is sought. This algorithm is the best all-around general numerical method for solving the linear system (33), but its performance can vary depending on the coefficient matrix A. In this section we will present some practical estimates for the error of the computed solution that will allow us to put some quality control guarantee on the answers that we obtain from (numerical) Gaussian elimination. We need to begin with a practical way to measure the "sizes" of vectors and matrices. We have already used the Euclidean length of a vector v to measure its size, and norms will be a generalization of this concept. We will introduce norms for vectors and matrices in this section, as well as the so-called condition numbers for square matrices. Shortly we will use norms and condition numbers to give precise estimates for the error of the computed solution of (33) (using Gaussian elimination). We will also explain some ideas to try when a system to be solved is poorly conditioned. The theory on modified algorithms that can deal with poorly conditioned systems contains an assortment of algorithms that can perform well if the (poorly conditioned) matrix takes on a special form. If one has the Student Version of MATLAB (or has the Symbolic Toolbox) there is always the option of working in exact arithmetic or with a fixed but greater number of significant digits. The main (and only) disadvantage of working in such arithmetic is that computations move a lot slower, so we will present some concrete criteria that will help us to decide when such a route might be needed. The whole subject of error analysis and refinements for numerically solving linear systems is quite vast and we will not be delving too deeply into it. For more details and additional results, the interested reader is advised to consult one of the following references (listed in order of increasing mathematical sophistication): [Atk-89], [Ort-90], [GoVL-83].

The Euclidean "length" of an n-dimensional vector $x = [x_1 \ x_2 \ \cdots \ x_n]$ is defined by:

$$\text{len}(x) = \sqrt{x_1^2 + x_2^2 + \cdots + x_n^2} \ . \tag{34}$$

For this definition it is immaterial whether x is a row or column vector. For example, if we are working in two dimensions and if the vector is drawn in the xy-plane from its tail at $(x,y) = (0,0)$ to its tip $(x,y) = (x_1, x_2)$, then $\text{len}(x)$ is (in most cases) the hypotenuse of a right triangle with legs having length $|x_1|$ and $|x_2|$, and so the formula (34) becomes the Pythagorean theorem. In the remaining cases where one of x_1 or x_2 is zero, then $\text{len}(x)$ is simply the absolute value of the other coordinate (in this case also the length of the vector x that will lie on either the x- or y-axis.) From what we know about plane geometry, we can deduce that $\text{len}(x)$ has the following properties:

$$\text{len}(x) \geq 0, \text{ and } \text{len}(x) = 0 \text{ if and only if } x = 0 \text{(vector)} \qquad (35A)$$

$$\text{len}(cx) = |c| \text{len}(x) \text{ for any scalar } c \qquad (35B)$$

$$\text{len}(x+y) \leq \text{len}(x) + \text{len}(y) \quad \text{(Triangle Inequality)} \qquad (35C)$$

Property (35A) is clear (even in n dimensions). Property (35B) corresponds to the geometric fact that when a vector is multiplied by a scalar, the length gets multiplied by the absolute value of the scalar (we learned this early in the chapter). The triangle inequality (35C) corresponds to the geometric fact (in two dimensions) that the length of any side of any triangle can never exceed the sum of the lengths of the other two sides. These properties remain true for general n-dimensional vectors (see Exercise 11 for a more general result).

A **vector norm** for n-dimensional (row or column) vectors $x = [x_1 \ x_2 \ \cdots \ x_n]$ is a way to associate a nonnegative number (default notation: $\|x\|$) with the vector x such that the following three properties hold:

$$\|x\| \geq 0, \|x\| = 0 \text{ if and only if } x = 0 \text{(vector)} \qquad (36A)$$

$$\|cx\| = |c| \|x\| \text{ for any scalar } c \qquad (36B)$$

$$\|x+y\| \leq \|x\| + \|y\| \quad \text{(Triangle Inequality)} \qquad (36C)$$

We have merely transcribed the properties (35) to obtain these three axioms (36) for a norm. It turns out that there is an assortment of useful norms, the aforementioned Euclidean norm being one of them. The one we will use mostly in this section is the so-called **max norm** (also known as the **infinity norm**) and this is defined as follows:

$$\|x\| = \|x\|_\infty = \max\{|x_i|, 1 \leq i \leq n\} = \max\{|x_1|, |x_2|, \cdots, |x_n|\}. \qquad (37)$$

The proper mathematical notation for this vector norm is $\|x\|_\infty$, but since it will be our default vector norm we will often denote it by $\|x\|$ for convenience. The max norm is the most simple of all vector norms, so working with it will allow the complicated general concepts from error analysis to be understood in the simplest possible setting. The price paid for this simplicity will be that some of the resulting error estimates that we obtain using the max norm may be somewhat more liberal than those obtained with other more complicated norms. Both the max and Euclidean norms are easy to compute on MATLAB (e.g., for the max norm of x we could simply type `max(abs(x))`, but (of course) MATLAB has built-in functions for both of these vector norms and many others.

`norm(x)` $\rightarrow$	Computes the length norm $\text{len}(x)$ of a (row or column) vector x.

norm(x, inf) $\rightarrow$	Computes the max norm $\|x\|$ of a (row or column) vector x.

EXAMPLE 7.17: For the two four-dimensional vectors $x = [1, 0, -4, 6]$ and $y = [3, -4, 1, -3]$ find the following:

(a) $\text{len}(x)$, $\text{len}(y)$, $\text{len}(x+y)$

(b) $\|x\|$, $\|y\|$, $\|x+y\|$

SOLUTION: First we do these computations by hand, and then redo them using MATLAB.

Part (a):   Using (34)   and   since $x = [4, -4, -3, 3]$ we   get   that

$\text{len}(x) = \sqrt{1^2 + 0^2 + (-4)^2 + 6^2} = \sqrt{53} = 7.2801...$, $\text{len}(y) = \sqrt{3^2 + (-4)^2 + 1^2 + (-3)^2}$

$= \sqrt{35} = 5.9160...$, and $\text{len}(x+y) = \sqrt{4^2 + (-4)^2 + (-3)^2 + 3^2} = \sqrt{50} = 7.0710...$

Part (b):   Using (37), we   compute: $\|x\| = \max\{|1|, |0|, |-4|, |6|\} = 6$, $\|y\| = \max\{|3|, |-4|, |1|, |-3|\} = 4$, and $\|x+y\| = \max\{|4|, |-4|, |-3|, |3|\} = 4$.

These computations give experimental evidence of the validity of the triangle inequality in this special case. We now repeat these same computations using MATLAB:

```
>> x=[1 0 -4 6]; y=[3 -4 1 -3];
>> norm(x), norm(y), norm(x+y)    →ans =  7.2801    5.9161    7.0711
>> norm(x,inf), norm(y,inf),  norm(x+y,inf) →ans =  6    4    4
```

EXERCISE FOR THE READER 7.23: Show that the max norm as defined by (37) is indeed a vector norm by verifying the three vector norm axioms of (36).

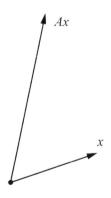

Given any vector norm, we define an **associated matrix norm** by the following:

$$\|A\| \equiv \max\left\{\frac{\|Ax\|}{\|x\|}, \ x \neq 0 \,(\text{vector})\right\}. \qquad (38)$$

For any nonzero vector x, its norm $\|x\|$ will be a positive number (by (36A)), the transformed vector Ax will be another vector and so will have a norm $\|Ax\|$. The norm of the matrix A can be thought of as the maximum magnification factor by which the transformed vector Ax's norm will have changed from the original vector x's norm, see Figure 7.35.

FIGURE 7.35: Graphic for the matrix norm definition (38). The matrix A will transform x into another vector Ax (of same dimension if A is square). The norm of A is the maximum magnification that the transformed vector Ax norm will have in terms of the norm of x.

It is interesting that matrix norms, despite the daunting definition (38), are often easily computed from other formulas. For the max vector norm, it can be shown that the corresponding matrix norm (38), often called the **"infinity" matrix norm**, is given by the following formula:

$$\|A\| \equiv \max_{1 \le i \le n} \left\{ \sum_{j=1}^{n} |a_{ij}| \right\} = \max_{1 \le i \le n} \{ |a_{i1}| + |a_{i2}| + \cdots + |a_{in}| \} . \tag{39}$$

This more practical definition is simple to compute: We take the sum of each of the absolute values of the entries in each row of the matrix A, and $\|A\|$ will equal the maximum of these "row sums." MATLAB has a command that will do this computation for us:

`norm(A,inf)` →	Computes the infinity norm $\|A\|$ of a matrix A.

One simple but very important consequence of the definition (38) is the following inequality:

$$\|Ax\| \le \|A\| \|x\| \text{ (for any matrix } A \text{ and vector } x \text{ of compatible size).} \tag{40}$$

To see why (40) is true is easy: First if x is the zero vector, then so is Ax and so both sides of (40) are equal to zero. If x is not the zero vector then by (38) we have $\|Ax\|/\|x\| \le \|A\|$, so we can multiply both sides of this inequality by the positive number $\|x\|$ to produce (40).

EXAMPLE 7.18: Let $A = \begin{bmatrix} 1 & 2 & -1 \\ 0 & 3 & -1 \\ 5 & -1 & 1 \end{bmatrix}$ and $x = \begin{bmatrix} 1 \\ 0 \\ -2 \end{bmatrix}$. Compute $\|x\|$, $\|Ax\|$, and $\|A\|$ and check the validity of (40).

SOLUTION: Since $Ax = \begin{bmatrix} 3 \\ 2 \\ 3 \end{bmatrix}$, we obtain: $\|x\| = 2$, $\|Ax\| = 3$, and using (39),

$\|A\| = \max\{1 + 2 + |-1|, \ 0 + 3 + |-1|, \ 5 + |-1| + 1\} = \max\{4, 4, 7\} = 7$. Certainly $\|Ax\| \le \|A\| \|x\|$ holds here ($3 \le 7 \cdot 2$).

EXERCISE FOR THE READER 7.24: Prove the following two facts about matrix norms: For two $n \times n$ matrices A and B:

(a) $\|AB\| \le \|A\| \|B\|$.

(b) If A is nonsingular, then $\|A^{-1}\| = \left(\min_{x \ne 0} \frac{\|Ax\|}{\|x\|} \right)^{-1}$.

With matrix norms introduced, we are now in a position to define the condition number of a nonsingular (square) matrix. For such a matrix A, the **condition number** of A, denoted by $\kappa(A)$, is the product of the norm of A and the norm of A^{-1}, i.e.,

$$\kappa(A) = \text{condition number of } A \equiv \| A \| \, \| A^{-1} \|. \tag{41}$$

By convention, for a singular matrix A, we define $\kappa(A) = \infty$.[14] Unlike the determinant, a large condition number is a reliable indicator that a square matrix is **nearly singular** (or **poorly conditioned**); and condition numbers will be a cornerstone in many of the error estimates for linear systems that we give later in this section. Of course, the condition number depends on the vector norm that is being used (which determines the matrix norm), but unless explicitly stated otherwise, we will always use the infinity vector norm (and the associated matrix norm and condition numbers). To compute the condition number directly is an expensive computation in general, since it involves computing the inverse A^{-1}. There are good algorithms to estimate condition numbers relatively quickly to any degree of accuracy. We will forgo presenting such algorithms, but will take the liberty of using the following MATLAB built-in function for computing condition numbers:

`cond(A, inf)` $\rightarrow$	Computes and outputs the condition number (with respect to the infinity vector norm) of the square matrix A.

The condition number has the following general properties (actually valid for condition numbers arising from any vector norm):

$$\kappa(A) \geq 1, \quad \text{for any square matrix } A. \tag{42}$$

If D is a diagonal matrix with nonzero diagonal entries: $d_1, d_2, \cdots, d_n$, then

$$\kappa(D) = \frac{\max\{| d_i |\}}{\min\{| d_i |\}}. \tag{43}$$

If A is a square matrix and c is a nonzero scalar, then $\kappa(cA) = \kappa(A)$. $\tag{44}$

In particular, from (43) it follows that $\kappa(I) = 1$. The proofs of these identities will be left to the exercises. Before giving our error analysis results (for linear systems), we state here a theorem that shows, quite quantitatively, that nonsingular matrices with large condition numbers are truly very close to being singular. Recall that the singular square matrices are precisely those whose determinant is

---

[14] Sometimes this condition number is denoted $\kappa_\infty(A)$ to emphasize that it derives from the infinity vector and matrix norm. Since this will be the only condition number that we use, no ambiguity should arise by our adopting this abbreviated notation.

zero. For a given $n \times n$ nonsingular matrix A, we think of the distance from A to the set of all $n \times n$ singular matrices to be $\min\{\|S - A\| : \det(S) = 0\}$. (Just as with absolute values, the norm of a difference of matrices is taken to be the distance between the matrices.) We point out that $\min_{\det(S)=0} \|S - A\|$ can be thought of as the distance from A to the set of singular matrices. (See Figure 7.36.)

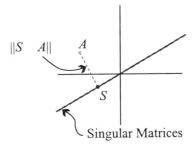

FIGURE 7.36: Heuristic diagram showing the distance from a nonsingular matrix A to the set of all singular matrices (line).

THEOREM 7.6: *(Geometric Characterization of Condition Numbers)* If A is any $n \times n$ nonsingular matrix, then we have:

$$\frac{1}{\kappa(A)} = \frac{1}{\|A\|} \cdot \min_{\det(S)=0} \|S - A\| = \frac{1}{\|A\|} . \tag{45}$$

Like all of the results we state involving matrix norms and condition numbers, this one is true, in general, for whichever matrix norm (and resulting condition number) we would like to use. A proof can be found in the paper [Kah-66]. This theorem suggests some of the difficulties in trying to numerically solve systems having large condition numbers. Gaussian elimination involves many computations and each time we modify our matrix, because of roundoff errors, we are actually dealing with matrices that are close to but not the same as the actual (mathematically exact) matrices. The theorem shows that for poorly conditioned matrices (i.e., ones with large condition numbers), this process is extremely sensitive since even a small change in a poorly conditioned matrix could result in one that is singular!

We close with an example that will review some of the concepts about norms and condition numbers that have been introduced.

EXAMPLE 7.19: Consider the matrix: $A = \begin{bmatrix} 7 & -4 \\ -5 & 3 \end{bmatrix}$

(a) Is there a (2×1) vector x such that: $\|Ax\| > 8\|x\|$? If yes, find one; otherwise explain why one does not exist.

(b) Is there a nonzero vector x such that $\|Ax\| \geq 12\|x\|$? If so, find one; otherwise explain why one does not exist.

(c) Is there a singular matrix $S = \begin{bmatrix} a & b \\ c & d \end{bmatrix}$ (i.e., $ad - bc = 0$) such that $\|S - A\| \leq 0.2$? If so find one; otherwise explain why one does not exist.

(d) Is there a singular matrix $S = \begin{bmatrix} a & b \\ c & d \end{bmatrix}$ (i.e., $ad - bc = 0$) such that $\|S - A\| \leq 0.05$? If so, find one; otherwise explain why one does not exist.

SOLUTION: Parts (a) and (b): Since $\|A\| = 7 + 4 = 11$, it follows from (38) that there exist (nonzero) vectors x with $\|Ax\|/\|x\| = 11$ or, put differently, (multiply by $\|x\|$) $\|Ax\| = 11\|x\|$, but there will not be any nonzero vectors x that will make this equation work if 11 gets replaced by any larger number. (The maximum amount that matrix multiplication by A can magnify the norm of any nonzero vector x is 11 times.) Thus part (a) will have a vector solution but part (b) will not. To find an explicit vector x that will solve part (a), we will actually do more and find one that undergoes the maximum possible magnification $\|Ax\| = 11\|x\|$. The procedure is quite simple (and general). The vector x will have entries being either 1 or -1. To find such an appropriate vector x, we simply identify the row of A that gives rise to its norm being 11; this would be the first row (in general if more than one row gives the norm, we can choose either one). We simply choose the signs of the x-entries so that when they are multiplied in order by the corresponding entries in the just-identified row of A, all products are positive. In other words, if an entry in the special row of A is positive, take the corresponding component of x to be 1; if the special row entry of A is negative, take the corresponding component of x to be -1. In our case the special row of A is (the first row) $[7\ -4]$, and so in accordance we take $x = [1\ -1]'$. The first entry of the vector Ax is $[7\ -4]\cdot[1\ -1]' = 7(1) - 4(-1) = 7 + 4 = 11$, so $\|x\| = 1$ and $\|Ax\| = 11$ (actually, this shows only $\|Ax\| \geq 11$ since we have not yet computed the other component of Ax, but from what was already said, we know $\|Ax\| \leq 11$, so that indeed $\|Ax\| = 11$). This procedure easily extends to any matrices of any size.

Parts (c) and (d): We rewrite equation (45) to isolate the distance from A to the singular matrices (simply multiply both sides by $\|A\|$):

$$\min_{\det(S)=0} \|S - A\| = \frac{\|A\|}{\kappa(A)}.$$

Appealing to MATLAB to compute the right side (and hence the distance from A to singulars):

```
>> A=[7 -4;-5 3];
>> norm(A,inf)/cond(A,inf)
→ ans =          0.0833
```

Since this distance is less than 0.2, the theorem tells us that there is a singular matrix satisfying the requirement of part (c) (the theorem unfortunately does not help us to find one), but there is no singular matrix satisfying the more stringent requirements of part (d). We use *ad hoc* methods to find a specific matrix S that satisfies the requirements of part (c). Note that $\det A = 7 \cdot 3 - (-5) \cdot (-4) = 1$. We will try to tweak the entries of A into those of a singular matrix $S = \begin{bmatrix} a & b \\ c & d \end{bmatrix}$ with determinant $ad - bc = 0$. The requirement of the distance being less than 0.2 means that our perturbations in each row must add up (in absolute value) to at most 0.2. Let's try tweaking 7 to $a = 6.9$ and 3 to $d = 2.9$ (motive for this move: right now A has $ad = 21$, which is one more than $bc = 20$; we need to tweak things so that ad is brought down a bit and bc is brought up to meet it). Now we have $ad = 20.01$, and we still have a perturabtion allowance of 0.1 for both entries b and c and we need only bring bc up from its current value of 20 to 20.01. This is easy— there are many ways to do it. For example, keep $c = -5$ and solve $bc = 20.01$, which gives $c = 20.01/-5 = -4.002$ (well within the remaining perturbation allowance). In summary, the matrix $S = \begin{bmatrix} 6.9 & -4.002 \\ -5 & 2.9 \end{bmatrix}$ meets the requirements that were asked for in part (c). Indeed S is singular (its determinant was arranged to be zero), and the distance from this matrix to A is

$$\| S - A \| = \left\| \begin{bmatrix} 6.9 & -4.002 \\ -5 & 2.9 \end{bmatrix} - \begin{bmatrix} 7 & -4 \\ -5 & 3 \end{bmatrix} \right\| = \left\| \begin{bmatrix} -.1 & -.002 \\ 0 & -.1 \end{bmatrix} \right\| = .102 < 0.2.$$

NOTE: The matrix S that we found was actually quite a bit closer to A than what was asked for. Of course, the closer that we wish to find a singular matrix to the ultimate distance, the harder we will have to work with such *ad hoc* methods. Also, the idea used to construct the "extremal" vector x can be modified to give a proof of identity (39); this task will be left to the interested reader as Exercise 10.

When we use Gaussian elimination to solve a nonsingular linear system (33): $Ax = b$, we will get a **computed solution** vector z that will, in general, differ from the **exact (mathematical) solution** x by the **error term** Δx :

$$z = x + \Delta x \quad \longleftarrow \quad \text{Error Term}$$

Computed Solution   Exact Solution

The main goal for the error analysis is to derive estimates for the size of the error (vector) term: $\| \Delta x \|$. Such estimates will give us quality control on the computed solution z to the linear system.

Caution: It may seem that a good way to measure the quality of the computed solution is to look at the size (norm) of the so-called **residual vector**:

$$r = \text{residual vector} \equiv b - Az . \tag{46}$$

Indeed, if z were the exact solution x then the residual would equal the zero vector. We note the following different ways to write the residual vector:

$$r \equiv b - Az = Ax - Az = A(x - z) = A(x - (x + \Delta x)) = A(-\Delta x) = -A(\Delta x) ; \qquad (47)$$

in particular, the residual is simply the (negative of) the matrix A multiplied by the error term vector. The matrix A may distort a large error term into a much smaller vector thus making the residual much smaller than the actual error term. The following example illustrates this phenomenon (see Figure 7.37).

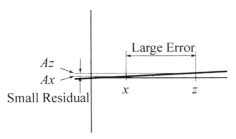

FIGURE 7.37: Heuristic illustration showing the unreliability of the residual as a gauge to measure the error. This phenomenon is a special case of the general principle that a function having a very small derivative can have very close outputs resulting from different inputs spread far apart.

EXAMPLE 7.20: Consider the following linear system $Ax = b$:

$$\begin{bmatrix} 1 & 2 \\ 1.0001 & 2 \end{bmatrix} \begin{bmatrix} x_1 \\ x_2 \end{bmatrix} = \begin{bmatrix} 3 \\ 3.0001 \end{bmatrix}.$$

This system has (unique) exact solution $x = [1, 1]'$. Let's consider the (poor) approximation $z = [3, 0]'$. The (norm of the) error of this approximation is $\|x - z\| = \|[-2, 1]'\| = 2$, but the residual vector,

$$r = b - Az = \begin{bmatrix} 3 \\ 3.0001 \end{bmatrix} - \begin{bmatrix} 1 & 2 \\ 1.0001 & 2 \end{bmatrix} \begin{bmatrix} 3 \\ 0 \end{bmatrix} = \begin{bmatrix} 3 \\ 3.0001 \end{bmatrix} - \begin{bmatrix} 3 \\ 3.0003 \end{bmatrix} = \begin{bmatrix} 0 \\ -.0002 \end{bmatrix},$$

has a much smaller norm of only 0.0002. This phenomenon is also depicted in Figure 7.37.

Despite this drawback about the residual itself, it can be manipulated to indeed give us a useful error estimate. Indeed, from (47), we may multiply left sides by A^{-1} to obtain:

$$r = A(-\Delta x) \implies -\Delta x = A^{-1} r .$$

If we now take norms of both sides and apply (40), we can conclude that:

$$\|\Delta x\| = \|-\Delta x\| = \|A^{-1} r\| \le \|A^{-1}\| \|r\| .$$

(We have used the fact that $\|-v\| = \|v\|$ for any vector v; this follows from norm axiom (36B) using $c = -1$.) We now summarize this simple yet important result in the following theorem.

THEOREM 7.7: (*Error Bound via Residual*) If z is an approximate solution to the exact solution x of the linear system (33) $Ax = b$, with A nonsingular, and $r = b - Az$ is the residual vector, then

$$\text{error} \equiv \|x - z\| \le \|A^{-1}\| \|r\| . \tag{48}$$

Remark: Using the condition number $\kappa(A) = \|A\| \|A^{-1}\|$, Theorem 7.7 can be reformulated as

$$\|x - z\| \le \kappa(A) \frac{\|r\|}{\|A\|} . \tag{49}$$

EXAMPLE 7.21: Consider once again the linear system $Ax = b$ of the preceding example:

$$\begin{bmatrix} 1 & 2 \\ 1.0001 & 2 \end{bmatrix} \begin{bmatrix} x_1 \\ x_2 \end{bmatrix} = \begin{bmatrix} 3 \\ 3.0001 \end{bmatrix} .$$

If we again use the vector $z = [3, 0]'$ as the approximate solution, then, as we saw in the last example, the error = 2 and the residual is $r = [0, -0.0002]'$. The estimate for the error provided in Theorem 9.2, is (with MATLAB's help) found to be 4:

```
>>A=[1 2; 1.0001 2]; r=[0 -.0002];
>> norm(inv(A),inf)*norm(r,inf)     →ans = 4.0000
```

Although this estimate for the error is about as far off from the actual error as the approximation z is from the actual solution, as far as an estimate for the error is concerned, it is considered a decent estimate. An estimate for the error is considered good if it has approximately the same order of magnitude (power of 10 in scientific notation) as the actual error.

Using the previous theorem on the error, we obtain the following analogous result for the relative error.

THEOREM 7.8: (*Relative Error Bound via Residual*) If z is an approximate solution to the exact solution x of the linear system (33) $Ax = b$, with A nonsingular, $b \ne 0$ (vector), and $r = b - Az$ is the residual vector, then

$$\text{relative error} \equiv \frac{\|x - z\|}{\|x\|} \le \frac{\|A\| \|A^{-1}\|}{\|b\|} \|r\| . \tag{50}$$

Remark: In terms of condition numbers, Theorem 7.8 takes on the more appealing form:

$$\frac{\|x-z\|}{\|x\|} \le \kappa(A)\frac{\|r\|}{\|b\|}.$$ (51)

Proof of Theorem 7.8: We first point out that $x \ne 0$ since $b \ne 0$ (and $Ax = b$ with A nonsingular). Using identity (40), we deduce that:

$$\|b\| = \|Ax\| \le \|A\|\|x\| \implies \frac{1}{\|x\|} \le \frac{\|A\|}{\|b\|}.$$

We need only multiply both sides of this latter inequality by $\|x-z\|$ and then apply (48) to arrive at the desired inequality:

$$\frac{\|x-z\|}{\|x\|} \le \frac{\|A\|}{\|b\|} \cdot \|x-z\| \le \frac{\|A\|}{\|b\|} \cdot \|A^{-1}\|\|r\|.$$

EXAMPLE 7.21: (cont.) Using MATLAB to compute the right side of (50),

```
>> cond(A,inf)*norm(r,inf)/norm([3   3.0001]',inf)
→ans = 4.0000
```

Once again, this compares favorably with the true value of the relative error whose explicit value is $\|x-z\|/\|x\| = 2/1 = 2$ (see Example 7.20).

EXAMPLE 7.22: Consider the following (large) linear system $Ax = b$, with

$$A = \begin{bmatrix} 4 & -1 & 0 & 0 & -1 & 0 & 0 & 0 & \cdots & 0 & 0 \\ -1 & 4 & -1 & 0 & 0 & -1 & 0 & 0 & \cdots & 0 & 0 \\ 0 & -1 & 4 & -1 & 0 & 0 & -1 & 0 & \cdots & & 0 \\ 0 & 0 & -1 & 4 & 0 & 0 & 0 & -1 & \ddots & & 0 \\ -1 & 0 & 0 & 0 & 4 & -1 & 0 & 0 & \ddots & & 0 \\ 0 & -1 & 0 & 0 & -1 & 4 & -1 & 0 & \ddots & & 0 \\ \vdots & & \ddots & \ddots & \ddots & \ddots & \ddots & \ddots & \ddots & & \\ & \vdots & & & & & & & & & \\ 0 & & & & \ddots & \ddots & \ddots & \ddots & \ddots & \ddots & 0 \\ 0 & 0 & \cdots & & \ddots & -1 & 0 & 0 & -1 & 4 & -1 \\ 0 & 0 & 0 & \cdots & & 0 & -1 & 0 & 0 & -1 & 4 \end{bmatrix}, \quad b = \begin{bmatrix} 1 \\ 2 \\ 3 \\ 4 \\ 5 \\ 6 \\ \vdots \\ \\ 798 \\ 799 \\ 800 \end{bmatrix}.$$

The 800×800 coefficient matrix A is diagonally banded with a string of 4's down the main diagonal, a string of -1's down each of the diagonals 4 below and 4 above the main diagonal, and each of the diagonals directly above and below the main diagonal consist of the vector that starts off with $[-1\ -1\ -1]$, and repeatedly tacks the sequence $[0\ -1\ -1\ -1]$ onto this until the diagonal fills. Such banded coefficient matrices are very common in finite difference methods for solving (ordinary and partial) differential equations. Despite the intimidating size of this system, MATLAB's "left divide" can take advantage of its special structure and will produce solutions with very decent accuracy as we will see below:

(a) Compute the condition number of the matrix A.

(b) Use the left divide (Gaussian elimination) to solve this system, and call the computed solution z. Use Theorem 7.7 to estimate its error and Theorem 7.8 to estimate the relative error. Do not print out the vector z!

(c) Obtain a second numerical solution $z2$, this time by left multiplying the equation by the inverse A^{-1}. Use Theorem 7.7 to estimate its error and Theorem 7.8 to estimate the relative error. Do not print out the vector $z2$!

SOLUTION: We first enter the matrix A, making use of MATLAB's useful `diag` function:

```
>> A=diag(4*ones(1,800));
>> a1=[-1 -1 -1]; vrep=[0 -1 -1 -1];
>> for i=1:199, a1=[a1, vrep]; end %this is level +1/-1 diagonal
>> v4=-1*ones(1,796); %this is level +4/-4 diagonal
>> A=A+diag(a1,1)+diag(a1,-1)+diag(v4,4)+diag(v4,-4);
>> A(1:8,1:8) %we make a quick check to see how A looks
```
→ans =

```
    4   -3    0    0   -1    0    0    0
   -3    4   -3    0    0   -1    0    0
    0   -3    4   -3    0    0   -1    0
    0    0   -3    4    0    0    0   -1
   -1    0    0    0    4   -3    0    0
    0   -1    0    0   -3    4   -3    0
    0    0   -1    0    0   -3    4   -3
    0    0    0   -1    0    0   -3    4
```

The matrix A looks as it should. The vector b is, of course, easily constructed.

```
>>b=1:800; b=b'; %needed to take transpose to make b a column vector
```

Part (a):
```
>> c= cond(A,inf)    → c = 2.6257e + 003
```

With a condition number under 3000, considering its size, the matrix A is rather well conditioned.

Part (b): Here and in part (c), we use the condition number formulations (49) and (51) for the error estimates of Theorems 7.7 and 7.8.

```
>> z=A\b; r=b-A*z; errest=c*norm(r,inf)/norm(A,inf)
```
→ errest =   2.4875e - 010

```
>> relerrest=c*norm(r,inf)/norm(b,inf)
```
→ relerrest =   3.7313e - 012

Part (c): `>> z2=inv(A)*b; r2=b-A*z2; errest2=c*norm(r2,inf)/norm(A,inf)`
→ errest2 =   6.8656e - 009

```
>> relerrest2=c*norm(r2,inf)/norm(b,inf)
```
→ relerrest2 =   1.0298e - 010

Both methods have produced solutions of very decent accuracy. All of the computations here were done with lightning speed. Thus even larger such systems (that are decently conditioned) can be dealt with safely with MATLAB's "left divide." The matrix in the above problem had a very high percentage of its entries being zeros. Such matrices are called sparse matrices, and MATLAB has efficient ways to store and manipulate such matrices. We will discuss this topic in the next section.

For (even moderately sized) poorly conditioned linear systems, quality control of computed solutions becomes a serious issue. The estimates provided in Theorems 7.7 and 7.8 are just that, estimates that give a guarantee of the closeness of the computed solution to the actual solution. The actual errors may be a lot smaller than the estimates that are provided. Another more insidious problem is that computation of the error bounds of these theorems is expensive, since it either involves the norm of A^{-1} directly or the condition number of A (which implicitly requires computing the norm of A^{-1}). Computer errors can lead to inaccurate computation of these error bounds that we would like to use to give us confidence in our numerical solutions. The next example will demonstrate and attempt to put into perspective some of these difficulties. The example will involve the very poorly conditioned Hilbert matrix that we introduced in Section 7.4. We will solve the system exactly (using MATLAB's symbolic toolbox),[15] and thus be able to compare estimated errors (using Theorems 7.7 and 7.8) with the actual errors. We warn the reader that some of the results of this example may be shocking, but we hasten to add that the Hilbert matrix is notorious for being extremely poorly conditioned.

EXAMPLE 7.23: Consider the linear system $Ax = b$ with

$$A = \begin{bmatrix} 1 & \frac{1}{2} & \frac{1}{3} & \cdots & \frac{1}{48} & \frac{1}{49} & \frac{1}{50} \\ \frac{1}{2} & \frac{1}{3} & \frac{1}{4} & \cdots & \frac{1}{49} & \frac{1}{50} & \frac{1}{51} \\ \frac{1}{3} & \frac{1}{4} & \frac{1}{5} & \cdots & \frac{1}{50} & \frac{1}{51} & \frac{1}{52} \\ \vdots & & & \ddots & & & \vdots \\ \frac{1}{48} & \frac{1}{49} & \frac{1}{50} & \cdots & \frac{1}{96} & \frac{1}{97} & \frac{1}{98} \\ \frac{1}{49} & \frac{1}{50} & \frac{1}{51} & \cdots & \frac{1}{97} & \frac{1}{98} & \frac{1}{99} \\ \frac{1}{50} & \frac{1}{51} & \frac{1}{52} & \cdots & \frac{1}{98} & \frac{1}{99} & \frac{1}{100} \end{bmatrix}, \qquad b = \begin{bmatrix} 1 \\ 2 \\ 3 \\ \vdots \\ 48 \\ 49 \\ 50 \end{bmatrix}.$$

Using MATLAB, perform the following computations.
(a) Compute the condition number of the 50×50 Hilbert matrix A (on MATLAB using the usual floating point arithmetic).
(b) Compute the same condition number using symbolic (exact) arithmetic on MATLAB.

---

[15] For more on the Symbolic Toolbox, see Appendix A. This toolbox may or may not be in the version of MATLAB that you are using. A reduced version of it comes with the student edition. It is not necessary to have it to understand this example.

(c) Use MATLAB's left divide to solve this system and label the computed solution as z, then use Theorem 7.7 to estimate the error. (Do not actually print the solution.)

(d) Solve the system by (numerically) left multiplying both sides by A^{-1} and label the computed solution as z2; then use Theorem 7.7 to estimate the error. (Do not actually print the solution.)

(e) Solve the system exactly using MATLAB's symbolic capabilities, label this exact solution as x, and compute the norm of this solution vector. Then use this exact solution to compute the exact errors of the two approximate solutions in parts (a) and (b).

SOLUTION: Since MATLAB has a built-in function for the generating Hilbert matrices, we may very quickly enter the data A and b:

```
>> A=hilb(50);   b=1:50;b=b';
```

Part (a): We invoke MATLAB's built-in function for computing condition numbers:

```
>> c1=cond(A,inf)
→Warning: Matrix is close to singular or badly scaled.
      Results may be inaccurate. RCOND = 3.615845e - 020.
> In C:\MATLABR11\toolbox\matlab\matfun\cond.m at line 44
→c1 = 5.9243e + 019
```

This is certainly very large, but it came with a warning that it may be an inaccurate answer due to the poor conditioning of the Hilbert matrix. Let's now see what happens when we use exact arithmetic.

Part (b): Several of MATLAB's built-in functions are not defined for symbolic objects; and this is true for the norm and condition number functions. The way around this is to work directly with the definition (41) of the condition number: $\kappa(A) = \left\| A \right\| \left\| A^{-1} \right\|$, compute the norm of A directly (no computational difficulties here), and compute the norm of A^{-1} by first computing A^{-1} in exact arithmetic, then using the double command to put the answer from "symbolic" form into floating point form, so we can take its norm as usual (the computational difficulty is in computing the inverse, not in finding the norm).

```
>> c=norm(double(inv(sym(A))),inf)*norm(A,inf)   % "sym" declares A as
a symbolic variable, so inv is calculated exactly; double switches
the symbolic answer back into floating point form.
→c1 = 4.3303e + 074
```

The difference here is astounding! This condition number means that, although the Hilbert matrix A has its largest entry being 1 (and smallest being 1/99), the inverse matrix will have some entries having absolute values at least $4.33 \times 10^{74} / 50 = 8.66 \times 10^{72}$ (why?). With floating point arithmetic, however,

MATLAB's computed inverse has all entries less than 10^{20} in absolute value, so that MATLAB's inverse is totally out in left field!

Part (c):
```
>> z=A\b;  r=b-A*z;
→Warning: Matrix is close to singular or badly scaled.
    Results may be inaccurate. RCOND = 3.615845e - 020.
```

As expected, we get a flag about our poorly conditioned matrix.

```
>> norm(r,inf)      →ans = 6.2943e - 005
```

Thus the residual of the computed solution is somewhat small. But the extremely large condition number of the matrix will overpower this residual to render the following useless error estimate (see (49)):

```
>> errest=c1*norm(r,inf)/norm(A,inf)      →errest = 9.5437e+014
```

Since
```
>> norm(z,inf)      →ans = 5.0466e+012
```

this error estimate is over 100 times as large as the largest component of the numerical solution. Things get even worse (if you can believe this is possible) with the inverse multiplication method that we look at next.

Part (d):
```
>> z2=inv(A)*b;  r2=b-A*z2;
→Warning: Matrix is close to singular or badly scaled.
    Results may be inaccurate. RCOND = 3.615845e - 020.
>>norm(r2,inf)      →ans = 1.6189e + 004
```

Here, even the norm of the residual is unacceptably large.

```
>> errest2=c1*norm(r2,inf)/norm(A,inf)      →errest = 2.2078e+023
```

Part (e):
```
>> S=sym(A);  %declares A as a symbolic matrix
>> x=S\b;   %Computes exact solution of system
>> x=double(x);  %Converts x back to a floating point vector
>>  norm(x,inf)       →ans = 7.4601e + 040
```

We see that the solution vector has some extremely large entries.

```
>> norm(x-z,inf)        →ans= 7.4601e + 040
>> norm(x-z2,inf)       →ans = 7.4601e + 040
>> norm(z-z2,inf)       →ans = 3.8429e + 004
```

Comparing all of these norms, we see that the two approximations are closer to each other than to the exact solution (by far). The errors certainly met the estimates provided for in the theorem, but not by much. The (exact arithmetic) computation of x took only a few seconds for MATLAB to do. Some comments are in order. The reader may wonder why one should not always work using exact arithmetic, since it is so much more reliable. The reasons are that it is often not necessary to do this—floating point arithmetic usually provides acceptable (and usually decent) accuracy, and exact arithmetic is much more expensive. However, when we get such a warning from MATLAB about near singularity of a matrix, we must discard the answers, or at least do some further analysis. Another option (again using the Symbolic Toolbox of MATLAB) would be to use variable precision arithmetic rather than exact arithmetic. This is less expensive than exact arithmetic and allows us to declare how many significant digits with which we would like to compute. We will give some examples of this arithmetic in a few rare cases where MATLAB's floating point arithmetic is not sufficient to attain the desired accuracy (see also Appendix A).

EXERCISE FOR THE READER 7.25: Repeat all parts of the previous example to the following linear system $Ax = b$, with:

$$A = \begin{bmatrix} 1 & 1 & 1 & \cdots & 1 & 1 & 1 \\ 2^{11} & 2^{10} & 2^9 & \cdots & 4 & 2 & 1 \\ 3^{11} & 3^{10} & 3^9 & \cdots & 9 & 3 & 1 \\ \vdots & & & \ddots & & \vdots & \vdots \\ 10^{11} & 10^{10} & 10^9 & \cdots & 100 & 10 & 1 \\ 11^{11} & 11^{10} & 11^9 & \cdots & 121 & 11 & 1 \\ 12^{11} & 12^{10} & 12^9 & \cdots & 144 & 12 & 1 \end{bmatrix}, \quad b = \begin{bmatrix} 1 \\ -2 \\ 3 \\ \vdots \\ -10 \\ 11 \\ -12 \end{bmatrix}.$$

This coefficient matrix is the 12×12 Vandermonde matrix that was introduced in Section 7.4 with polynomial interpolation.

We next move on to the concepts of eigenvalues and eigenvectors of a matrix. These concepts are most easily motivated geometrically in two dimensions, so let us begin with a 2×2 matrix $A = \begin{bmatrix} a & b \\ c & d \end{bmatrix}$, and a nonzero column vector $x = \begin{bmatrix} x_1 \\ x_2 \end{bmatrix}$ $\neq \begin{bmatrix} 0 \\ 0 \end{bmatrix}$. We view A (as in Section 7.1) as a linear transformation acting on the vector x. The vector x will have a positive length given by the Pythagorean formula:

$$\text{len}(x) = \sqrt{x_1^2 + x_2^2}. \tag{52}$$

Thus A transforms the two-dimensional vector x into another vector $y = \begin{bmatrix} y_1 \\ y_2 \end{bmatrix}$ by matrix multiplication $y = Ax$. We consider the case where y is also not the zero vector; this will always happen if A is nonsingular. In general, when we graph the vectors x and y together in the same plane, they can have different lengths as well as different directions (see Figure 7.38a).

Sometimes, however, there will exist vectors x for which y will be **parallel** to x (meaning that y will point in either the same direction or the opposite direction as

x, see Figure 7.38b). In symbols, this would mean that we could write $y = \lambda x$ for some **scalar** (number) λ. Such a vector x is called an eigenvector for the matrix A, and the number λ is called an associated eigenvalue. Note that if λ is positive, then x and $y = \lambda x$ point in the same direction and $\text{len}(y) = \lambda \cdot \text{len}(x)$, so that λ acts as a magnification factor. If λ is negative, then (as in Figure 7.38b) y points in the opposite direction as x, and finally if $\lambda = 0$, then y must be the zero vector (so has no direction). By convention, the zero vector is parallel to any vector. This is permissible, as long as x is not the zero vector. This definition generalizes to square matrices of any size.

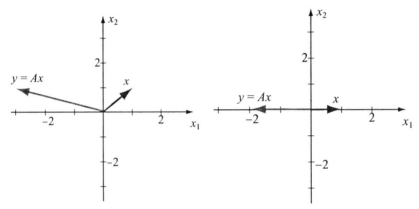

FIGURE 7.38: Actions of the matrix $A = \begin{bmatrix} -2 & -1 \\ 0 & 1 \end{bmatrix}$ on a pair of vectors: (a) (left) The (shorter) vector $x = \begin{bmatrix} 1 \\ 1 \end{bmatrix}$ of length $\sqrt{2}$ gets transformed to the vector $y = Ax = \begin{bmatrix} -3 \\ 1 \end{bmatrix}$ of length $\sqrt{10}$. Since the two vectors are not parallel, x is not an eigenvector of A. (b) (right) The (shorter) unit vector $x = \begin{bmatrix} 1 \\ 0 \end{bmatrix}$ gets transformed to the vector $y = Ax = \begin{bmatrix} -2 \\ 0 \end{bmatrix}$ (red) of length 2, which is parallel to x, therefore x is an eigenvector for A.

DEFINITION: Let A be an $n \times n$ matrix. An $n \times 1$ nonzero column vector x is called an **eigenvector** for the matrix A if for some scalar λ we have

$$Ax = \lambda x . \tag{53}$$

The scalar λ is called the **eigenvalue** associated with the eigenvector x.

Finding all eigenvalues and associated eigenvectors for a given square matrix is an important problem that has been extensively studied and there are numerous algorithms devoted to this and related problems. It turns out to be useful to know this **eigendata** for a matrix A for an assortment of applications. It is actually quite easy to look at eigendata in a different way that will give an immediate method for finding it. We can rewrite equation (53) as follows:

$$Ax = \lambda x \Leftrightarrow Ax = \lambda Ix \Leftrightarrow \lambda Ix - Ax = 0 \Leftrightarrow (\lambda I - A)x = 0 \ .$$

Thus, using what we know about solving linear equations, we can restate the eigenvalue definition in several equivalent ways as follows:

λ is an eigenvalue of A $\;\Leftrightarrow\;$ $(\lambda I - A)x = 0$ has a nonzero solution x

(which will be an eigenvector)

$\Leftrightarrow \quad \det(\lambda I - A) = 0$

Thus the eigenvalues of A are precisely the roots λ of the equation $\det(\lambda I - A) = 0$. If we write out this determinant with a bit more detail,

$$\det(\lambda I - A) = \det \begin{bmatrix} \lambda - a_{11} & -a_{12} & -a_{13} & \cdots & -a_{1,n-1} & -a_{1n} \\ -a_{21} & \lambda - a_{22} & -a_{23} & & & \\ -a_{31} & -a_{32} & \lambda - a_{33} & & \ddots & \vdots \\ \vdots & & & & & \\ -a_{n-1,1} & -a_{n-1,2} & -a_{n-1,3} & \cdots & \lambda - a_{n-1,n-1} & -a_{n-1,n} \\ -a_{n1} & -a_{n2} & -a_{n3} & \cdots & -a_{n,n-1} & \lambda - a_{nn} \end{bmatrix},$$

it can be seen that this expression will always be a polynomial of degree n in the variable λ , for any particular matrix of numbers $A = [a_{ij}]$ (see Exercise 30). This polynomial, because of its importance for the matrix A, is called the **characteristic polynomial** of the matrix A, and will be denoted as $p_A(\lambda)$. Thus $p_A(\lambda) = \det(\lambda I - A)$, and in summary:

The eigenvalues of a matrix A are the roots of the characteristic polynomial, i.e., the solutions of the equation $p_A(\lambda) = 0$.

Finding eigenvalues is an algebraic problem that can be easily solved with MATLAB; more on this shortly. Each eigenvalue will always have associated eignvectors. Indeed, the matrix equation $(\lambda I - A)x = 0$ has a singular coefficient matrix when λ is an eigenvalue (since its determinant is zero). We know from our work on solving linear equations that a singular ($n \times n$) linear system of form $Cx = 0$ can have either no solutions or infinitely many solutions, but $x = 0$ is (obviously) always a solution of such a linear system, and consequently such a singular system must have infinitely many solutions. To find eigenvectors associated with a particular eigenvalue λ , we could compute them numerically by applying `rref` rather than Gaussian elimination to the augmented matrix $[\lambda I - A \mid 0]$. Although theoretically sound, this approach is not a very effective numerical method. Soon we will describe MATLAB's relevant built-in functions that are based on more sophisticated and effective numerical methods.

EXAMPLE 7.24: For the matrix $A = \begin{bmatrix} -2 & -1 \\ 1 & 1 \end{bmatrix}$, do the following:

(a) Find the characteristic polynomial $p_A(\lambda)$.

(b) Find all roots of the characteristic polynomial (i.e., the eigenvalues of A).
(c) For each eigenvalue, find an associated eigenvector.

SOLUTION: Part (a): $p_A(\lambda) = \det(\lambda I - A) =$

$$\det\left(\begin{bmatrix} \lambda+2 & 1 \\ -1 & \lambda-1 \end{bmatrix}\right) = (\lambda+2)(\lambda-1)+1 = \lambda^2 + \lambda - 1 .$$

Part (b): The roots of $p_A(\lambda) = 0$ are easily obtained from the quadratic formula:

$$\lambda = \frac{-1 \pm \sqrt{1 - 4 \cdot 1 \cdot (-1)}}{2} = \frac{-1 \pm \sqrt{5}}{2} \approx -1.6180, .6180 .$$

Part (c): For each of these two eigenvalues, let's use rref to find (all) associated eigenvectors.

Case: $\lambda = (-1-\sqrt{5})/2$

```
>> A=[-2 -1;1 1];   lambda=(-1-sqrt(5))/2;
>> C=lambda*eye(2)-A;  C(:,3)=zeros(2,1);
>> rref(C)
           1.0000   2.6180      0
→ ans =        0        0       0
```

From the last matrix, we can read off the general solution of the system $(\lambda I - A)x = 0$ (written out to four decimals)

$$\begin{cases} x_1 + 2.6180x_2 = 0 \\ (= \text{any number}) \end{cases} \Rightarrow \begin{cases} x_1 = -2.6180t \\ x_2 = t \end{cases}, \; t = \text{any number}$$

These give, for all choices of the parameter t except $t = 0$, all of the associated eigenvectors.   For a specific example, if we take $t = 1$, this will give us the eigenvector $x = \begin{bmatrix} -2.6180 \\ 1 \end{bmatrix}$.   We can verify this geometrically by plotting the vector x along with the vector $y = Ax$ to see that they are parallel.

Case: $\lambda = (-1+\sqrt{5})/2$

```
>> lambda=(-1+sqrt(5))/2;  C=lambda*eye(2)-A;  rref(C)
→ ans =    1.0000  0.3820      0
               0       0        0
```

As in the preceding case, we can get all of the associated eigenvectors.   We consider the eigenvector $x = \begin{bmatrix} -.3820 \\ 1 \end{bmatrix}$ for this second eigenvalue.   Since the eigenvalue is postive, x and Ax will point in the same directions, as can be checked. Of course, each of these eigenvectors have been written in format short; if we wanted we could have displayed them in format long and thus written our eigenvectors with more significant figures, up to about 15 (MATLAB's accuracy limit).

Before discussing MATLAB's relevant built-in functions for eigendata, we state a theorem detailing some useful facts about eigendata of a matrix. First we give a definition. Since an eigenvalue λ of a matrix A is a root of the characteristic polynomial $p_A(x)$, we know that $(x - \lambda)$ must be a factor of $p_A(x)$. Recall that the **algebraic multiplicity** of the root λ is the highest exponent m such that $(x - \lambda)^m$ is still a factor of $p_A(x)$, i.e., $p_A(x) = (x - \lambda)^m q(x)$ where $q(x)$ is a polynomial (of degree $n - m$) such that $q(\lambda) \neq 0$.

THEOREM 7.9: (*Facts about Eigenvalues and Eigenvectors*): Let $A = [a_{ij}]$ be an $n \times n$ matrix.
(i) The matrix A has at most n (real) eigenvalues, and their algebraic multiplicities add up to at most n.
(ii) If u, w are both eigenvectors of A corresponding to the same eigenvalue λ, then $u + w$ (if nonzero) is also an eigenvector of A corresponding to λ, and if c is a nonzero constant, then cu is also an eigenvector of A.[16]   The set of all such eigenvectors associated with λ, together with the zero vector, is called the **eigenspace** of A **associated with the eigenvalue** λ.
(iii)   The dimension of the eigenspace of A associated with the eigenvalue λ, called the **geometric multiplicity of the eigenvalue** λ, is always less than or equal to the algebraic multiplicity of the eigenvalue λ.
(iv)   In general a matrix A need not have any (real) eigenvalues,[17] but if A is a **symmetric matrix** (meaning: A coincides with its transpose matrix), then A will always have a full set of n real eigenvalues, provided each eigenvalue is repeated according to its geometric multiplicity.

The proofs of (i) and (ii) are rather easy; they will be left as exercises. The proofs of (iii) and (iv) are more difficult; we refer the interested reader to a good linear algebra textbook, such as [HoKu-71], [Kol-99] or [Ant-00]. There is an extensive theory and several factorizations associated with eigenvalue problems. We should also point out a couple of more advanced texts. The book [GoVL-83] has become the standard reference for numerical analysis of matrix computations. The book [Wil-88] is a massive treatise entirely dedicated to the eigenvalue problem; it remains the standard reference on the subject. Due to space limitations, we will

---

[16] Thus when we throw all eigenvectors associated with a particular eigenvalue λ of a matrix A together with the zero vector, we get a set of vectors that is closed under the two linear operations: vector additon and scalar multiplication. Readers who have studied linear algebra, will recognize such a set as a vector space; this one is called the **eigenspace** associated with the eigenvalue λ of the matrix A. Geometrically the eigenspace will be either a line through the origin (one-dimensional), a plane through the origin (two-dimensional), or in general, any k-dimensional hyperplane through the origin ($k \leq n$).

[17] This is reasonable since, as we have seen before, a polynomial need not have any real roots (e.g., $x^2 + 1$). If complex numbers are considered, however, a polynomial will always have a complex root (this is the so-called "Fundamental Theorem of Algebra") and so any matrix will always have at least a complex eigenvalue. Apart from this fact, the theory for complex eigenvalues and eigenvectors parallels that for real eigendata.

not be getting into comprehensive developments of eigenalgorithms; we will merely give a few more examples to showcase MATLAB's relevant built-in functions.

EXAMPLE 7.25: Find a matrix A that has no real eigenvalues (and hence no eigenvectors), as indicated in part (iv) of Theorem 7.9.

SOLUTION: We should begin to look for a 2×2 matrix A. We need to find one for which its characteristic polynomial $p_A(\lambda)$ has no real root. One approach would be to take a simple second-degree polynomial, that we know does not have any real roots, like $\lambda^2 + 1$, and try to build a matrix $A = \begin{bmatrix} a & b \\ c & d \end{bmatrix}$ which has this as its characteristic polynomial. Thus we want to choose a, b, c, and d such that:

$$\det\begin{pmatrix} \lambda - a & b \\ c & \lambda - d \end{pmatrix} = \lambda^2 + 1.$$

If we put $a = d = 0$, and compute the determinant we get $\lambda^2 - bc = \lambda^2 + 1$, so we are okay if $bc = -1$. For example, if we take $b = 1$ and $c = -1$, we get that the matrix $A = \begin{bmatrix} 0 & 1 \\ -1 & 0 \end{bmatrix}$ has no real eigenvalues.

   MATLAB has the built-in function `eig` that can find eigenvalues and eigenvectors of a matrix.   The two possible syntaxes of this function are as follows:

`eig(A)` →	If A is a square matrix, this command produces a vector containing the eigenvalues of A. Both real and complex eigenvalues are given.
`[V, D]=eig(A)` →	If A is an $n \times n$ matrix, this command will create two $n \times n$ matrices. D is a diagonal matrix whose diagonal entries are the eigenvalues of A, and V is matrix whose columns are corresponding eigenvectors for A. For complex eigenvalues, the corresponding eigenvectors will also be complex.

Since, by Theorem 7.9(ii), any nonzero scalar multiple of an eigenvector is again an eigenvector, MATLAB's `eig` chooses its eigenvectors to have length = 1.

For example, let's use these commands to find eigendata for the matrix of Example 7.24:

```
>> [V,D]=eig([-2 -1;1 1])
→ V =           -0.9342   0.3568
                 0.3568  -0.9342

→ D =           -1.6180      0
                    0     0.6180
```

The diagonal entries in the matrix D are indeed the eigenvalues (in short format) that we found in the example. The corresponding eigenvectors (from the columns of V) $\begin{bmatrix} -0.9342 \\ 0.3568 \end{bmatrix}$ and $\begin{bmatrix} 0.3568 \\ -0.9342 \end{bmatrix}$ are different from the two we gave in that example, but can be obtained from the general form of eigenvectors that was found in that example. Also, unlike those that we gave in the example, it can be checked that these two eigenvectors have length equal to 1.

In the case of an eigenvalue with geometric multiplicity greater than 1, eig will find (whenever possible) corresponding eigenvectors that are linearly independent.[18] Watch what happens when we apply the eig function to the matrix that we constructed in Example 7.25:

```
>> [V,D]=eig([0 1;-1 0])
→ V =                 0.7071          0.7071
                      0 + 0.7071i     0 - 0.7071i

→ D =                 0 + 1.0000i     0
                      0               0 - 1.0000i
```

We get the eigenvalues (from D) to be the complex numbers $\pm i$ (where $i = \sqrt{-1}$) and the two corresponding eigenvectors also have complex numbers in them. Since we are interested only in real eigendata, we would simply conclude from such an output that the matrix has no real eigenvalues.

If you are interested in finding the characteristic polynomial $p_A(\lambda) = a_n \lambda^n + a_{n-1} \lambda^{n-1} + \cdots + a_1 \lambda + a_0$ of an $n \times n$ matrix A, MATLAB has a function poly that works as follows:

	For an $n \times n$ matrix A, this command will produce the vector $v = [a_n \ a_{n-1} \cdots a_1 \ a_0]$ of the $n + 1$ coefficients of the nth-degree characteristic polynomial $p_A(\lambda) = \det(\lambda I - A) = a_n \lambda^n + a_{n-1} \lambda^{n-1} + \cdots + a_1 \lambda + a_0$ of the matrix A.
poly(A) →	

For example, for the matrix we constructed in Example 7.25, we could use this command to check its characteristic polynomial:

```
>> poly([0 1;-1 0])   →ans = 1   0   1
```

which translates to the polynomial $1 \cdot \lambda^2 + 0 \cdot \lambda + 1 = \lambda^2 + 1$, as was desired. Of course, this command is particularly useful for larger matrices where computation of determinants by hand is not feasible. The MATLAB function roots will find the roots of any polynomial:

---

[18] Linear independence is a concept from linear algebra. What is relevant for the concept at hand is that any other eigenvector associated with the same eigenvalue will be expressible as a linear combination of eigenvectors that MATLAB produces. In the parlance of linear algebra, MATLAB will produce eigenvectors that form a basis of the corresponding eigenspaces.

roots(v) →	For a vector $\begin{bmatrix} a_n & a_{n-1} & \cdots a_1 & a_0 \end{bmatrix}$ of the $n + 1$ coefficients of the nth-degree polynomial $p(x) = a_n x^n + a_{n-1} x^{n-1} + \cdots + a_1 x + a_0$ this command will produce a vector of length n containing all of the roots of $p(x)$ (real and complex) repeated according to algebraic multiplicity.

Thus, another way to get the eigenvalues of the matrix of Example 7.24 would be as follows: [19]

```
>> roots(poly([-2 -1;1 1]))
→ ans        -1.6180
              0.6180
```

EXERCISE FOR THE READER 7.26: For the matrix $A = \begin{bmatrix} 2 & 1 & 0 & 0 \\ 0 & 2 & 0 & 0 \\ 0 & 0 & 1 & 0 \\ 0 & 0 & 0 & 1 \end{bmatrix}$, do the

following:
(a) By hand, compute $p_A(\lambda)$, the characteristic polynomial of A, in factored form, and find the eigenvalues and the algebraic multiplicity of each one.
(b) Either by hand, or using MATLAB, find all the corresponding eigenvectors for each eigenvalue and find the geometric multiplicity of each eigenvalue.
(c) After doing part (a), can you figure out a general rule for finding the eigenvalues of an upper triangular matrix?

EXERCISES 7.6:

1. For each of the following vectors x, find $\text{len}(x)$ and find $\|x\|$.

    $x = [2, -6, 0, 3]$

    $x = [\cos(n), \sin(n), 3^n]$ (n = a positive integer)

    $x = [1, -1, 1, -1, \cdots, 1, -1]$ (vector has $2n$ components)

2. For each of the following matrices A, find the infinity norm $\|A\|$.

    (a) $A = \begin{bmatrix} 2 & -3 \\ 1 & 6 \end{bmatrix}$

    (b) $A = \begin{bmatrix} 4 & -5 & -2 \\ 1 & 2 & 3 \\ -2 & -4 & -6 \end{bmatrix}$

    (c) $A = \begin{bmatrix} \cos(\pi/4) & -\sin(\pi/4) & 0 \\ \sin(\pi/4) & \cos(\pi/4) & 0 \\ 0 & 0 & 1 \end{bmatrix}$

    (d) $A = H_n$ the $n \times n$ Hilbert matrix (introduced and defined in Example 7.8)

3. For each of the matrices A (parts (a) through (d)) of Exercise 2, find a nonzero vector x such that $\|Ax\| = \|A\|\|x\|$.

---

[19] We mention, as an examination of the M-file will show (enter type roots), that the roots of a polynomial are found using the eig command on an associated matrix.

4. For the matrix $A = \begin{bmatrix} 2 & -3 \\ -2 & 4 \end{bmatrix}$, calculate by hand the following: $\|A\|$, $\|A^{-1}\|$, $\kappa(A)$, and then verify your calculations with MATLAB. If possible, find a singular matrix S such that $\|S - A\| = 1/3$.

5. For the matrix $A = \begin{bmatrix} 1 & -1 \\ -1 & 1.001 \end{bmatrix}$, calculate by hand the following: $\|A\|$, $\|A^{-1}\|$, $\kappa(A)$, and then verify your calculations with MATLAB. Is there a singular matrix S such that $\|S - A\| = 1/1000$? Explain.

6. Consider the matrix $B = \begin{bmatrix} 2.6 & 0 & -3.2 \\ 3 & -8 & -4 \\ 1 & 2 & -1 \end{bmatrix}$.

   (a) Is there a nonzero (3×1) vector x such that $\|Bx\| \geq 13\|x\|$? If so, find one; otherwise explain why one does not exist.

   (b) Is there a singular 3×3 matrix S such that $\|S - B\| \leq 0.2$? If so, find one; otherwise explain why one does not exist.

7. Consider the matrices: $A = \begin{bmatrix} 2 & -6 \\ 11 & -5 \end{bmatrix}$, $B = \begin{bmatrix} 7 & 1 & -4 \\ 5 & -8 & -5 \\ 4 & 4 & 4 \end{bmatrix}$.

   (a) Is there a (2×1) vector X such that: $\|AX\| > 12\|X\|$?

   (b) Is there a nonzero vector X such that $\|AX\| \geq 16\|X\|$? If so, find one; otherwise explain why one does not exist.

   (c) Is there a nonzero (3×1) vector X such that $\|BX\| \geq 20\|X\|$? If so, find one; otherwise explain why one does not exist.

   (d) Is there a singular matrix $S = \begin{bmatrix} a & b \\ c & d \end{bmatrix}$ (i.e., $ad - bc = 0$) such that $\|S - A\| \leq 4.5$? If yes, find one; otherwise explain why one does not exist.

   (e) Is there a singular 3×3 matrix S such that $\|S - B\| \leq 2.25$? If so find one; otherwise explain why one does not exist.

8. Prove identities (42), (43), and (44) for condition numbers.
   Suggestion: The identity that was established in Exercise for the Reader 7.24 can be helpful.

9. (*True/False*) For each statement below, either explain why it is always true or provide a counterexample of a single situation where it is false:

   (a) If A is a square matrix with $\|A\| = 0$, then $A = 0$(matrix), i.e., all the entries of A are zero.

   (b) If A is any square matrix, then $|\det(A)| \leq \kappa(A)$.

   (c) If A is a nonsingular square matrix, then $\kappa(A^{-1}) = \kappa(A)$.

   (d) If A is a square matrix, then $\kappa(A') = \kappa(A)$.

   (e) If A and B are same-sized square matrices, then $\|AB\| \leq \|A\|\|B\|$.

   Suggestion: As is always recommended, unless you are sure about any of these identities, run a bunch of experiments on MATLAB (using randomly generated matrices).

10. Prove identity (39).
    Suggestion: Reread Example 7.19 and the note that follows it for a useful idea.

11.   (*A General Class of Norms*) For any real number $p \geq 1$, the p-norm $\|\cdot\|_p$ defined for an n-dimensional vector x by the equation

$$\|x\|_p = \left(\sum_{i=1}^{n} |x_i|^p\right)^{1/p} = \left(|x_1|^p + |x_2|^p + \cdots + |x_n|^p\right)^{1/p}$$

turns out to satisfy the norm axioms ($36A - C$). In the general setting, the proof of the triangle inequality is a bit involved (we refer to one of the more advanced texts on error analysis cited in the section for details).

(a)   Show that $\mathrm{len}(x) = \|x\|_2$ (this is why the length norm is sometimes called the 2-norm).

(b)   Verify the norm axioms ($36A - C$) for the 1-norm $\|\cdot\|_1$.

(c)   For a vector x, is it always true that $\|x\|_\infty \leq \|x\|_1$? Either prove it is always true or give a counterexample of an instance of a certain vector x for which it fails.

(d)   For a vector x, is it always true that $\|x\|_\infty \leq \|x\|_2$? Either prove it is always true or give a counterexample of an instance of a certain vector x for which it fails.

(e)   How are the norms $\|\cdot\|_1$ and $\|\cdot\|_2$ related? Does one always seem to be at least as large as the other? Do (lots of) experiments with some randomly generated vectors of different sizes.

Note: Experiments will probably convince you of a relationship (and inequality) for part (e), but it might be difficult to prove, depending on your background; the interested reader can find a proof in one of the more advanced references listed in the section. The reason that the infinity norm got this name is that for any fixed vector x, we have

$$\lim_{p \to \infty} \|x\|_p = \|x\|_\infty.$$

As the careful reader might have predicted, the MATLAB built-in function for the p-norm of a vector x is `norm(x,p)`.

12.   Let $A = \begin{bmatrix} -11 & 3 \\ 4 & -1 \end{bmatrix}$, $b = \begin{bmatrix} 2 \\ 0 \end{bmatrix}$ $z = \begin{bmatrix} 1.2 \\ 7.8 \end{bmatrix}$.

(a) Let z be an approximate solution of the system $Ax = b$; find the residual vector r.
(b) Use Theorem 7.7 to give an estimate for the error of the approximation in part (a).
(c) Give an estimate for the relative error of the approximation in part (a).
(d) Find the norm of the exact error $\|\Delta x\|$ of the approximate solution in part (a).

13.   Repeat all parts of Exercise 12 for the following matrices:

$$A = \begin{bmatrix} -0.1 & -9 \\ 11 & 1000 \end{bmatrix}, \quad b = \begin{bmatrix} -0.1 \\ 10 \end{bmatrix} \quad z = \begin{bmatrix} 11 \\ -0.2 \end{bmatrix}.$$

14.   Let $A = \begin{bmatrix} 1 & 0 & 1 \\ -1 & 1 & 1 \\ -1 & -1 & 1 \end{bmatrix}$, $b = \begin{bmatrix} 1 \\ 2 \\ 3 \end{bmatrix}$.

(a) Use MATLAB's "left divide" to solve the system $Ax = b$.
(b) Use Theorems 7.7 and 7.8 to estimate the error and relative error of the solution obtained in part (a).
(c) Use MATLAB to solve the system $Ax = b$ by left multiplying the equation by the `inv(A)`.
(d) Use Theorems 7.7 and 7.8 to estimate the error and relative error of the solution obtained in part (c).
(e) Solve the system using MATLAB's symbolic capabilities and compute the actual errors of the solutions obtained in parts (a) and (c).

15.   Let A be the 60×60 matrix whose entries are 1's across the main diagonal and the last column,

-1's below the main diagonal, and whose remaining entries (above the main diagonal) are zeros, and let $b = [1 \quad 2 \quad 3 \quad \cdots \quad 58 \quad 59 \quad 60]'$.

(a) Use MATLAB's left divide to solve the system $Ax = b$ and label this computed solution as z. Print out only z (37).

(b) Use Theorems 7.7 and 7.8 to estimate the error and relative error of the solution obtained in part (a).

(c) Use MATLAB to solve the system $Ax = b$ by left multiplying the equation by the inv(A) and label this computed solution as z2. Print out only z2 (37).

(d) Use Theorems 7.7 and 7.8 to estimate the error and relative error of the solution obtained in part (c).

(e) Solve the system using MATLAB's symbolic capabilities, and print out x (37) of the exact solution. Then compute the norms of the actual errors of the solutions obtained in parts (a) and (c).

16.   (*Iterative Refinement*) Let z_0 be the computer solution of Exercise 15(a), and let r_0 denote the corresponding residual vector. Now use the Gaussian program to solve the system $Ax = r_0$, and call the computer solution z_1, and the corresponding residual vector r_1. Next use the Gaussian program once again to solve $Ax = r_1$, and let z_2 and r_2 denote the corresponding approximate solution and residual vector. Now let

$$z = z_0, \quad z' = z + z_1, \quad \text{and} \quad z'' = z' + z_2.$$

Viewing these three vectors as solutions of the original system $Ax = b$ (of Exercise 15), use the error estimate Theorem 7.8 to estimate the relative error of each of these three vectors. Then compute the norm of the actual errors by comparing with the exact solution of the system as obtained in part (e) of Exercise 15. See the next exercise for more on this topic.

17.   In theory, the iterative technique of the previous exercise can be useful to improving accuracy of approximate solutions in certain circumstances. In practice, however, roundoff errors and poorly conditioned matrices can lead to unimpressive results. This exercise explores the effect that additional digits of precision can have on this scheme.

(a) Using variable precision arithmetic (see Appendix A) with 30 digits of accuracy, redo the previous exercise, starting with the computed solution of Exercise 15(a) done in MATLAB's default floating point arithmetic.

(b) Using the same arithmetic of part (a), solve the original system using MATLAB's left divide.

(c) Compare the norms of the actual errors of the three approximate solutions of part (a) and the one of part (b) by using symbolic arithmetic to get MATLAB to compute the exact solution of the system.

Note: We will learn about different iterative methods in the next section.

18.   This exercise will examine the benefits of variable precision arithmetic over MATLAB's default floating point arithmetic and over MATLAB's more costly symbolic arithmetic. As in Section 7.4, we let $H_n = \left[1/(i+j-1) \right]$ denote the $n \times n$ Hilbert matrix. Recall that it can be generated in MATLAB using the command hilb(n).

(a)   For the values $n = 5, 10, 15, \cdots, 100$ create the corresponding Hilbert matrices H_n in MATLAB as symbolic matrices and compute symbolically the inverses of each. Use tic/toc to record the computation times (these times will be machine dependent; see Chapter 4). Go as far as you can until your cumulative MATLAB computation time exceeds one hour. Next compute the corresponding condition numbers of each of these Hilbert matrices.

(b) Starting with MATLAB's default floating point arithmetic (which is roughly 15 digits of variable precision arithmetic), and then using variable precision arithmetic starting with 20 digits and then moving up in increments of 5 (25, 30, 35, ...) continue to compute the inverses of each of the Hilbert matrices of part (a) until you get a computed inverse whose norm differs from the norm of the exact inverse in part (a) by no more than 0.000001. Record (using tic/toc) the computation time for the final variable precision arithmetically computed inverse, along with the number of digits used and compare it to the corresponding computation time for the exact inverse that was done in part (a).

19. Prove the following inequality $\|Ax\| \geq \|x\|/\|A^{-1}\|$, where A is any invertible $n \times n$ matrix and x is any column vector with n entries.

20. Suppose that A is a 2×2 matrix with norm $\|A\| = 0.5$ and x and y are 2×1 vectors with $\|x - y\| \leq 0.8$. Show that: $\|A^2x - A^2y\| \leq 0.2$.

21. (*Another Error Bound for Computed Solutions of Linear Systems*) For a nonsingular matrix A and a computed inverse matrix C for A^{-1}, we define the resulting **residual matrix** as $R = I - CA$. If z is an approximate solution to $Ax = b$, and as usual $r = b - Az$ is the residual vector, show that

$$\text{error} \equiv \|x - z\| \leq \frac{\|CR\|}{1 - \|R\|},$$

provided that $\|R\| < 1$.

Hint: For part (a), first use the equation $I - R = CA$ to get that $(I - R)^{-1} = A^{-1}C^{-1}$ and so $A^{-1} = (I - R)^{-1}C$. (Recall that the inverse of a product of invertible matrices equals the reverse order product of the inverses.)

22. For each of the matrices A below, find the following:
    (a) The characteristic polynomial $p_A(\lambda)$.
    (b) All eigenvalues and all of their associated eigenvectors.
    (c) The algebraic and geometric multiplicity of each eigenvalue.
    (i) $A = \begin{bmatrix} 1 & 2 \\ 2 & 1 \end{bmatrix}$ (ii) $A = \begin{bmatrix} 1 & 1 \\ 2 & 2 \end{bmatrix}$ (iii) $A = \begin{bmatrix} 1 & 2 \\ 2 & 2 \end{bmatrix}$ (iv) $A = \begin{bmatrix} -1 & 0 \\ 2 & 3 \end{bmatrix}$

23. Repeat all parts of Exercise 22 for the following matrices.
    (i) $A = \begin{bmatrix} 1 & 2 & 2 \\ 2 & 1 & 2 \\ 2 & 2 & 1 \end{bmatrix}$ (ii) $A = \begin{bmatrix} 1 & 2 & 0 \\ 2 & 1 & 2 \\ 0 & 2 & 1 \end{bmatrix}$

    (iii) $A = \begin{bmatrix} 1 & 2 & 2 \\ 2 & 1 & 2 \\ 0 & 0 & 1 \end{bmatrix}$ (iv) $A = \begin{bmatrix} 1 & 2 & 2 \\ -2 & 1 & 2 \\ -2 & -2 & 1 \end{bmatrix}$

24. Consider the matrix $A = \begin{bmatrix} 11 & 11 & 4 \\ 7 & 7 & -4 \\ -7 & -11 & 0 \end{bmatrix}$.

    (a) Find all eigenvalues of A, and for each find just one eigenvector (give your eigenvectors as many integer components as possible).
    (b) For each of the eigenvectors x that you found in part (a), evaluate $y = (2A)x$. Is it possible to write $y = \lambda x$ for some scalar λ? In other words, is x also an eigenvector of the matrix $2A$?
    (c) Find all eigenvalues of $2A$. How are these related to those of the matrix A?
    (d) For each of your eigenvectors x from part (a), evaluate $y = (-5A)x$. Is it possible to write $y = \lambda x$ for some scalar λ? In other words, is x also an eigenvector of the matrix $-5A$?
    (e) Find all eigenvalues of $-5A$; how are these related to those of the matrix A?

(f) Based on your work in these above examples, without picking up your pencil or typing anything on the computer, what do you think the eigenvalues of the matrix $23A$ would be? Could you guess also some associated eigenvectors for each eigenvalue? Check your conclusions on MATLAB.

25.   Consider the matrix $A = \begin{bmatrix} 2 & 0 & 1 \\ 1 & -4 & 1 \\ 1 & 0 & 2 \end{bmatrix}$.

(a) Find all eigenvalues of A, and for each find just one eigenvector (give your eigenvectors as many integer components as possible).

(b) For each of the eigenvectors x that you found in part (a), evaluate $y = A^2 x$. Is it possible to write $y = \lambda x$ for some scalar λ? In other words, is x also an eigenvector of the matrix A^2?

(c) Find all eigenvalues of A^2; how are these related to those of the matrix A?

(d) For each of your eigenvectors x from part (a), evaluate $y = A^3 x$. Is it possible to write $y = \lambda x$ for some scalar λ? In other words, is x also an eigenvector of the matrix A^3?

(e) Find all eigenvalues of A^3; how are these related to those of the matrix A?

(f) Based on your work in the above examples, without picking up your pencil or typing anything on the computer, what do you think the eigenvalues of the matrix A^8 would be? Could you guess also some associated eigenvectors for each eigenvalue? Check your conclusions on MATLAB.

26.   Find the characteristic polynomial (factored form is okay) as well as all eigenvalues for the $n \times n$ identity matrix I. What are (all) of the corresponding eigenvectors (for each eigenvalue)?

27.   Consider the matrix $A = \begin{bmatrix} 3 & 3 & 3 \\ 3 & 2 & 4 \\ 3 & 4 & 2 \end{bmatrix}$.

(a) Find all eigenvalues of A, and for each find just one eigenvector (give your eigenvectors as many integer components as possible).

(b) For each of the eigenvectors x that you found in part (a), evaluate $y = (A^2 + 2A)x$. Is it possible to write $y = \lambda x$ for some scalar λ? In other words, is x also an eigenvector of the matrix $A^2 + 2A$?

(c) Find all eigenvalues of $A^2 + 2A$; how are these related to those of the matrix A?

(d) For each of your eigenvectors x from Part (a), evaluate $y = (A^3 - 4A^2 + I)x$. Is it possible to write $y = \lambda x$ for some scalar λ? In other words, is x also an eigenvector of the matrix $A^3 - 4A^2 + I$?

(e) Find all eigenvalues of $A^3 - 4A^2 + I$; how are these related to those of the matrix A?

(f) Based on your work in the above examples, without picking up your pencil or typing anything on the computer, what do you think the eigenvalues of the matrix $A^5 - 4A^3 + 2A - 4I$ would be? Could you guess also some associated eigenvectors for each eigenvalue? Check your conclusions on MATLAB.

NOTE: The **spectrum** of a matrix A, denoted $\sigma(A)$, is the set of all eigenvalues of the matrix A. The next exercise generalizes some of the results discovered in the previous four exercises.

28.   For a square matrix A and any polynomial $p(x) = a_m x^m + a_{m-1} x^{m-1} + \cdots + a_1 x + a_0$, we define a new matrix $p(A)$ as follows:

$$p(A) = a_m A^m + a_{m-1} A^{m-1} + \cdots + a_1 A + a_0 I.$$

(We simply substituted A for x in the formula for the polynomial; we also had to replace the constant term a_0 by this constant times the identity matrix—the matrix analogue of the number

1.) Prove the following appealing formula;
$$\sigma(p(A)) = p(\sigma(A)),$$
which states that the spectrum of the matrix $p(A)$ equals the set
$$\{p(\lambda): \lambda \text{ is an eigenvalue of } A\}.$$

29. Prove parts (i) and (ii) of Theorem 7.9.

30. Show that the characteristic polynomial of any $n \times n$ matrix is always a polynomial of degree n in the variable λ.
     Suggestion: Use induction and cofactor expansion.

31. (a) Use the basic Gaussian elimination algorithm (Program 7.6) to solve the linear systems of Exercise for the Reader 7.16, and compare with results obtained therein.
     (b) Use the Symbolic Toolbox to compute the condition numbers of the Hilbert matrices that came up in part (a). Are the estimates provided by Theorem 7.8 accurate or useful?
     (c) Explain why the algorithm performs so well with this problem despite the large condition numbers of A.
     Suggestion: For part (c), examine what happens after the first pivot operation.

7.7: ITERATIVE METHODS

As mentioned earlier in this chapter, Gaussian elimination is the best all-around solver for nonsingular linear systems $Ax = b$. Being a universal method, however, there are often more economical methods that can be used for particular forms of the coefficient matrix. We have already seen the tremendous savings, both in storage and in computations that can be realized in case A is tridiagonal, by using the Thomas method. All methods considered thus far have been **direct methods** in that, mathematically, they compute the exact solution and the only errors that arise are numerical. In this section we will introduce a very different type of method called an **iterative method**. Iterative methods begin with an initial guess at the (vector) solution $x^{(0)}$, and produce a sequence of vectors, $x^{(1)}, x^{(2)}, x^{(3)}, \cdots,$ which, under certain circumstances, will converge to the exact solution. Of course, in any floating point arithmetic system, a solution from an iterative method (if the method converges) can be made just as accurate as that of a direct method.

In solving differential equations with so-called finite difference methods, the key numerical step will be to solve a linear system $Ax = b$ having a large and sparse coefficient matrix A (a small percentage of nonzero entries) that will have a special form. The large size of the matrix will often make Gaussian elimination too slow. On the other hand, the special structure and sparsity of A can make the system amenable to a much more efficient iterative method. We have seen that in general Gaussian elimination for solving an n variable linear system performs in $O(n^3)$-time. We take this as the ceiling performance time for any linear system solver. The Thomas method, on the other hand, for the very special triadiagonal systems, performed in only $O(n)$-time. Since just solving n independent linear equations (i.e., with A being a diagonal matrix) will also take this amount, this is the theoretical floor performance time for any linear system solver. Most of iterative methods today perform theoretically in $O(n^2)$ -time, but in practice can perform in times closer to the theoretical floor $O(n)$ -time. In recent years, iterative methods have become increasingly important and have a promising future, as increasing computer performance will make the improvements over Gaussian elimination more and more dramatic.

We will describe three common iterative methods: Jacobi, Gauss-Seidel, and SOR iteration. After giving some simple examples showing the sensitivity of these methods to the particular form of A, we give some theoretical results that will guarantee convergence. We then make some comparisons among these three methods in flop counts and computation times for larger systems and then with Gaussian elimination. The theory of iterative methods is a very exciting and interesting area of numerical analysis. The Jacobi, Gauss-Seidel, and SOR iterative methods are quite intuitive and easy to develop. Some of the more state-of-the-art methods such as conjugate gradient methods and GMRES (generalized minimum residual method) are more advanced and would take a lot more work to develop and understand, so we refer the interested reader to some references for more details on this interesting subject: [Gre-97], [TrBa-97], and [GoVL-83]. MATLAB, however, does have some built-in functions for performing such more advanced iterative methods. We introduce these MATLAB functions and do some performance comparisons involving some (very large and) typical coefficient matrices that arise in finite difference schemes.

We begin with a nonsingular linear system:

$$Ax = b . \tag{54}$$

In scalar form, it looks like this:

$$a_{i1}x_1 + a_{i2}x_2 + \cdots a_{in}x_n = b_i \quad (1 \le i \le n) . \tag{55}$$

Now, if we assume that each of the diagonal entries of A are nonzero, then each of the equations in (55) can be solved for x_i to arrive at:

$$x_i = \frac{1}{a_{ii}} \left[b_i - a_{i1}x_1 - a_{i2}x_2 - \cdots - a_{i,i-1}x_{i-1} - a_{i,i+1}x_{i+1} - \cdots a_{in}x_n \right] (1 \le i \le n) . \tag{56}$$

The **Jacobi iteration** scheme is obtained from using formula (56) with the values of current iteration vector $x^{(k)}$ on the right to create, on the left, the values of the next iteration vector $x^{(k+1)}$. We record the simple formula:

Jacobi Iteration:

$$x_i^{(k+1)} = \frac{1}{a_{ii}} \left[b_i - \sum_{\substack{j=1, \\ j \ne i}}^{n} a_{ij}x_j^{(k)} \right] \quad (1 \le i \le n) . \tag{57}$$

Let us give a (very) simple example illustrating this scheme on a small linear system and compare with the exact solution.

EXAMPLE 7.26: Consider the following linear system:

$$\begin{array}{rcrcrcl} 3x_1 & + & x_2 & - & x_3 & = & -3 \\ 4x_1 & - & 10x_2 & + & x_3 & = & 28 \\ 2x_1 & + & x_2 & + & 5x_3 & = & 20 \end{array}.$$

(a) Starting with the vector $x^{(0)} = [0\ 0\ 0]'$, apply the Jacobi iteration scheme with up to 30 iterations until (if ever) the 2-norm of the differences $x^{(k+1)} - x^{(k)}$ is less than 10^{-6}. Plot the norms of these differences as a function of the iteration. If convergence occurs, record the number of iterations and the actual 2-norm error of the final iterant with the exact solution.
(b) Repeat part (a) on the equivalent system obtained by switching the first two equations.

SOLUTION: Part (a): The Jacobi iteration scheme (57) becomes:

$$\begin{aligned} x_1^{(k+1)} &= (-3 - x_2^{(k)} + x_3^{(k)})/3 \\ x_2^{(k+1)} &= (28 - 4x_1^{(k)} - x_3^{(k)})/(-10) \\ x_3^{(k+1)} &= (20 - 2x_1^{(k)} - x_2^{(k)})/5 \end{aligned}.$$

The following MATLAB code will perform the required tasks:

```
xold = [0 0 0]';   xnew=xold;
for k=1:30
  xnew(1)=(-3-xold(2)+xold(3))/3;
  xnew(2)=(28-4*xold(1)-xold(3))/(-10);
  xnew(3)=(20-2*xold(1)-xold(2))/5;
  diff(k)=norm(xnew-xold,2);
  if diff(k)<1e-6
    fprintf('Jacobi iteration has converged in %d iterations', k)
    return
  end
  xold=xnew;
end
→ Jacobi iteration has converged in 26 iterations
```

The exact solution is easily seen to be $[1\ -2\ 4]'$. The exact 2-norm error is thus given by:

```
>>norm(xnew-[1 -2 4]',2)      → ans = 3.9913e - 007
```

which compares favorably with the norm of the last difference of the iterates (i.e., the actual error is smaller):

```
>> diff(k)                     →ans = 8.9241e - 007
```

We will see later in this section that finite difference methods typically exhibit linear convergence (if they indeed converge); the quality of convergence will thus depend on the asymptotic error constant (see Section 6.5 for the terminology).

Due to this speed of the decay of errors, an ordinary plot will not be so useful (as the reader should verify), so we use a log scale on the y-axis. This is accomplished by the following MATLAB command:

`semilogy(x,y)` →	If x and y are two vectors of the same size, this will produce a plot where the y-axis numbers are logarithmically spaced rather than equally spaced as with `plot(x,y)`.
`semilogx(x,y)` →	Works as the above command, but now the x-axis numbers are logarithmically spaced.

The required plot is now created with the following command and the result is shown in Figure 7.39(a).

```
>>semilogy(1:k,diff(1:k))
```

Part (b): Switching the first two equations of the given system leads to the following modified Jacobi iteration scheme:

$$x_1^{(k+1)} = (28 + 10x_2^{(k)} - x_3^{(k)})/4$$
$$x_2^{(k+1)} = -3 - 3x_1^{(k)} + x_3^{(k)}$$
$$x_3^{(k+1)} = (20 - 2x_1^{(k)} - x_2^{(k)})/5.$$

In the above MATLAB code, we need only change the two lines for xnew(54) and xnew(55) accordingly:

```
xnew(1)=(28+10*xold(2)-xold(3))/4;
xnew(2)=-3-3*xold(1)+xold(3);
```

Running the code, we see that this time we do not get convergence. In fact, a semilog plot will show that quite the opposite is true, the iterates badly diverge. The plot, obtained just as before, is shown in Figure 7.39(b). We will soon show how such sensitivities of iterative methods depend on the form of the coefficient matrix.

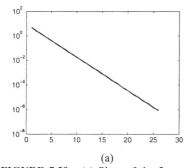
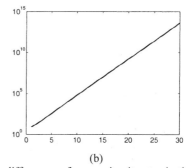

(a)   (b)

FIGURE 7.39: (a) Plots of the 2-norms of the differences of successive iterates in the Jacobi scheme for the linear system of Example 7.26, using the zero vector as the initial iterate. The convergence is exponential. (b) The corresponding errors when the same scheme is applied to the equivalent linear system with the first two equations being permuted. The sequence now badly diverges, showing the sensitivity of iterative methods to the particular form of the coefficient matrix.

The code given in the above example can be easily generalized into a MATLAB M-file for performing the Jacobi iteration on a general system. This task will be delegated to the next exercise for the reader.

EXERCISE FOR THE READER 7.27:   (a) Write a function M-file, [x,k,diff]= jacobi(A,b,x0,tol,kmax), that performs the Jacobi iteration on the linear system $Ax = b$. The inputs are the coefficient matrix A, the inhomogeneity (column) vector b, the seed (column) vector x0 for the iteration process, the tolerance tol which will cause the iteration to stop if the 2-norms of successive iterates become smaller than tol, and kmax, the maximum number of iterations to perform. The outputs are the final iterate x, the number of iterations performed k, and a vector diff that records the 2-norms of successive differences of iterates. If the last three input variables are not specified, default values of x0 = the zero column vector, tol = 1e-10, and kmax = 100 are used.
(b) Apply the program to recover the data obtained in part (a) of Example 7.26. If we reset the tolerance for accuracy to 1e-10 in that example, how many iterations would the Jacobi iteration need to converge?

If we compute the values of $x^{(k+1)}$ in order, it seems reasonable to update the values used on the right side of (57) sequentially, as they become available. This modification in the scheme gives the **Gauss-Seidel iteration**. Notice that the Gauss-Seidel scheme can be implemented so as to roughly cut in half the storage requirements for the iterates of the solution vector x. Although the M-file we present below does not take advantage of such a scheme, the interested reader can easily modify it to do so. Futhermore, as we shall see, the Gauss-Seidel scheme almost always outperforms the Jacobi scheme.

Gauss-Seidel Iteration:

$$x_i^{(k+1)} = \frac{1}{a_{ii}}\left[b_i - \sum_{j=1}^{i-1} a_{ij}x_j^{(k+1)} - \sum_{j=i+1}^{n} a_{ij}x_j^{(k)} \right] \ (1 \le i \le n). \tag{58}$$

We proceed to write an M-file that will apply the Gauss-Seidel scheme to solving the nonsingular linear system (54).

PROGRAM 7.7: A function M-file,

[x,k,diff]=gaussseidel(A,b,x0,tol,kmax)

that performs the Gauss-Seidel iteration on the linear system $Ax = b$. The inputs are the coefficient matrix A, the inhomogeneity (column) vector b, the seed (column) vector x0 for the iteration process, the tolerance tol, which will cause the iteration to stop if the 2-norms of successive iterates become smaller than tol, and kmax, the maximum number of iterations to perform. The outputs are the final iterate x, the number of iterations performed k, and a vector diff that records the 2-norms of successive differences of iterates. If the last two input variables are not specified, default values of tol = 1e-10 and kmax = 100 are used.

```
function [x, k, diff] = gaussseidel(A,b,x0,tol,kmax)
% performs the Gauss-Seidel iteration on the linear system Ax = b.
% Inputs:  the coefficient matrix 'A', the inhomogeneity (column)
% vector 'b',the seed (column) vector 'x0' for the iteration process,
% the tolerance 'tol' which will cause the iteration to stop if the
% 2-norms of differences of successive iterates becomes smaller than
% 'tol', and 'kmax' that is the maximum number of iterations to
% perform.
% Outputs:  the final iterate 'x', the number of iterations performed
% 'k', and a vector 'diff' that records the 2-norms of successive
% differences of iterates.
% If either of the  last three input variables are not specified,
% default values of x0= zero column vector, tol=1e-10 and kmax=100
% are used.

%assign default input variables, as necessary
if nargin<3, x0=zeros(size(b)); end
if nargin<4, tol=1e-10; end
if nargin<5, kmax=100; end

if min(abs(diag(A)))<eps
    error('Coefficient matrix has zero diagonal entries, iteration
                          cannot be performed.\r')
end

[n m]=size(A);
x=x0;
k=1; diff=[];

while k<=kmax
    norm=0;
    for i=1:n
        oldxi=x(i); x(i)=b(i);
        for j=[1:i-1 i+1:n]
            x(i)=x(i)-A(i,j)*x(j);
        end
        x(i)=x(i)/A(i,i);
        norm=norm+(oldxi-x(i))^2;
    end
    diff(k)=sqrt(norm);
    if diff(k)<tol
            fprintf('Gauss-Seidel iteration has converged in %d
                              iterations/r', k)
            return
    end
    k=k+1;
end
fprintf('Gauss-Seidel iteration failed to converge./r')
```

EXAMPLE 7.27: For the linear system of the last example, apply Gauss-Seidel iteration with initial iterate being the zero vector and the same tolerance as that used for the last example. Find the number of iterations that are now required for convergence and compare the absolute 2-norm error of the final iterate with that for the last example.

SOLUTION: Reentering, if necessary, the data from the last example, create corresponding data for the Gauss-Seidel iteration using the preceding M-file:

```
>> [xGS, kGS, diffGS] = gaussseidel(A,b,zeros(size(b))),1e-6);
→Gauss-Seidel iteration has converged in 17 iterations
```

Thus with the same amount of work per iteration, Gauss-Seidel has done the job in only 17 versus 26 iterations for Jacobi.

Looking at the absolute error of the Gauss-Seidel approximation,
```
>> norm(xGS-[1 -2 4]',2)              →ans = 1.4177e - 007
```

we see it certainly meets our tolerance goal of 1e-6 (and, in fact, is smaller than that for the Jacobi iteration).

The Gauss-Seidel scheme can be extended to include a new parameter, ω, that will allow the next iterate $x^{(k+1)}$ to be expressed as a linear combination of the current iterate $x^{(k)}$ and the Gauss-Seidel values given by (58). This gives a family of iteration schemes, collectively known as **SOR (successive over relaxation)** whose iteration schemes are given by the following formula:

SOR Iteration:

$$x_i^{(k+1)} = \frac{\omega}{a_{ii}}\left[b_i - \sum_{j=1}^{i-1}a_{ij}x_j^{(k+1)} - \sum_{j=i+1}^{n}a_{ij}x_j^{(k)}\right] + (1-\omega)x_i^{(k)} \quad (1 \le i \le n) \tag{59}$$

The parameter ω, called the **relaxation parameter**, controls the proportion of the Gauss-Seidel update versus the current iterate to use in forming the next iterate. We will soon see that for SOR to converge, we will need the relaxation parameter to satisfy $0 < \omega < 2$. Notice that when $\omega = 1$, SOR reduces to Gauss-Seidel. For certain values of ω, SOR can accelerate the convergence realized in Gauss-Seidel.

With a few changes to the Program 7.7, a corresponding M-file for SOR is easily created. We leave this for the next exercise for the reader.

EXERCISE FOR THE READER 7.28: (a) Write a function M-file, `[x,k,diff]=sorit(A,b,omega,x0,tol,kmax)`, that performs the SOR iteration on the linear system $Ax = b$. The inputs are the coefficient matrix A, the inhomogeneity (column) vector b, the relaxation parameter omega, the seed (column) vector x0 for the iteration process, the tolerance tol, which will cause the iteration to stop if the 2-norms of successive iterates become smaller than tol, and kmax, the maximum number of iterations to perform. The outputs are the final iterate x, the number of iterations performed k, and a vector diff that records the 2-norms of successive differences of iterates. If the last three input variables are not specified, default values of x0 = the zero column vector, tol = 1e-10, and kmax = 100 are used.
(b) Apply the program to recover the solution obtained in Example 7.27.

(b) Apply the program to recover the solution obtained in Example 7.27.

(c) If we use $\omega = 0.9$, how many iterations would the SOR iteration need to converge?

EXAMPLE 7.28: Run a set of SOR iterations by letting the relaxation parameter run from 0.05 to 1.95 in increments of 0.5. Use a tolerance for error = 1e-6, but set kmax = 1000. Record the number of iterations needed for convergence (if there is convergence) for each value of the ω (up to 1000) and plot this number as a function of ω.

SOLUTION: We can use the M-file `sorit` of Exercise for the Reader 7.28 in conjunction with a loop to easily obtain the needed data.

```
>> omega=0.05:.05:1.95;
>> length(omega)      →ans = 39
>> for i=1:39
[xSOR, kSOR(i), diffSOR] = sorit(A,b,omega(i),zeros(size(b))),...
                   1e-6,1000);
end
```

The above loop has overwritten all but the iteration counters, which were recorded as a vector. We use this vector to locate the best value (from among those in our vector omega) to use in SOR.

```
>> [mink ind]=min(kSOR)      →mink = 9, ind = 18
>> omega(18)                 →ans = 0.9000
```

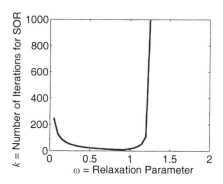

Thus we see that the best value of ω to use (from those we tested) is $\omega = 0.9$, which requires only nine iterations in the SOR scheme, nearly a 50% savings over Gauss-Seidel. The next two commands will produce the desired plot of the required number of iterations needed in SOR versus the value of the parameter ω. The resulting plot is shown in Figure 7.40.

```
>> plot(omega, kSOR),
>> axis([0 2 0 1000])
```

FIGURE 7.40: Graph of the number of iterations required for convergence (to a tolerance of 1e-6) using SOR iteration as a function of the relaxation parameter ω. The *k*-values are truncated at 1000. Notice from the graph that the convergence for Gauss-Seidel ($\omega = 1$) can be improved.

Figure 7.41 gives a plot that compares the convergences of three methods: Jacobi, Gauss-Seidel, and SOR (with our pseudo-optimal value of ω); the next exercise for the reader will ask to reproduce this plot.

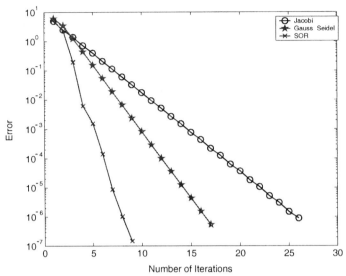

FIGURE 7.41: Comparison of the errors versus the number of iterations for each of the three iteration methods: Jacobi (o), Gauss-Seidel (*), and SOR (x).

EXERCISE FOR THE READER 7.29: Use MATLAB to reproduce the plot of Figure 7.41. The key in the upper-right corner can be obtained by using the "Data Statistics" tool from the "Tools" menu of the MATLAB graphics window once the three plots are created.

Of course, even though the last example has shown that SOR can converge faster than Gauss-Seidel, the amount of work required to locate a good value of the parameter greatly exceeded the actual savings in solving the linear system of Example 7.26. In the SOR iteration the value of $\omega = 0.9$ was used as the relaxation parameter.

There is some interesting research involved in determining the optimal value of ω to use based on the form of the coefficient matrix. What is needed to prove such results is to get a nice formula for the eigenvalues of the matrix (in general an impossible problem, but for special types of matrices one can get lucky) and then compute the value of ω for which the corresponding maximum absolute value of the eigenvalues is as small as possible . A good survey of the SOR method is given in [You-71]. A sample result will be given a bit later in this section (see Proposition 7.14).

We now present a general way to view iteration schemes in matrix form. From this point of view it will be a simple matter to specialize to the three forms we gave above. More importantly, the matrix notation will allow a much more natural way to perform error analysis and other important theoretical tasks.

To cast iteration schemes into matrix form, we begin by breaking the coefficient matrix A into three pieces:

$$A = D - L - U, \tag{60}$$

where D is the diagonal part of A, L is (strictly) lower triangular and U is the (strictly) upper triangular. In long form this (60) looks like:

$$A = \begin{bmatrix} a_{11} & & & \\ & a_{22} & & 0 \\ & & a_{33} & \\ 0 & & & \ddots \\ & & & & a_{nn} \end{bmatrix} - \begin{bmatrix} 0 & & & \\ -a_{21} & 0 & & 0 \\ -a_{31} & -a_{32} & 0 & \\ & & & \ddots \\ -a_{n1} & -a_{n2} & -a_{n3} & \cdots & 0 \end{bmatrix} - \begin{bmatrix} 0 & -a_{12} & -a_{13} & \cdots & -a_{1n} \\ & 0 & -a_{23} & \cdots & -a_{2n} \\ & & 0 & \ddots & \vdots \\ & 0 & & \ddots & -a_{n-1,n} \\ & & & & 0 \end{bmatrix}$$

This decomposition is actually quite simple. Just take D to be the diagonal matrix with the diagonal entries equal to those of A, and take L/U to be, respectively, the strictly lower/upper triangular matrix whose nonzero entries are the opposites of the corresponding entries of A.

Next, we will examine the following general (matrix form) iteration scheme for solving the system (54) $Ax = b$:

$$Bx^{(k+1)} = (B - A)x^{(k)} + b, \tag{61}$$

where B is an invertible matrix that is to be determined. Notice that if B is chosen so that this iteration scheme produces a convergent sequence of iterates: $x^{(k)} \to \tilde{x}$, then the limiting vector $\tilde{x}$ must solve (54). (Proof: Take the limit in (61) as $k \to \infty$ to get $B\tilde{x} = (B - A)\tilde{x} + b = B\tilde{x} - A\tilde{x} + b \Rightarrow A\tilde{x} = b$.) The matrix B should be chosen so that the linear system is easy to solve for $x^{(k+1)}$ (in fact, much easier than our original system $Ax = b$ lest this iterative scheme would be of little value) and so that the convergence is fast.

To get some idea of what sort of matrix B we should be looking for, we perform the following error analysis on the iterative scheme (61). We mathematically solve (61) for $x^{(k+1)}$ by left multiplying by B^{-1} to obtain:

$$x^{(k+1)} = B^{-1}(B - A)x^{(k)} + B^{-1}b = (I - B^{-1}A)x^{(k)} + B^{-1}b.$$

Let x denote the exact solution of $Ax = b$ and $e^{(k)} = x^{(k)} - x$ denote the error vector of the kth iterate. Note that $-(I - B^{-1}A)x = -x + B^{-1}b$. Using this in conjunction with the last equation, we can write:

$$\begin{aligned} e^{(k+1)} = x^{(k+1)} - x &= (I - B^{-1}A)x^{(k)} + B^{-1}b - x \\ &= (I - B^{-1}A)x^{(k)} - (I - B^{-1}A)x \\ &= (I - B^{-1}A)(x^{(k)} - x) \\ &= (I - B^{-1}A)e^{(k)} \end{aligned}$$

We summarize this important error estimate:

$$e^{(k+1)} = (I - B^{-1}A)e^{(k)}. \tag{62}$$

From (62), we see that if the matrix $(I - B^{-1}A)$ is "small" (in some matrix norm), then the errors will decay as the iterations progress.[20]   But this matrix will be small if $B^{-1}A$ is "close" to I, which in turn will happen if B^{-1} is "close" to A^{-1} and this translates to B being close to A.

Table 7.1 summarizes the form of the matrix B for each of our three iteration schemes introduced earlier. We leave it as an exercise to show that with the matrices given in Table 7.1, (61) indeed is equivalent to each of the three iteration schemes presented earlier (Exercise 20).

TABLE 7.1: Summary of matrix formulations of each of the three iteration schemes: Jacobi, Gauss-Seidel, and SOR.

Iteration Scheme:	Matrix B in the corresponding formulation (61) $Bx^{(k+1)} = (B-A)x^{(k)} + b$ in terms of (60) $A = D - L - U$
Jacobi (see formula (57))	$B = D$
Gauss-Seidel (see formula (58))	$B = D - L$
SOR with relaxation parameter ω (see formula (59))	$B = \frac{1}{\omega}D - L$

Thus far we have done only experiments with iterations. Now we turn to some of the theory.

THEOREM 7.10:   (*Convergence Theorem*) Assume that A is a nonsingular (square) matrix, and that B is any nonsingular matrix of the same size as A. The (real and complex) eigenvalues of the matrix $I - B^{-1}A$ all have absolute value less than one if and only if the iteration scheme (61) converges (to the solution of $Ax = b$) for any initial seed vector $x^{(0)}$.

For a proof of this and the subsequent theorems in this section, we refer the interested reader to the references [Atk-89] or to [GoVL-83].   MATLAB's eig function is designed to produce the eigenvalues of a matrix. Since, as we have pointed out (and seen in examples), the Gauss-Seidel iteration usually converges faster than the Jacobi iteration, it is not surprising that there are examples where

---

[20] It is helpful to think of the one-dimensional case, where everything in (62) is a number. If $(I - B^{-1}A)$ is less than one in absolute value, then we have exponential decay, and furthermore, the decay is faster when absolute values of the matrix are smaller. This idea can be made to carry over to matrix situations. The corresponding needed fact is that all of the eigenvalues (real or complex) of the matrix $(I - B^{-1}A)$ are less than one in absolute value. In this case, it can be shown that we have exponential decay also for the iterative scheme, regardless of the initial iterate. For complete details, we refer to [Atk-89] or to [GoVL-83].

the former will converge but not the latter (see Exercise 8). It turns out that there are examples where the Jacobi iteration will converge even though the Gauss-Seidel iteration will fail to converge.

EXAMPLE 7.29: Using MATLAB's `eig` function for finding eigenvalues of a matrix, apply Theorem 7.10 to check to see if it tells us that the linear system of Example 7.26 will always lead to a convergent iteration method with each of the three schemes: Jacobi, Gauss-Seidel, and SOR with $\omega = 0.9$. Then create a plot of the maximum absolute value of the eigenvalues of $(I - B^{-1}A)$ for the SOR method as ω ranges from 0.05 to 1.95 in increments of 0.5, and interpret.

SOLUTION: We enter the relevant matrices into a MATLAB session:

```
>> A=[3 1 -1;4 -10 1;2 1 5];  D=diag(diag(A));
>> L=[0 0 0;-4 0 0;-2 -1 0];  U=D-L-A;  I = eye(3);
>> % Jacobi
>> max(abs(eig(I-inv(D)*A)))          →ans = 0.5374
>> % Gauss Seidel
>> max(abs(eig(I-inv(D-L)*A)))        →ans = 0.3513
>> % SOR omega = 0.9
>> max(abs(eig(I-inv(D/.9-L)*A)))     →ans = 0.1301
```

The three computations give the maximum absolute values of the eigenvalues of $(I - B^{-1}A)$ for each of the three iteration methods. Each maximum is less than one, so the theorem tells us that, whatever initial iteration vector we choose for any of the three schemes, the iterations will always converge to the solution of $Ax = b$. Note that the faster converging methods tend to correspond to smaller maximum absolute values of the eigenvalues of $(I - B^{-1}A)$; we will see a corroboration of this in the next part of this solution.

The next set of commands produces a plot of the maximum value of the eigenvalues of $(I - B^{-1}A)$ for the various values of ω, which is shown in Figure 7.42.

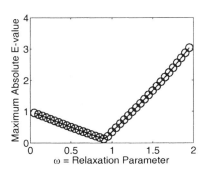

```
>> omega=0.05:.05:1.95;
>> for i=1:length(omega)
   rad(i)=max(abs(eig(I-...
   inv(D/omega(i)-L)*A)));
end
>> plot(omega, rad, 'o-')
```

FIGURE 7.42: Illustration of the maximum absolute value of the eigenvalues of the matrix $I - B^{-1}A$ of Theorem 7.10 for the SOR method (see Table 7.1) for various values of the relaxation parameter ω. Compare with the corresponding number of iterations needed for convergence (Figure 7.40).

The above theorem is quite universal in that it applies to all situations. The drawback is that it relies on the determination of eigenvalues of a matrix. This eigenvalue problem can be quite a difficult numerical problem, especially for very large matrices (the type that we would want to apply iteration methods to). MATLAB's `eig` function may perform unacceptably slowly in such cases and/or may produce inaccurate results. Thus, the theorem has limited practical value. We next give a more practical result that gives a sufficient condition for convergence of both the Jacobi and Gauss-Seidel iterative methods.

Recall that an $n \times n$ matrix A is **strictly diagonally dominant (by rows)** if the absolute values of each diagonal entry is greater than the sum of the absolute values of all other entries in the same row, i.e.,

$$| a_{ii} | > \sum_{\substack{j=i \\ j \neq i}}^{n} | a_{ij} |, \text{ for } i = 1, 2, ..., n .$$

THEOREM 7.11: (*Jacobi and Gauss-Seidel Convergence Theorem*) Assume that A is a nonsingular (square matrix). If A is strictly diagonally dominant, then the Jacobi and Gauss-Seidel iterations will converge (to the solution of $Ax = b$) for any initial seed vector $x^{(0)}$.

Usually, the more diagonally dominant A is, the faster the rate of convergence will be. Note that the coefficient matrix of Example 7.26 is strictly diagonally dominant, so that Theorem 7.11 tells us that no matter what initial seed vector $x^{(0)}$ we started with, both the Jacobi and Gauss-Seidel iteration schemes would produce a sequence that converges to the solution. Although we knew this already from Theorem 7.10, note that, unlike the eigenvalue condition, the strict diagonal dominance was trivial to verify (by inspection). There are matrices A that are not strictly diagonally dominant but for which both Jacobi and Gauss-Seidel schemes will always converge. For an outline of a proof of the Jacobi part of the above result, see Exercise 23.

For the SOR method (for other values of ω than 1) there does not seem to be such a simple useful criterion. There is, however, another equivalent condition in the case of a symmetric coefficient matrix A with positive diagonal entries. Recall that an $n \times n$ matrix A is symmetric if $A = A'$; also A is **positive definite** provided that $x'Ax > 0$ for any nonzero $n \times 1$ vector x.

THEOREM 7.12: (*SOR Convergence Theorem*) Assume that A is a nonsingular (square matrix). Assume that A is symmetric and has positive diagonal entries. For any choice of relaxation parameter $0 < \omega < 2$, the SOR iteration will converge (to the solution of $Ax = b$) for any initial seed vector $x^{(0)}$, if and only if A is positive definite.

Matrices that satisfy the hypotheses of Theorems 7.11 and 7.12 are actually quite common in numerical solutions of differential equations. We give a typical example of such a matrix shortly.   For reference, we collect in the following theorem two equivalent formulations for a symmetric matrix to be positive definite, along with some necessary conditions for a matrix to be positive definite. Proofs can be found in [Str-88], p. 331 (for the equivalences) and [BuFa-01], p. 401 (for the necessary conditions).

THEOREM 7.13: (*Positive Definite Matrices*)   Suppose that A is a symmetric $n \times n$ matrix.
(a) The following two conditions are each equivalent to A being positive definite:
    (i) All eigenvalues of A are positive, or
    (ii) The determinants of all upper-left submatrices of A have positive determinants.
(b) If A is positive definite, then each of the following conditions must hold:
    (i) A is nonsingular.
    (ii) $a_{ii} > 0$ for $i = 1, 2, ..., n$.
    (iii) $a_{ii} a_{jj} > a_{ij}^2$ whenever $i \neq j$.

We will be working next with a certain class of sparse matrices, which is typical of those that arise in finite difference methods for solving partial differential equations.   We study such problems and concepts in detail in [Sta-05]; here we only very briefly outline the connection.

The matrix we will analyze arises in solving the so-called Poisson boundary value problem on the two-dimensional unit square $\{(x, y): 0 \leq x, y \leq 1\}$, which asks for the determination of a function $u = u(x, y)$ that satisfies the following partial differential equation and boundary conditions:

$$\begin{cases} -\Delta u = f(x, y), \text{ inside the square: } 0 < x, y < 1 \\ u(x, y) = 0, \text{ on the boundary: } x = 0, 1 \text{ or } y = 0, 1 \end{cases}$$

Here Δu denotes the Laplace differential operator $\Delta u = u_{xx} + u_{yy}$.   The finite difference method "discretizes" the problem into a linear system.   If we use the same number N of grid points both on the x- and the y-axis, the linear system $Ax = b$ that arises will have the $N^2 \times N^2$ coefficient matrix shown in (63).

In our notation, the partition lines break the $N^2 \times N^2$ matrix up into smaller $N \times N$ block matrices (N^2 of them).   The only entries indicated are the nonzero entries that occur on the five diagonals shown.   Because of their importance in applications, matrices such as the one in (63) have been extensively studied in the context of iterative methods.   For example, the following result contains some very practical and interesting results about this matrix.

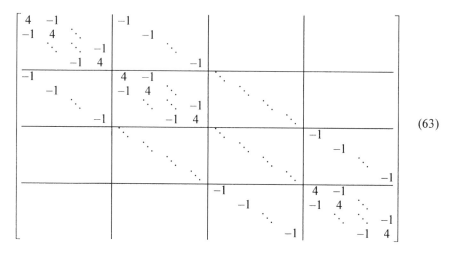

$$(63)$$

PROPOSITION 7.14: Let A be the $N^2 \times N^2$ matrix (63).

(a) A is positive definite (so SOR will converge by Theorem 7.12) and the optimal relaxation parameter ω for an SOR iteration scheme for a linear system $Ax = b$ is as follows:

$$\omega = \frac{2}{1 + \sin\left(\frac{\pi}{N+1}\right)}.$$

(b) With this optimal relaxation parameter, the SOR iteration scheme works on order of N times as fast as either the Jacobi or Gauss-Seidel iteration schemes. More precisely, the following quantities R_J, R_{GS}, R_{SOR} indicate the approximate number of iterations that each of these three schemes would need, respectively, to reduce the error by a factor of $1/10$:

$$R_J \approx 0.467(N+1)^2, \ R_{GS} = \tfrac{1}{2}R_J \approx 0.234(N+1)^2, \text{ and } R_{SOR} \approx 0.367(N+1).$$

In our next example we compare the different methods by solving a very large fictitious linear system $Ax = b$ involving the matrix (63). This will allow us to make exact error comparisons with the true solution.

A proof of the above proposition, as well as other related results, can be found in Section 8.4 of [StBu-93]. Note that since A is not (quite) strictly diagonally dominant, the Jacobi/Gauss-Seidel convergence theorem (Theorem 7.11) does not apply. It turns out that the Jacobi iteration scheme indeed converges, along with SOR (and, in particular, the Gauss-Seidel method); see Section 8.4 of [StBu-93].

Consider the matrix A shown in (63) with $N = 50$. The matrix A has size 2500×2500 so it has 6.25 million entries. But of these only about $5N^2 = 12,500$ are nonzero. This is about 0.2% of the entries, so A is quite sparse.

EXERCISE FOR THE READER 7.30: Consider the problem of multiplying the matrix A in (63) (using $N = 50$) by the vector $x = [1\ 2\ 1\ 2\ 1\ 2\ \cdots\ 1\ 2]'$.

(a) Compute (by hand) the vector $b \equiv Ax$ by noticing the patterns present in the multiplication.

(b) Get MATLAB to compute $b = Ax$ by first creating and storing the matrices A and x and performing a usual matrix multiplication. Use `tic/toc` to time the parts of this computation.

(c) Store only the five nonzero diagonals of A (as column vectors): d, a1, aN, b1, bN (d stands for main diagonal, a for above main diagonal, b for below main diagonal). Recompute b by suitably manipulating these 5 vectors in conjunction with x Use `tic/toc` to time the computation and compare with that in part (b).

(d) Compare all three answers. What happens to the three methods if we bump N up to 100?

Shortly we will give a general development on the approach hinted at in part (c) of the above Exercise for the Reader 7.30.

As long as the coefficient matrix A is not too large to be stored in a session, MATLAB's left divide is quite an intelligent linear system solver. It has special more advanced algorithms to deal with positive definite coefficient matrices, as well as with other special types of matrices. It can numerically solve systems about as large as can be stored; but the accuracy of the numerical solutions obtained depends on the condition number of the matrix, as explained earlier in this chapter.[21] The next example shows that even with all that we know about the optimal relaxation parameter for the special matrix (63), MATLAB's powerful left divide will still work more efficiently for a very large linear system than our SOR program. After the example we will remedy the situation by modifying the SOR program to make it more efficient for such banded sparse matrices.

EXAMPLE 7.30: In this example we do some comparisons in some trial runs of solving a linear system $Ax = b$ where A is the matrix of (63) with $N = 50$, and the vectors x and b are as in the preceding Exercise for the Reader 7.30. Having the

---

[21] Depending on the power of the computer on which you are running MATLAB's as well as the other processes being run, computation times and storage capacities can vary. At the time of writing this section on the author's 1.6 MHz, 256 MB RAM Pentium IV PC, some typical limits, for random (dense) matrices, are as follows: The basic Gaussian elimination (Program 7.6) starts taking too long (toward an hour) when the size of the coefficient matrix gets larger than 600×600; for it to take less than about one minute the size should be less than about 250×250. Before memory runs out, on the other hand, matrices of sizes up to about 6000×6000 can be stored, and MATLAB's left divide can usually (numerically) solve them in a reasonable amount of time (provided that the condition number is moderate). To avoid redundant storage problems, MATLAB does have capabilities of storing sparse matrices. Such functionality introduced at the end of this section. Taking advantage of the structure of sparse banded matrices (which are the most important ones in numerical differential equations) will enable us to solve many such linear systems that are quite large, say up to about $50,000 \times 50,000$. Such large systems often come up naturally in numerical differential equations.

exact solution will allow us to look at the exact errors resulting from any of the methods.

(a) Solve the linear system by using MATLAB's left divide (Gaussian elimination). Record the computation time and error of the computed solution.

(b) Solve the system using the Gauss-Seidel program `gaussseidel` (Program 7.7) with the default number of iterations and initial vector. Record the computation time and error. Repeat using 200 iterations.

(c) Solve again using the SOR program `sorit` (from Exercise for the Reader 7.28) with the optimal relaxation parameter ω given in Proposition 7.14. Record the computation time and error. Repeat using 200 iterations.

(d) Reconcile the data of parts (b) and (c) with the results of part (c) of Proposition 7.14.

SOLUTION: We first create and store the relevant matrices and vectors:

```
>> x=ones(2500,1); x(2:2:2500,1)=2;
>> A=4*eye(2500);
>> v1=-1*ones(49,1); v1=[v1;0]; %seed vector for sub/super diagonals
>> secdiag=v1;
>> for i=1:49
     if i<49
        secdiag=[secdiag;v1];
     else
        secdiag=[secdiag;v1(1:49)];
     end
end

>> A=A+diag(secdiag,1)+diag(secdiag,-1)-diag(ones(2450,1),50)...
            -diag(ones(2450,1),-50);
>> b=A*x;
```

Part (a):
```
>> tic, xMATLAB=A\b; toc      →elapsed_time = 9.2180
>> max(xMATLAB-x)             →ans = 6.2172e - 015
```

Part (b):
```
>> tic, [xGS, k, diff]=gaussseidel(A,b); toc
→Gauss-Seidel iteration failed to converge.→elapsed_time = 181.6090
>> max(abs(xGS-x))                →ans = 1.4353

>> tic, [xGS2, k, diff] = gaussseidel(A,b,zeros(size(b)), 1e-10,200);
toc
→Gauss-Seidel iteration failed to converge.→elapsed_time = 374.5780
>> max(abs(xGS2-x))→ans = 1.1027
```

Part (c):
```
>> tic, [xSOR, k, diff]=sorit(A,b, 2/(1+sin(pi/51))); toc
→SOR iteration failed to converge.   →elapsed_time = 186.7340
>> max(abs(xSOR-x))              →ans = 0.0031

>> tic, [xSOR2, k, diff]=sorit(A,b, 2/(1+sin(pi/51)),...
            zeros(size(b)), 1e-10,200); toc
→SOR iteration failed to converge.   →elapsed_time = 375.2650
```

```
>> max(abs(xSOR2-x))        →ans = 1.1885e - 008
```

Part (d): The above data shows that our iteration programs pale in performance when compared with MATLAB's left divide (both in time and accuracy). The attentive reader will realize that both iteration programs do not take advantage of the sparseness of the matrix (63). They basically run through all of the entries in this large matrix for each iteration. One other thing that should be pointed out is that the above comparisons are somewhat unfair because MATAB's left divide is a compiled code (built-into the system) whereas the other programs were interpreted codes (created as external programs—M-files). After this example, we will show a way how to make these programs perform more efficiently by taking advantage of the special banded structure of such matrices. The resulting modified programs will then perform more efficiently than MATLAB's left divide, at least for the linear system of this example.

Using $N = 50$ in part (b) of Proposition 7.14, we see that in order to cut errors by a factor of 10 with the Gauss-Seidel method, we need approximately $0.234 \cdot 51^2 \approx 609$ additional iterations, but for the SOR with optimal relaxation parameter the corresponding number is only $.367 \cdot 51 \approx 18.7$. This corroborates well with the experimental data in parts (b) and (c) above. For Gauss-Seidel, we first used 100 iterations and then used 200. The theory tells us we would need over 600 more iterations to reduce the error by 90%. Using 100 more iterations resulted in a reduction of error by about 23%. On the other hand, with SOR, the 100 additional iterations gave us approximately $100/18.7 \approx 5.3$ reductions in the errors each by factors of 1/10, which corresponds nicely to the (exact) error shrinking from about 3e-3 to 1e-8 (literally, about 5.3 decimal places!).

In order to take advantage of sparsely banded matrices in our iteration algorithms, we next record here some elementary observations regarding multiplications of such matrices by vectors. MATLAB is enabled with features to easily manipulate and store such matrices. We will now explore some of the underlying concepts and show how exploiting the special structure of some sparse matrices can greatly expand the sizes of linear systems that can be effectively solve. MATLAB has its own capability for storing and manipulating general sparse matrices; at the end of the section we will discuss how this works.

The following (nonstandard) mathematical notations will be convenient for the present purpose: For two vectors v, w that are both either row- or column-vectors, we let $v \otimes w$ denote their juxtaposition. For the pointwise product of two vectors of the same size, we use the notation $v \odot w = [v_i w_i]$. So, for example, if $v = [v_1 \ v_2 \ v_3]$, and $w = [w_1 \ w_2 \ w_3]$ are both 3×1 row vectors, then $v \otimes w$ is the 6×1 row vector $[v_1 \ v_2 \ v_3 \ w_1 \ w_2 \ w_3]$ and $v \odot w$ is the 3×1 row vector $[v_1 w_1 \ v_2 w_2 \ v_3 w_3]$. We also use the notation $\mathbf{0}_n$ to denote the zero vector with n components. We will be using this last notation only in the context of juxtapositions so that whether 0_n is meant to be a row vector or a column vector will be clear from the context.[22]

LEMMA 7.15: (a) Let S be an $n \times n$ matrix whose kth superdiagonal ($k \geq 1$) is made of the entries (in order) of the $(n-k) \times 1$ vector v, i.e.,

$$S = \begin{bmatrix} 0 & 0 & \cdots & 0 & v_1 & 0 & & \cdots & 0 \\ 0 & 0 & 0 & \cdots & 0 & v_2 & 0 & & \vdots \\ \vdots & \vdots & 0 & \ddots & \cdots & 0 & v_3 & \ddots & 0 \\ & \vdots & & 0 & & \ddots & \ddots & \ddots & 0 \\ & & & & \ddots & & & 0 & v_{n-k} \\ \vdots & & & & & \ddots & & & 0 \\ \vdots & \vdots & & & & & \ddots & \ddots & \vdots \\ 0 & \vdots & \vdots & & & & 0 & 0 & 0 \\ 0 & 0 & 0 & 0 & \cdots & & \cdots & 0 & 0 & 0 \end{bmatrix},$$

where $v = [v_1 \ v_2 \ \cdots \ v_{n-k}]$. (In MATLAB's notation, the matrix S could be entered as `diag(v,k)`, once the vector v has been stored.) Let $x = [x_1 \ x_2 \ x_3 \ \cdots \ x_n]'$ be any $n \times 1$ column vector. The following relation then holds:

$$Sx = (v \otimes 0_k) \odot ([x_{k+1} \ x_{k+2} \ \cdots \ x_n] \otimes 0_k). \tag{64}$$

(b) Analogously, if S is an $n \times n$ matrix whose kth subdiagonal ($k \geq 1$) is made of the entries (in order) of the $(n-k) \times 1$ vector v, i.e.,

$$S = \begin{bmatrix} 0 & 0 & \cdots & & & & \cdots & 0 & 0 \\ \vdots & 0 & 0 & \cdots & & & & & 0 \\ 0 & \vdots & 0 & 0 & \cdots & & & & \vdots \\ v_1 & 0 & \vdots & 0 & 0 & \cdots & & & \\ 0 & v_2 & 0 & & \ddots & \ddots & & & \\ \vdots & 0 & v_3 & 0 & & \ddots & \ddots & & \\ \vdots & & \ddots & \ddots & & \ddots & \ddots & \ddots & \vdots \\ 0 & \vdots & \vdots & 0 & v_{n-k-1} & 0 & 0 & \cdots & 0 \\ 0 & 0 & 0 & 0 & 0 & v_{n-k} & 0 & \cdots & 0 \end{bmatrix},$$

where $v = [v_1 \ v_2 \ \cdots \ v_{n-k}]$ and $x = [x_1 \ x_2 \ x_3 \ \cdots \ x_n]'$ is any $n \times 1$ column vector, then

$$Sx = (0_k \otimes v) \odot (0_k \otimes [x_1 \ x_2 \ \cdots \ x_k]). \tag{65}$$

The proof of the lemma is left as Exercise 21. The lemma can be easily applied to greatly streamline all of our iteration programs for sparse banded matrices. The

---

[22] We point out that these notations are not standard. The symbol " $\otimes$ " is usually reserved for the so-called *tensor product*.

next exercise for the reader will ask the reader to perform this task for the SOR iteration scheme.

EXERCISE FOR THE READER 7.31: (a) Write a function M-file with the following syntax:

```
[x,k,diff] = sorsparsediag(diags, inds, b, omega, x0, tol, kmax)
```

that will perform the SOR iteration to solve a nonsingular linear system $Ax = b$ in which the coefficient matrix A has entries only on a sparse set of diagonals. The first two input variables are diags, an $n \times j$ matrix where each column consists of the entries of A on one of its diagonals (with extra entries at the end of the column being zeros), and inds, a $1 \times j$ vector of the corresponding set of indices for the diagonals (index zero corresponds to the main diagonal and should be first). The remaining input and output variables will be exactly as in the M-file sorit of Exercise for the Reader 7.28.   The program should function just like the sorit M-file, with the only exceptions being that the stopping criterion for the norms of the difference of successive iterates should now be determined by the infinity norm[23] and the default number of iterations is now 1000. The algorithm should, of course, be based on formula (59) for the SOR iteration, but the sum need only be computed over the index set (inds) of the nonzero diagonals. To this end, the above lemma should be used in creating this M-file so that it will avoid unnecessary computations (with zero multiplications) as well as storage problems with large matrices.
(b) Apply the program to redo part (c) of Example 7.30.

EXAMPLE 7.31: (a) Invoke the M-file sorsparsediag of the preceding exercise for the reader to obtain SOR numerical solutions of the linear system of Example 7.30 with error goal 5e-15 (roughly MATLAB's machine epsilon) and compare the necessary runtime with that of MATLAB's left divide, which was recorded in that example.
(b) Next use the program to solve the linear system $Ax = b$ with A as in (63) with $N = 300$ and $b = [1\ 2\ 1\ 2\ ...\ 1\ 2]'$. Use the default tolerance 1e-10, then, looking at the last norm difference (estimate for the actual error) use Proposition 7.14 to help to see how much to increase the maximum number of iterations to ensure convergence of the method.   Record the runtimes.   The size of A is $90,000 \times 90,000$ and so it has over 8 billion entries. Storage of such a matrix would require a supercomputer.

SOLUTION: Part (a): We first need to store the appropriate data for the matrix A. Assuming the variables created in the last example are still in our workspace, this can be accomplished as follows:

---

[23] The infinity norm of a vector x is simply, in MATLAB's notation, max(abs(x)). This is a rather superficial change in the M-file, merely to allow easier performance comparisons with MATLAB's left divide system solver.

```
>> diags=zeros(2500,5);
>> diags(:,1)=4*ones(2500,1);
>> diags(1:2499,2:3)=[secdiag secdiag];
>> diags(1:2450,4:5)=-ones(2450,2);
>> inds=[0 1 -1 50 -50];

>> tic, [xSOR, k, diff]=sorsparsediag(diags, inds,b,...
            2/(1+sin(pi/51)), zeros(size(b)), 5e-15); toc
→SOR iteration has converged in 308 iterations  →elapsed_time = 1.3600
>> max(abs(xSOR-x))                              →ans = 2.3537e - 014
```

Our answer is quite close to machine precision (there were roundoff errors) and
the answer obtained by MATLAB's left divide. The runtime of the modified SOR
program is now, however, significantly smaller than that of the MATLAB solver.
We will see later, however, that when we store the matrix A as a sparse matrix (in
MATLAB's syntax), the left divide method will work at comparable speed to our
modified SOR program.

Part (b): We first create the input data by suitably modifying the code in part (a):

```
>> b=ones(90000,1); b(2:2:90000,1)=2;
>> v1=-1*ones(299,1); v1=[v1;0]; %seed vector for sub/super diagonals
>> secdiag=v1;
>> for i=1:299
       if i<299
          secdiag=[secdiag;v1];
       else
          secdiag=[secdiag;v1(1:299)];
       end
   end
>> diags=zeros(90000,5);
>> diags(:,1)=4*ones(90000,1);
>> diags(1:89999,2:3)=[secdiag secdiag];
>> diags(1:89700,4:5)= [-ones(89700,1) -ones(89700,1)];
>> inds=[0 1 -1 300 -300];
>>tic, [xSORbig, k, diff]=sorsparsediag(diags, inds,b,...
            2/(1+sin(pi/301))); toc
→SOR iteration failed to converge.   →elapsed_time = 167.0320
>> diff(k-1)                          →ans = 1.3845e - 005
```

We need to reduce the current error by a factor of 1e-5. By Proposition 7.14, this
means that we should bump up the number of iterations by a bit more than
$5R_{SOR} \approx 5 \cdot 0.367 \cdot 301 \approx 552$. Resetting the default number of iterations to be 1750
(from 1000) should be sufficient. Here is what transpires:

```
>> tic, [xSORbig, k, diff]=sorsparsediag(diags, inds,b,...
            2/(1+sin(pi/301)), zeros(size(b)), 1e-10, 1750); toc
→SOR iteration has converged in 1620 iterations  →elapsed_time = 290.1710
>> diff(k - 1)                                    →ans = 1.0550e - 010
```

We have thus solved this extremely large linear system, and it only took about
three minutes!

As promised, we now give a brief synopsis of some of MATLAB's built-in, state-of-the-art iterative solvers for linear systems $Ax = b$. The methods are based on more advanced concepts that we briefly indicated and referenced earlier in the section. Mathematical explanations of how these methods work would lie outside the focus of this book. We do, however, outline the basic concept of **preconditioning**. As seen early in this section, iterative methods are very sensitive to the particular form of the coefficient matrix (we gave an example where simply switching two rows of A resulted in the iterative method diverging when it originally converged). An invertible matrix (usually positive definite) M is used to precondition our linear system when we apply the iterative method instead to the equivalent system: $M^{-1}Ax = M^{-1}b$. Often, preconditioning a system can make it more suitable for iterative methods. For details on the practice and theory of preconditioning, we refer to part II of [Gre-97], which includes, in particular, preconditioning techniques appropriate for matrices that arise in solving numerical PDEs. See also Part IV of [TrBa-97].

Here is a detailed description of MATLAB's function for the so-called **preconditioned conjugate gradient method**, which assumes that the coefficient matrix A is symmetric positive definite.

x=pcg(A,b,tol,kmax,M1,M2,x0) →	Performs the preconditioned gradient method to solve the linear system $Ax = b$, where the $N \times N$ coefficient matrix A must be symmetric positive definite and the preconditioner $M = M1 * M2$. Only the first two input variables are required; any tail sequence of input variables can be omitted. The default values of the optional variables are as follows: tol = 1e-6, kmax = min(N,20), M1 = M2 = I (identity matrix), and x0 = the zero vector. Setting any of these optional input variables equal to [ ] gives them their default values.

[x, flag] =pcg(A,b,tol,kmax,M1,M2,x0) →	Works as above but returns additional output flag: flag = 0 means pcg converged to the desired tolerance tol within kmax iterations; flag = 1 means pcg iterated kmax times but did not converge. For a detailed explanation of other flag values, type help pcg.

We point out that with the default values $M1 = M2 = I$, there is no conditioning and the method is called the conjugate gradient method.

Another powerful and and more versatile method is the **generalized minimum residual method (GMRES)**. This method works well for general (nonsymmetric) linear systems. MATLAB's syntax for this function is similar to the above, but there is one additional (optional) input variable:

x=gmres(A,b,restart,tol, kmax,M1,M2,x0) →	Performs the generalized minimum residual method to solve the linear system $Ax = b$, with preconditioner M = M1*M2. Only the first two input variables are required, any tail sequence of input variables can be omitted. The default values of the optional variables are as follows: restart = [ ] (unrestarted method) tol = 1e-6, kmax = min(N,20), M1 = M2 = I (identity matrix), and x0 = the zero vector. Setting any of these optional input variables equal to [ ], gives them their default values. An optional second output variable flag will function in a similar fashion as with pcg.

EXAMPLE 7.32: (a) Use pcg to resolve the linear system of Example 7.30, with the default settings and flag. Repeat by resetting the tolerance at 1e-15 and the maximum number of iterations to be 100 and then 200. Record the runtimes and compare these and the errors to the results for the SOR program of the previous example.
(b) Repeat part (a) with gmres.

SOLUTION: Assume that the matrices and vectors remain in our workspace (or recreate them now if necessary); we need only follow the above syntax instructions for pcg:

Part (a):
```
>> tic, [xpcg, flagpcg]=pcg(A,b); toc        →elapsed_time = 3.2810
>> max(abs(xpcg-x))                           →ans = 1.5007
>> flagpcg                                    →flagpcg = 1
```

The flag being = 1 means after 20 iterations, pcg did not converge within tolerance (1e-5), a fact that we knew from the exact error estimate.

```
>> tic, [xpcg, flagpcg]=pcg(A,b,5e-15, 100); toc →elapsed_time = 15.8900
>> max(abs(xpcg-x))                               →ans = 4.5816e - 006
>> flagpcg                                        →flagpcg = 1

>> tic, [xpcg, flagpcg]=pcg(A,b,5e-15, 200); toc →elapsed_time = 29.7970
>> flagpcg                                        →flagpcg = 0
>> max(abs(xpcg-x))                               →ans = 3.2419e - 014
```

The flag being = 0 in this last run shows we have convergence. The max norm is a different one from the 2-norm used in the M-file; hence the slight discrepancy. Notice the unconditioned conjugate gradient method converged in fewer iterations than did the optimal SOR method, and in much less time than the original sorit program. The more efficient sorsparsediag program, however, got the solution in by far the shortest amount of real time. Later, we will get a more equitable comparison when we store A as a sparse matrix.

Part (b):
```
>> tic, [xgmres, flaggmres]=gmres(A,b, [],[], 200); toc
```

```
>> tic, [xgmres, flaggmres]=gmres(A,b); toc    →elapsed_time = 2.3280
>> max(abs(xgmres-x))                          →ans = 1.5002
>> flaggmres                                   →flaggmres = 1

>> tic, [xgmres, flaggmres]=gmres(A,b, [],5e-15, 100); toc
→elapsed_time = 17.2820
>> max(abs(xgmres-x))      →ans = 6.9104e - 006

>> tic, [xgmres, flaggmres]=gmres(A,b, [],5e-15, 200); toc
→elapsed_time = 37.1250
>> max(abs(xgmres-x))      →ans = 9.2037e - 013
```

The results for GMRES compare well with those for the preconditioned conjugate gradient method. The former method converges a bit more slowly in this situation. We remind the reader that the conjugate gradient method is ideally suited for positive definite matrices, like the one we are dealing with.

Figure 7.43 gives a nice graphical comparison of the relative speeds of convergence of the five iteration methods that have been introduced in this section. An exercise will ask the reader to reconstruct this MATLAB graphic.

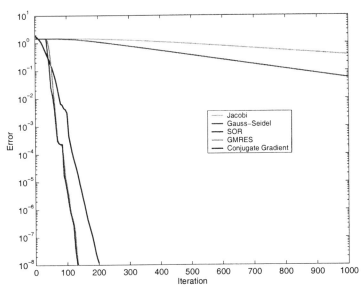

FIGURE 7.43: Comparison of the convergence speed of the various iteration methods in the solution of the linear system $Ax = b$ of Example 7.30 where the matrix A is the matrix (63) of size 2500×2500. In the SOR method the optimal relaxation parameter of Proposition 7.14 was used. Since we did not invoke any conditioning, the preconditioned conjugate gradient method is simply referred to as the conjugate gradient method. The errors were measured in the infinity norm.

In the construction of the above data, the program `sorsparsediag` was used to get the SOR data and despite the larger number of iterations than GMRES and the conjugate gradient method, the SOR data was computed more quickly. The sorsparsediag program is easily modified to construct similar programs for the Jacobi and Gauss-Seidel iterations (of course Gauss-Seidel could simply be done by setting $\omega = 0$ in the SOR program), and such programs were used to get the data for these iterations. Note that the GMRES and conjugate gradient methods take several iterations before errors start to decrease, unlike the SOR method, but they soon catch up. Note also the comparable efficiencies between the GMRES and conjugate gradient methods.

We close this chapter with a brief discussion of how to store and manipulate sparse matrices directly with MATLAB. Sparse matrices in MATLAB can be stored using three vectors: one for the nonzero entries, the other two for the corresponding row and column indices. Since in many applications sparse matrices will be banded, we will explain only a few commands useful for the creation and storage of such sparse matrices. Enter help sparse for more detailed information. To this end, suppose that we have an $n \times n$ banded matrix A and we wish to store it as a sparse matrix. Let the indices corresponding to the nonzero bands (diagonals) of A have numbers stored in a vector d (so the size of d is the number of bands, 0 corresponds to the main diagonal, positive numbers mean above the main diagonal, negatives numbers mean below). Letting p denote the length of the vector d we form a corresponding $n \times p$ matrix Diags, containing as its columns the corresponding bands (diagonals) of A. When columns are longer than the bands they replace (this will be the case except for main diagonal), super (above) diagonals should be put on the lower portion of Diags and sub (below) diagonals on the upper portion of Diags, with remaining entries on the column being set to zero.

S=spdiags(Diags,d,n,n) →	This command creates a sparse matrix data type S, of size $n \times n$ provided that d is a vector of diagonal indices (say there are p), and Diags is the corresponding $n \times p$ matrix whose columns are the diagonals of the matrix (arranged as explained above).
full(S) →	Converts a sparse matrix data type back to its usual "full" form. This command is rarely used in dealing with sparse matrices as it defeats their purpose.

A simple example will help shed some light on how MATLAB deals with sparse data types. Consider the matrix $A = \begin{bmatrix} 0 & 1 & 0 & 0 \\ 4 & 0 & 2 & 0 \\ 0 & 5 & 0 & 3 \\ 0 & 0 & 6 & 0 \end{bmatrix}$. The following MATLAB commands will store A as a sparse matrix:

```
>> d=[-1 1];  Diags=[4 5 6 0; 0 1 2 3]';
>> S=spdiags(Diags,d,4,4)
→S =      (2,1)    4       (2,3)    2
          (1,2)    1       (4,3)    6
          (3,2)    5       (3,4)    3
```

The display shows the storage scheme.   Let's compare with the usual form:

```
>> full(S)
→ ans =      0   1   0   0
             4   0   2   0
             0   5   0   3
             0   0   6   0
```

The key advantage of sparse matrix storage in MATLAB is that if A is stored as a sparse matrix S, then to solve a linear system $Ax = b$, MATLAB's left divide operation x=S\b takes advantage of sparsity and can greatly increase the size of (sparse) problems we can solve.   In fact, at most all of MATLAB's matrix functions are able to operate on sparse matrix data types.   This includes MATLAB's iterative solvers pcg, gmres, etc. We invite the interested reader to perform some experiments and discover the additional speed and capacity that taking advantage of sparsity can afford. We end with an example of a rematch of MATLAB's left divide against our sorsparsediag program, this time allowing the left divide method to accept a sparse matrix. The results will be quite illuminating.

EXAMPLE 7.33: We examine the large ($10,000 \times 10,000$) system $Ax = b$, where A is given by (63) with $N = 100$, and $x = (1\ 1\ 1\ \cdots\ 1)'$. By examining the matrix multiplication we see that

$$b = Ax = (2\ 1\ 1\ \cdots 1\ 2|1\ 0\ 0\ \cdots 0\ 1|\cdots|1\ 0\ 0\ \cdots 0\ 1|2\ 1\ 1\ \cdots 1\ 2)'.$$

We thus have a linear system with which we can easily obtain the exact error of any approximate solution.
(a) Solve this system using MATLAB's left divide and by storing A as a sparse matrix.   Use tic/toc to track the computation time (on your computer); compute the error as measured by the infinity norm (i.e., as the maximum difference of any component of the computed solution with the exact solution).
(b) Solve the system using the sorsparsediag M-file of Exercise for the Reader 7.31. Compute the time and errors as in part (a) and compare.

SOLUTION: Part (a): We begin by entering the parameters (for (63)), creating the needed inputs for spdiags, and then using the latter to store A as a sparse matrix.

```
>> N=200; n=N^2; d=[-N -1 0 1 N];, dia=4*ones(1,n);
>> seed1=-1*ones(1,N-1); v1=[seed1 0];
for i=1:N-1, if i<N-1, v1 = [v1 [seed1 0]];, else, v1 = [v1 seed1];
>> end, end
>> b1=[v1 0]; a1=[0 v1]; %below/above 1 unit diagonals
>> %Next here are the below/above N unit diagonals
>> bN=[-ones(1,n-N)  zeros(1,N)];
>> aN=[zeros(1,N)   -ones(1,n-N)  ];
>> %Now we can form the n by 5 Diags matrix.
>> Diags=[bN; b1; dia; a1; aN]';
>> S=spdiags(Diags,d,n,n); %S is the sparsely stored matrix A
```

```
>> %We use a simple iteration to contruct the inhomogeneity
>> %vector b.
>> bseed1=ones(1,N);, bseed1([1 N])=[2 2]; %2 end pieces
>> bseed2=bseed1-ones(1,N); %N-2 middle pieces
>> b=bseed1; for k=1:N-2, b=[b bseed2];, end, b=[b bseed1];
>> b=b';
>> tic, xLD=S\b;, toc      →Elapsed time is 0.250000 seconds.
>> x=ones(size(xLD));
>> max(x-xLD)              → ans =1.0947e-013 (Exact Error)
```

Part (b): The syntax and creation of input variables is just as we did in Example 7.31.

```
>> d=[0 -N N -1 1];,  diags=zeros(n,5);
>> diags(:,1)=dia; diags(:,2:3)=[bN' bN']; diags(:,4:5)=[b1' b1'];
>> tic, [xSOR, k, diff]=sorsparsediag(diags, d,b,...
              2/(1+sin(pi/101))); toc
→Elapsed time is 8.734000 seconds.
>> max(x-xSOR)            → 3.9102e-012 (Exact Error)
```

Thus, now that the inputted data structures are similar, MATLAB's left divide has transcended our `sorsparsediags` program both in performance time and in accuracy. The reader is invited to perform further experiments with sparse matrices and MATLAB's iterative solvers.

EXERCISES 7.7

1. For each of the following data for a linear system $Ax = b$, perform the following iterations using the zero vector as the initial vector.
   (a) Use Jacobi iteration until the error (as measured by the infinity norm of the difference of successive iterates) is less than 1e-10, if this is possible. In cases where the iteration does not converge, try rearranging the rows of the matrices to attain convergence (through all $n!$ rearrangements, if necessary). Find the norm of the exact error (use MATLAB's left divide to get the "exact" solutions of these small systems).
   (b) Repeat part (a) with the Gauss-Seidel iteration.

(i) $A = \begin{bmatrix} 6 & -1 \\ -1 & 6 \end{bmatrix}$, $b = \begin{bmatrix} 1 \\ 2 \end{bmatrix}$.
(ii) $A = \begin{bmatrix} 6 & -1 & 0 \\ -1 & 6 & -1 \\ 0 & -1 & 6 \end{bmatrix}$, $b = \begin{bmatrix} 1 \\ 2 \\ 1 \end{bmatrix}$.

(iii) $A = \begin{bmatrix} -2 & 5 & 4 \\ 6 & 2 & -3 \\ 1 & 1 & -1 \end{bmatrix}$, $b = \begin{bmatrix} 1 \\ 2 \\ 3 \end{bmatrix}$.
(iv) $A = \begin{bmatrix} 2 & 1 & 0 & 0 \\ 2 & 4 & 1 & 0 \\ 0 & 4 & 8 & 2 \\ 0 & 0 & 8 & 16 \end{bmatrix}$, $b = \begin{bmatrix} 4 \\ -2 \\ 1 \\ 3 \end{bmatrix}$.

2. For each of the following data for a linear system $Ax = b$, perform the following iterations using the zero vector as the initial vector. Determine if the Jacobi and Gauss-Seidel iterations converge. In cases of convergence, produce a graph of the errors (as measured by the infinity norm of the difference of successive iterates) versus the number of iterations, that contains both the Jacobi iteration data as well as the Gauss-Seidel data. Let the errors go down to 10^{-10}. .

(i) $A = \begin{bmatrix} 5 & 0 \\ 2 & 4 \end{bmatrix}$, $b = \begin{bmatrix} -1 \\ 3 \end{bmatrix}$.
(ii) $A = \begin{bmatrix} 10 & 2 & -1 \\ 2 & 10 & 2 \\ -1 & 2 & 10 \end{bmatrix}$, $b = \begin{bmatrix} 1 \\ 2 \\ 3 \end{bmatrix}$.

$$
\text{(iii) } A = \begin{bmatrix} 7 & 5 & 4 \\ 3 & 2 & 1 \\ 2 & 8 & 21 \end{bmatrix}, \quad b = \begin{bmatrix} 1 \\ 0 \\ 5 \end{bmatrix}. \quad \text{(iv) } A = \begin{bmatrix} 3 & 1 & 0 & 0 \\ 1 & 9 & 1 & 0 \\ 0 & 1 & 27 & 1 \\ 0 & 0 & 1 & 81 \end{bmatrix}, \quad b = \begin{bmatrix} 4 \\ 3 \\ 2 \\ 1 \end{bmatrix}.
$$

3.   (a) For each of the linear systems specified in Exercise 1, run a set of SOR iterations with initial vector the zero vector by letting the relaxation parameter run form 0.05 to 1.95 in increments of 0.5. Use a tolerance of 1e-6, but a maximum of 1000 iterations. Plot the number of iterations versus the relaxation parameter ω.
     (b) Using MATLAB's `eig` function, let the relaxation parameter ω run through the same range 0.05 to 1.95 in increments of 0.5, and compute the maximum absolute value of the eigenvalues of the matrix $I - B^{-1}A$ where the matrix B is as in Table 7.1 (for the SOR iteration). Plot these maximum versus ω, compare and comment on the relationship with the plot of part (a) and Theorem 7.10.

4.   Repeat both parts (a) and (b) for each of the linear systems $Ax = b$ of Exercise 2.

5.   For the linear system specified in Exercise 2 (iv), produce graphs of the exact errors of each component of the solution: x_1, x_2, x_3, x_4 as a function of the iteration. Use the zero vector as the initial iterate. Measure the errors as the absolute values of the differences with the corresponding components of the exact solution as determined using MATLAB's left divide. Continue with iterations until the errors are all less than 10^{-10}. Point out any observations.

6.   (a) For which of the linear systems specified in Exercise 1(i)–(iv) will the Jacobi iteration converge for all initial iterates?
     (b) For which of the linear systems specified in Exercise 1(i)–(iv) will the Gauss-Seidel iteration converge for all initial iterates?

7.   (a) For which of the linear systems specified in Exercise 2(i)–(iv) will the Jacobi iteration converge for all initial iterates?
     (b) For which of the linear systems specified in Exercise 2(i)–(iv) will the Gauss-Seidel iteration converge for all initial iterates?

8.   (*An Example Where Gauss-Seidel Iteration Converges, but Jacobi Diverges*)   Consider the following linear system:

$$
\begin{bmatrix} 5 & 3 & 4 \\ 3 & 6 & 4 \\ 4 & 4 & 5 \end{bmatrix} \begin{bmatrix} x_1 \\ x_2 \\ x_3 \end{bmatrix} = \begin{bmatrix} 12 \\ 13 \\ 13 \end{bmatrix}.
$$

(a) Show that if initial iterate $x^{(0)} = [0\ 0\ 0]'$, the Jacobi iteration converges to the exact solution $x = [1\ 1\ 1]'$. Show that the same holds true if we start with $x^{(0)} = [10\ 8\ -6]'$.

(b) Show that if initial iterate $x^{(0)} = [0\ 0\ 0]'$, the Gauss-Seidel iteration will diverge. Show that the same holds true if we start with $x^{(0)} = [10\ 8\ -6]'$.

(c) For what sort of general initial iterates $x^{(0)}$ do the phenomena in parts (a) and (b) continue to hold?
(d) Show that the coefficient matrix of this system is positive definite. What does the SOR convergence theorem (Theorem 7.12) allow us to conclude?
Suggestion: For all parts (especially part (c)) you should first do some MATLAB experiments, and then aim to establish the assertions mathematically.

9.   (*An Example Where Jacobi Iteration Converges, but Gauss-Seidel Diverges*)   Consider the following linear system:

$$\begin{bmatrix} 1 & 2 & -2 \\ 1 & 1 & 1 \\ 2 & 2 & 1 \end{bmatrix} \begin{bmatrix} x_1 \\ x_2 \\ x_3 \end{bmatrix} = \begin{bmatrix} 1 \\ 3 \\ 5 \end{bmatrix}.$$

(a) Show that if initial iterate $x^{(0)} = [0\ 0\ 0]'$, the Jacobi iteration will converge to the exact solution $x = [1\ 1\ 1]'$ in just four iterations. Show that the same holds true if we start with $x^{(0)} = [10\ 8\ -6]'$.

(b) Show that if initial iterate $x^{(0)} = [0\ 0\ 0]'$, the Gauss-Seidel iteration will diverge. Show that the same holds true if we start with $x^{(0)} = [10\ 8\ -6]'$.

(c) For what sort of general initial iterates $x^{(0)}$ do the phenomena in parts (a) and (b) continue to hold?

Suggestion: For all parts (especially part (c)) you should first do some MATLAB experiments, and then aim to establish the assertions mathematically.

Note: This example is due to Collatz [Col-42].

10. (a) Use the formulas of Lemma 7.15 to write a function M-file with the following syntax:

$$b\ =\ \texttt{sparsediag(diags, inds, x)}$$

The input variables are `diags`, an $n \times j$ matrix where each column consists of the entries of A on one of its diagonals and `inds`, a $1 \times j$ vector of the corresponding set of indices for the diagonals (index zero corresponds to the main diagonal). The last input x is the $n \times 1$ vector to be multiplied by A. The output is the corresponding product $b = Ax$.

(b) Apply this program to check that $x = [1\ 1\ 1]'$ solves the linear system of Exercise 9.

(c) Apply this program to compute the matrix products of Exercise for the Reader 7.30 and check the error against the exact solution obtained in the latter.

11. (a) Modify the program `sorsparsediag` of Exercise for the Reader 7.31 to construct an analogous M-file:

```
[x,k,diff] = jacbobisparsediag(diags, inds, b, x0, tol, kmax)
```

for the Jacobi method.

(b) Modify the program `sorsparsediag` of Exercise for the Reader 7.31 to construct an analogous M-file:

```
[x,k,diff]=gaussseidelsparsediag(diags, inds, b, x0, tol, kmax)
```

for the Gauss-Seidel method.

(c) Apply these programs to recover the results of Examples 7.26 and 7.27.

(d) Using the M-files of parts (a) and (b), along with `sorsparsediag`, recreate the MATLAB graphic that is shown in Figure 7.43.

12. (a) Find a 2×2 matrix A whose optimal relaxation parameter ω appears to be greater than 1.5 (as demonstrated by a MATLAB plot like the one in Figure 7.42) resulting from the solution of some linear system $Ax = b$.

(b) Repeat part (a), but this time try to make the optimal value of ω to be less than 0.5.

13. Repeat both parts of Exercise 12, this time working with 3×3 matrices.

14. (a) Find a 2×2 matrix A whose optimal relaxation parameter ω appears to be greater than 1.5 (as demonstrated by a MATLAB plot like the one in Figure 7.42) resulting from the solution of some linear system $Ax = b$.

(b) Repeat part (a), but this time try to make the optimal value of ω to be less than 0.5.

15. (*A Program to Estimate the Optimal SOR Parameter* ω) (a) Write a program that will aim to find the optimal relaxation parameter ω for the SOR method in the problem of solving a certain

linear system $Ax = b$ for which it is assumed that the SOR method will converge. (For example, by the SOR convergence theorem, if A is symmetric positive definite, this program is applicable.) The syntax is as follows:

```
omega = optimalomega(A,b,tol,iter)
```

Of the input and output variables, only the last two input variables need explanation. The input variable `tol` is simply the accuracy goal that we wish to approximate `omega`. The variable `iter` denotes the number of iterations to use on each trial run. The default value for `tol` is 1e-3 and for `iter` it is 10. (For very large matrices a larger value may be needed for `iter`, and likewise for very small matrices a smaller value should be used.) Once this tolerance is met, the program terminates. The program should work as follows: First run through a set of SOR iterations with the values of ω running from 0.05 to 1.95 in increments of 0.05. For each value of ω we run through `iter` iterations. For each of these we keep track of the infinity norm of the difference of the final iterate and the immediately preceding iterate. For each tested value of $\omega = \omega_0$ for which this norm is minimal, we next run the tests on the values of ω running from $\omega_0 - .05$ to $\omega_0 + .05$ in increments of 0.005 (omit the values $\omega = 0$ or $\omega = 2$ should these occur as endpoints). In the next iteration, we single out those new values of ω for which the new error estimate is minimized. For each new corresponding value $\omega = \omega_0$ for which the norm is minimized, we will next run tests on the set of values from $\omega_0 - .005$ to $\omega_0 + .005$ in increments of 0.0005. At each iteration, the minimizing values of $\omega = \omega_0$ should be unique; if they are not, the program should deliver an error message to this effect, and recommend to try running the program again with a larger value of `iter`. When the increment size is less than tol, the program terminates and outputs the resulting value of $\omega = \omega_0$.

(b) Apply the above program to aim to determine the optimal value of the SOR parameter ω for the linear system of Example 7.26 with default tolerances. Does the resulting output change if we change `iter` to 5? To 20?

(c) Repeat part (b), but now change the default tolerance to 1e-6.

(d) Run the SOR iteration on the linear system using the values of the relaxation parameter computed in parts (a) and (b) and compare the rate of convergences with each other and with that seen in the text when $\omega = 0.9$ (Figure 7.41).

(e) Is the program in part (a) practical to run on the large matrix such as the 2500×2500 matrix of Example 7.30 (perhaps using a small value for `iter`)? If yes, run the program and compare with the result of Proposition 7.14.

16. *(A Program to Estimate the Optimal SOR Parameter ω for Sparse Banded Systems)* (a) Write a program that will aim to find the optimal relaxation parameter ω for the SOR method in the problem of solving a certain linear system $Ax = b$ for which it is assumed that the SOR method will converge. The functionality of the program will be similar to that of the preceding exercise, except that now the program should be specially designed to deal with sparsely banded systems, as did the program `sorsparsediag` of Exercise for the Reader 7.31 (in fact, the present program should call on this previous program). The syntax is as follows:

```
omega = optimalomegasparsediag(diags,inds,b,tol,iter)
```

The first three input variables are as explained in Exercise for the Reader 7.31 for the program `sorsparsediag`. The remaining variables and functionality of the program are as explained in the preceding exercise.

(b) Apply the above program to aim to determine the optimal value of the SOR parameter ω for the linear system of Example 7.26 with default tolerances. Does the resulting output change if we change `iter` to 5? To 20?

(c) With default tolerances, run the program on the linear system of Example 7.30 and compare with the exact result of Proposition 7.14. You may need to experiment with different values of `iter` to attain a successful approximation. Run SOR on the system with this computed value for the optimal relaxation parameter, and 308 iterations. Compute the exact error and compare with the results of Example 7.31.

(d) Repeat part (c) but now with `tol` reset to 1e-6.

NOTE: For tridiagonal matrices that are positive definite, the following formula gives the optimal value of the relaxation parameter for the SOR iteration:

$$\omega = \frac{2}{1 + \sqrt{1 - \rho(D - L)^2}}, \tag{66}$$

where the matrices D and L are as in (60) $A = D - L - U$,[24] and $\rho(D - L)$ denotes the spectral radius of the matrix $D - L$.

17. We consider tridiagonal square $n \times n$ matrices of the following form:

$$F = \begin{bmatrix} 2 & a & & & & \\ a & 2 & a & & \text{\Large 0} & \\ & a & 2 & a & & \\ & & a & 2 & a & \\ & \text{\Large 0} & & \ddots & \ddots & a \\ & & & & a & 2 \end{bmatrix}$$

(a) With $a = -1$ and $n = 10$, show that F is positive definite.

(b) What does formula (17) give for the optimal SOR parameter for the linear system?

(c) Run the SOR iteration with the value of ω obtained in part (b) for the linear system $Fx = b$ where the exact solution is $x = [1\ 2\ 1\ 2\ ...\ 1\ 2]'$. How many iterations are needed to get the exact error to be less than 1e-10?

(d) Create a graph comparing the performance of the SOR of part (c) along with the corresponding Jacobi and Gauss-Seidel iterations.

18. Repeat all parts of Exercise 17, but change a to –0.5.

19. Repeat all parts of Exercise 17, but change n to 100. Can you prove that with $a = -1$, the matrix F in Exercise 17 is always positive definite? If you cannot, do some MATLAB experiments (with different values of n) and conjecture whether you think this is a true statement.

20. (a) Show that the Jacobi iteration scheme is represented in matrix form (61) by the matrix B indicated in Table 7.1.

(b) Repeat part (a) for the Gauss-Seidel iteration.

(c) Repeat part (a) for the SOR iteration.

21. Prove Lemma 7.15.

22. (a) Given a nonsingular matrix A, find a corresponding matrix T so that the Jacobi iteration can be expressed in the form $x^{(k+1)} = x^{(k)} + Tr^{(k)}$, where $r^{(k)} = b - Ax^{(k)}$ is the residual vector for the kth iterate.

(b) Repeat part (a) for the Gauss-Seidel iteration.

(c) Can the result of part (b) be generalized for the SOR iteration?

23. (*Proof of Jacobi Convergence Theorem*) Complete the following outline for a proof of the Jacobi Convergence Theorem (part of Theorem 7.11): As in the text, we let $e^{(k)} = x^{(k)} - x$ denote the error vector of the kth iterate $x^{(k)}$ for the Jacobi method for solving a linear system $Ax = b$, where the $n \times n$ matrix A is assumed to be strictly diagonally dominant. For each

---

[24] $D - L$ is just the iteration matrix for the Gauss-Seidel scheme; see Table 7.1.

(row) index i, we let $\mu_i = \dfrac{1}{|a_{ii}|} \displaystyle\sum_{\substack{j=1 \\ j \neq i}}^{n} |a_{ij}|$ ($1 \leq i \leq n$). For any vector v, , we let $\|v\|$ denote its

infinity norm: $\|v\|_{\infty} = \max(|v_i|)$.

For each iteration index k and component index i, use the triangle inequality to show that

$$| e_i^{(k)} | \leq \frac{1}{|a_{ii}|} \sum_{\substack{j=1 \\ j \neq i}}^{n} |a_{ij}| \, | e_i^{(k-1)} | \ \leq \ \mu_i \left\| e^{(k-1)} \right\| ,$$

and conclude that

$$\left\| e^{(k)} \right\| \leq \| \mu \| \left\| e^{(k-1)} \right\| ,$$

and, in turn, that the Jacobi iteration converges.

Appendix A: MATLAB's Symbolic Toolbox

A.1: WHAT ARE SYMBOLIC COMPUTATIONS?

This appendix is meant as a quick reference for occasions in which exact mathematical calculations or manipulations are needed and are too arduous to expediently do by hand. Examples include the following:

1. Computing the (formula) for the derivative or antiderivative of a function.
2. Simplifying or combining algebraic expressions.
3. Computing a definite integral exactly and expressing the answer in terms of known functions and constants such as π, e, $\sqrt{7}$ (if possible).
4. Finding analytical solutions of differential equations (if possible).
5. Solving algebraic or matrix equations exactly (if possible).

Such exact arithmetic computations are known collectively as **symbolic computations**. MATLAB is unable to perform symbolic computations but the *Symbolic Math Toolbox* is available (or included with the Student Version), which uses the MATLAB interface to communicate with MAPLE, a symbolic computing system. Thus, MATLAB has essentially subcontracted symbolic computations to MAPLE, and acts as a middleman so that it is not necessary to use two separate softwares while working on problems. Invoking such symbolic capabilities needs specific actions on the user's part, such as declaring certain variables to be symbolic variables. This is a safety device since symbolic calculations are usually much more expensive than the default floating point calculations and are usually not called for (see Chapter 5). It is important to point out that symbolic expressions are different data types than the other sorts of data types that MATLAB uses. Consequently, care needs to be taken when passing data from one type of data to the other. Moreover, most mathematical problems have answers that cannot be expressed in terms of well-known functions (e.g., $\ln(x)$, $\sqrt{x}$, $\arcsin(x)$) and/or constants (e.g., $e, \pi, \sqrt{2}$), and therefore cannot be solved symbolically.

There are also circumstances where the precision of MATLAB's floating point arithmetic is not good enough for a given computation and we might wish to work in more than the 15 (or so) significant digits that MATLAB uses as a default. As a middle ground between this and exact arithmetic, the Symbolic Toolbox also offers what is called **variable precision arithmetic**, where the user can specify how many significant digits to work with. We point out that there are a few special occasions where symbolic calculations have been used in the text.

The remainder of this appendix will present a brief survey of some of the functionality and features of the Symbolic Toolbox that will be useful for our needs. All of the MATLAB code and output given in a particular section results from a new MATLAB session having been started at the beginning of that section.

A.2: ANALYTICAL MANIPULATIONS AND CALCULATIONS

To begin a symbolic calculation, we need to declare the relevant variables as symbolic. To declare x, y as symbolic variables we enter:

```
>> syms x y
```

Let's now do a few algebraic manipulations. The basic algebra manipulation commands that MAPLE has are as follows: expand, factor, simplify; they work on algebraic expressions just as anyone who knows algebra would expect. The next examples will showcase their functionality. We point out that any new variable introduced whose formula depends on a symbolic variable will also be symbolic.

```
>> p2=(x+2*y)^2;,   p4=(x+2*y)^4;
>> expand(p2) %Multiplies out the binomial product.
→ans = x^2+4*x*y+4*y^2
>> expand(p4)
→ans =x^4+8*x^3*y+24*x^2*y^2+32*x*y^3+16*y^4
>> pretty(ans) %Puts the answer in a prettier form.
   →        4    3      2 2      3      4
            x + 8 x y + 24 x y + 32 x y + 16 y
```

In general, for any sort of analytic expression exp, the command expand(exp) will use known analytical identities to try and rewrite exp in a form in which sums and products are expanded whenever possible.

```
>> pretty(expand(tan(x+2*y))) →
```

$$\frac{\tan(x) + 2 \dfrac{\tan(y)}{1 - \tan(y)^2}}{1 - 2 \dfrac{\tan(x)\,\tan(y)}{1 - \tan(y)^2}}$$

To clean up (simplify) any sort of analytical expression (involving powers, radicals, trig functions, exponential functions, logs, etc.), the simplify function is extremely useful.

```
>> simplify(log(2*sin(x)^2+cos(2*x)))     →ans =0
>> h=x^6-x^5-12*x^4-2*x^3+41*x^2+51*x+18;
>> pretty(factor(h))
   →            2       3
        (x + 2) (x - 3) (x + 1)
```

This function will also factor positive integers into primes. This brings up an important point. MATLAB also has a function `factor` that (only) does this latter task. Due to the limitations of floating point arithmetic, MATLAB's version is more restrictive than MAPLE's; it is programmed to give an error if the input exceeds $2^{32} \approx 4.2950\text{e}+009$.

```
>> factor(3^101-1)
??? Error using ==> factor
The maximum value of n allowed is 2^32.
>> factor(sym(3^101-1)) %declaring the integer input as symbolic
%brings forth the MAPLE version this command.
→ans = (2)^110*(43)*(47)*(89)*(6622026029)
```

Compare the ways of getting help from functions of the two systems:

`mhelp(<maplefunction>)` →	Gives information on a Symbolic Toolbox function (`<maplefunction>`).
`help(<MATLABfunction>)` →	Gives information on a MATLAB function (`<MATLABfunction>`).

Just like the `help` feature is useful in learning about MATLAB functions, much can be learned about MAPLE functions by using `mhelp`.

The `factor` function is programmed to look only for real rational factors, so it will not perform factorizations such as $x^2 - 3 = (x+\sqrt{3})(x-\sqrt{3})$ or $x^2 + 1 = (x+i)(x-i)$. Recall (Chapter 6) that it is not always possible to find explicit expressions for all roots/factors of a polynomial, but nevertheless, by the fundamental theorem of algebra, any degree n polynomial always has n roots (counted according to multiplicity) that can be real or complex numbers.   In cases where it is possible, the `solve` command can find them for us; otherwise, it produces decimal approximations.

`solve(exp,var)` →	If `exp` is a symbolic expression that involves the symbolic variable `var`, this command asks MAPLE to find all real and complex roots of the equation `exp=0`. In cases where they cannot be found exactly (symbolically), numerical (decimal) approximations are found. If there are additional symbolic variables, MAPLE solves for `var` in terms of them.

To solve the equation $x^5 - 5x^4 + 8x^3 - 40x^2 + 16x - 80 = 0$, we simply enter:

```
>>solve(x^5-5*x^4+8*x^3-40*x^2+16*x-80)
>> %shorter syntax if only one var
→ ans =    [ 2*i]     [ 2*i]     [ 5]
           [-2*i]      [-2*i]
```

The slightly perturbed polynomial equation $x^5 - 5x^4 + 8x^3 - 40x^2 + 16x - 78 = 0$ also has five different roots, but they cannot be expressed exactly, so MAPLE will give us numerical approximations, in its default 32 digits:

```
>> solve(x^5-5*x^4+8*x^3-40*x^2+16*x-78)
```

→ ans =   [-.28237724125630031806612784925449e-1 -
            2.1432362125064684675126753513414*i]
            [-.28237724125630031806612784925449e-1 +
            2.1432362125064684675126753513414*i]
            [.29428740076409528006464576345708e-1 -
            1.8429038593310837866143850920505*i]
            [.29428740076409528006464576345708e-1 +
            1.8429038593310837866143850920505*i]
            [4.9976179680984410076002964171595]

We can get the quadratic formula for the solutions of $ax^2 + bx + c = 0$ with the following commands:

```
>> syms a b c, solve(a*x^2+b*x+c,x)
→ ans =     [ 1/2/a*(-b+(b^2-4*a*c)^(1/2))]    [ 1/2/a*(-b-(b^2-4*a*c)^(1/2))]
```

Similarly, the Tartaglia formulas for the three solutions of the general cubic $ax^3 + bx^2 + cx + d = 0$, could be obtained.

A.3: CALCULUS

Table A.1 summarizes the Symbolic Toolbox commands needed to perform the most common "clerical" tasks in calculus: differentiation and integration.

TABLE A.1: Differentiation and integration using the Symbolic Toolbox.

Assume that f has been stored as a symbolic function of symbolic variables: $f(x)$ (or $f(x,y,...)$), if we have a function of several variables.	
`diff(f,x)` →	Computes $f'(x) = \dfrac{df}{dx}$ $\left(\text{or } \dfrac{\partial f}{\partial x}\right)$.
`diff(f,x,2)` →	Computes $f''(x) = \dfrac{d^2 f}{dx^2}$ $\left(\text{or } \dfrac{\partial^2 f}{\partial x^2}\right)$.
`int(f,x)` →	Calculates (if possible) an antiderivative of $f(x)$: $\int f(x)dx$ (does not add on integration constant). If there are other variables, they are treated as constant parameters.
`int(f,x,a,b)` →	Calculates (exactly, if possible) the definite integral: $\int_a^b f(x)dx$ (does not add on integration constant). If there are other variables, they are treated as constant parameters.

EXAMPLE A.1: Use the Symbolic Toolbox to compute the following:

(a) $\dfrac{d}{dx}x^x$

(b) $\dfrac{\partial^3}{\partial x \partial y^2}\left(\dfrac{\cos(x + y^2 + z^3)}{1 + x^2 + y^2}\right)$

(c) $\int \ln(x)\, dx$

(d) $\int \sin(x^2)dx$

(e) $\int_0^1 \sin(x^2)dx$

(f) $\int_{-\infty}^{-\infty} e^{-x^2} dx$

SOLUTION: Part (a):
```
>> syms x y z
>> diff(x^x)      →ans =x^x*(log(x)+1)
```
So the answer is $x^x(\ln x+1)$.

Part (b):
```
>> f=cos(x+y^2+z^3)/(1+x^2+y^2);
>> pdf=diff(diff(f,y,2),x)      →pdf =
4*sin(x+y^2+z^3)*y^2/(x^2+1+y^2)+8*cos(x+y^2+z^3)*y^2/(x^2+1+y^2)^2*x-
2*cos(x+y^2+z^3)/(x^2+1+y^2)+4*sin(x+y^2+z^3)/(x^2+1+y^2)^2*x+8*cos(x+y^2+z^3)*y^2/(x^
2+1+y^2)^2-32*sin(x+y^2+z^3)*y^2/(x^2+1+y^2)^3*x-8*sin(x+y^2+z^3)/(x^2+1+y^2)^3*y^2-
48*cos(x+y^2+z^3)/(x^2+1+y^2)^4*y^2*x+2*sin(x+y^2+z^3)/(x^2+1+y^2)^2+8*cos(x+y^2+z^3)
/(x^2+1+y^2)^3*x
```

We shall refrain from putting this mess in usual mathematical notation, but we will
do something else with it later (which is why we gave it a name).

Part (c): `>>int(log(x))`   → ans =x*log(x)-x

Part (d): `>> int(sin(x^2),x)` →ans =1/2*2^(1/2)*pi^(1/2)*FresnelS(2^(1/2)/pi^(1/2)*x)

This answer to part (d) needs a bit of explanation. Most indefinite integrals cannot
be expressed in terms of the elementary functions. Using some additional **special
functions** (e.g., Bessel functions, hypergeometric functions, the Error function,
and the above Fresnel Sine function), additional integrals can be computed (but
still only relatively few); thus MAPLE has found an antiderivative for us, but for
most practical purposes this answer by itself is not so interesting. A similar result
turns up (by the fundamental theorem of calculus) for the corresponding definite
integral.

Part (e): `>> int(sin(x^2),x,0,1)` →ans =1/2*FresnelS(2^(1/2)/pi^(1/2))*2^(1/2)*pi^(1/2)

The following commands show how to get a more useful decimal answer out of
this or any answer to a symbolic computation:

`vpa(a,d)` →	If a is a symbolic answer representing a number, d is a nonnegative number this command will convert the number a to decimal form with d significant digits. vpa stands for variable precision arithmetic. The default value is d=32[1]
`digits(d)` `vpa(a)` →	Has the same result as above, but now the default value of d=32 digits of MAPLE's arithmetic is reset to d in subsequent calculations.

`>> vpa(ans)`   →ans =.31026830172338110180815242316540

---
[1] Thus, MAPLE uses approximately a 32-digit floating point arithmetic system in cases where exact
answers are not possible. This is about double of what MATLAB uses and for many computations is
overkill since large-scale calculations would proceed much more slowly. Thus, generally speaking, use
of the Symbolic Toolbox should be limited to symbolic computations, except in the occasional
instances where, say, the problem being solved is very ill-conditioned and roundoff errors run out of
control with IEEE floating point arithmetic (see Chapter 5).

If we (for whatever reason) wanted to see the first 100 digits of π, we could simply enter:

```
>>vpa(pi,100) →ans=3.1415926535897932384626433832795028841971693993751 0582
                    0974944592307816406286208998628034825342117068
```

Part (f): Improper integrals are done with the same syntax as proper integrals.

```
>> int(exp(-x^2),x,-Inf, Inf)    →ans = pi^(1/2)
```

Thus we get that $\int_{-\infty}^{\infty} e^{-x^2} dx = \sqrt{\pi}$.

Often, we need to evaluate a symbolic expression or substitute some of its variables with other variables or expressions. The following command subs is very useful in this respect:

subs(S,old,new) →	If S is a symbolic expression, old is a symbolic variable appearing in S (or a vector of variables), new is symbolic number or symbolic expression (or a vector of such things having the same size as old), this command will produce the symbolic expression resulting from substituting in S each occurrence of old by the corresponding expression in new.

For example, suppose (in the setting of Example A.1) we wanted to compute

$$\left. \frac{\partial^3}{\partial x \partial y^2} \left(\frac{\cos(x + y^2 + z^3)}{1 + x^2 + y^2} \right) \right|_{\substack{x=\pi \\ y=\pi/2 \\ z=0}} .$$

From what we have already computed, we could simply enter:

```
>> subs(pdf,[x y z], [pi pi/2 0])    → ans =-0.2016
```

Since all symbolic variables were substituted with nonsymbolic (ordinary MATLAB floating point) numbers, the result is now a regular MATLAB floating point number. To retain the accuracy of symbolic computation in the substitution, we could instead enter:

```
>> exact=subs(pdf,[x y z], sym([pi pi/2 0])); %suppress messy output
>> vpa(exact) %could specify more or less digits here.
→ans = -.2016360958581 1087949860391144560
```

Note that the main difference is that in the latter we declared the numbers to be symbolic (exact):

fpn=double(sbn) →	If sbn is a (MAPLE) symbolic number, this command creates a (MATLAB) floating point number fpn from it essentially by rounding it off to about 16 digits of accuracy.

sbn=sym(fpn) →	If fpn is a (MATLAB) floating point number, this command creates a (MAPLE) symbolic number sbn from it by treating it as an exact number.

The Symbolic Toolbox has a simple way for computing Taylor series:

taylor(<fun>,n,a) →	If <fun> is a symbolic expression representing a function of a (previously declared) symbolic variable (say x), n is a positive integer, and a is a real number, this command will produce the Taylor polynomial of the function centered at $x = a$ of order (degree at most) n-1. The last input a is optional, the default value is a = 0.

EXAMPLE A.2: Obtain the 15th-order Taylor polynomial of $f(x) = x^2 \tan(x^3)$ centered at $x = 0$.

SOLUTION:

```
>> taylor(x^3*tan(x^2),16)    →ans =x^5+1/3*x^9+2/15*x^13
```

In the notation of Chapter 2, we can thus write $p_{15}(x) = x^5 + \dfrac{x^9}{3} + \dfrac{2x^{13}}{15}$.

Appendix B: Solutions to All Exercises for the Reader

NOTE: All of the M-files of this appendix (like the M-files of the text) are downloadable as text files from the ftp site for this text:

`ftp://ftp.wiley.com/public/sci_tech_med/numerical_preliminaries/`

Occasionally, for space considerations, we may refer a particular M-file to this site. Also, in cases where a long MATLAB command does not fit on a single line (in this appendix), it will be continued on the next line. In an actual MATLAB session, (long) compound commands should either be put on a single line, or three periods (...) should be entered after a line to hold off MATLAB's execution until the rest of the command is entered on subsequent lines and the ENTER key is pressed. The text explains these and other related concepts in greater detail.

CHAPTER 1: MATLAB Basics

EFR 1.1: `linspace(-2,3,11)`

EFR 1.2: `t = 0:.01:10*pi;,   x = 5*cos(t/5)+cos(2*t);,`
`y = 5*sin(t/5)+sin(3*t);, plot(x,y), axis('equal')`

EFR 1.3: Simply run the code through MATLAB to see if you analyzed it correctly.

CHAPTER 2: Basic Concepts Of Numerical Analysis With Taylor's Theorem

EFR 2.1: `x=-10:.05:10;,  y=cos(x);,  p2=1-x.^2/2;,  p4=1-`
`x.^2/2+x.^4/gamma(5); p6=1-x.^2/2+x.^4/gamma(5)-x.^6/gamma(7);, p8=1-`
`x.^2/2+x.^4/gamma(5)-x.^6/gamma(7)+x.^8/gamma(9);, p10=p8-`
`x.^10/gamma(11); hold on, plot(x,p10,'k:'), axis([-2*pi  2*pi  -1.5`
`1.5]), plot(x,p8,'c:'), plot(x,p6,'r-.'), plot(x,p4,'k--'),`
`plot(x,p2,'g'), plot(x,y,'+')`

EFR 2.2: Computing the first few derivatives of :

$$f(x) = x^{1/2}, \; f'(x) = \frac{1}{2}x^{-1/2}, \; f''(x) = -\frac{1 \cdot 1}{2 \cdot 2}x^{-3/2}, \; f'''(x) = \frac{1 \cdot 1 \cdot 3}{2 \cdot 2 \cdot 2}x^{-5/2},$$

$$f^{(4)}(x) = -\frac{1 \cdot 1 \cdot 3 \cdot 5}{2 \cdot 2 \cdot 2 \cdot 2}x^{-7/2} \; ..., \text{ leads us to discover the general pattern:}$$

$$f^{(n)}(x) = \; (-1)^{n+1}\frac{1 \cdot 3 \cdot 5 \cdots (2[n-1]-1)}{2^n}x^{-(2n-1)/2} \text{ (for } n \geq 2 \text{). Applying Taylor's theorem (with } a = $$

16, $x = 17$), we estimate the error of this approximation:

$$|R_n(17)| = \left|\frac{f^{(n+1)}(c)}{(n+1)!}1^{n+1}\right| = \left|\frac{1\cdot3\cdot5\cdots(2n-1)}{2^n(n+1)!}c^{-(2n+1)/2}\right| \le \frac{1\cdot3\cdot5\cdots(2n-1)}{2^n(n+1)!}16^{-(2n+1)/2} =$$

$$\frac{1\cdot3\cdot5\cdots(2n-1)}{2^n(n+1)!\cdot4\cdot16^n} = \frac{1\cdot3\cdot5\cdots(2n-1)}{2^{5n+2}(n+1)!}.$$ We use MATLAB to find the smallest n for which this last

expression is less than 10^{-10}; then Taylor's theorem will assure us that the Taylor polynomial of this order will provide us with the desired approximation.

```
>> n=2; ErrorEst=1*3/gamma(n+2)/2^(5*n+2);
>> while ErrorEst>1e-10, n=n+1;, ErrorEst=ErrorEst*(2*n-
1)/(n+1)/2^5;, end
>> n→n =7
>> ErrorEst→ErrorEst = 2.4386e-011 %this checks out.
```

So $p_7(17) = \sum_{k=0}^{7}\frac{1}{k!}f^{(k)}(16)\cdot1^k$ will give the desired approximation. We use MATLAB to perform

and check it:
```
>> sum=16^(1/2)+16^(-1/2)/2; %first order Taylor Polynomial
term = 16^(-1/2)/2; %first order term
for k=2:7, term = -term*(2*(k-1)-1)/2/16/k;, sum=sum+term;, end,
format long
>> sum→sum = 4.12310562562925 (approximation)
>>abs(sum-sqrt(17))→ans =1.1590e-011 %actual error excels goal
```

EFR 2.3:   Using ordinary polynomial substitution, subtraction and multiplication (and ignoring terms in the individual Maclaurin series that give rise to terms of order higher than 10), we use (9) and (10) to obtain: (a) $\sin(x^2) - \cos(x^3) =$

$$\left(x^2 - \frac{(x^2)^3}{3!} + \frac{(x^2)^5}{5!} - \cdots\right) - \left(1 - \frac{(x^3)^2}{2!} + \cdots\right) = -1 + x^2 - \left(\frac{1}{2!} - \frac{1}{3!}\right)x^6 + \frac{x^{10}}{5!}\cdots$$

(b) $\sin^2(x^2) = \left(x^2 - \frac{(x^2)^3}{3!} + \cdots\right)\cdot\left(x^2 - \frac{(x^2)^3}{3!} + \cdots\right) = x^4 - \frac{2}{3!}x^8 + \cdots$

In each case, $p_{10}(x)$ consists of all of the terms listed on the right-hand sides.

CHAPTER 3: Introduction to M-Files

EFR 3.1: In the left box we give the stored M-file; in the right we give the subsequent MATLAB session.

`% script file for EFR 3.1:`
`listp2`
`power =2;`
`while power <= n`
`    power`
`    power=2*power;`
`end`
	`>> n=5;, listp2   → power = 2, power = 4`
	`>> n=264;,listp2`
	`→ power = 2, power = 4, power = 8, power = 16,`
	`power = 32, power = 64, power =128, power = 256,`
	`>>n=2917;,listp2`
	`→ power = 2, power = 4, power = 8, power = 16,`
	`power = 32, power = 64, power =128, power = 256,`
	`power = 1024, power = 2048`

Note: If we wanted the output to be just a single vector of the powers of 2, the following modified script would do the job:

```
% script file for EFR 3.1:  listp2ver2
power =2; vector = [ ]; %start off with empty vector
```

```
while power <= n
    vector = [vector power];
    power=2*power;
end, vector
```
For example, with this file stored, if we enter >> n=264; listp2ver2, we get the following vector output:
→vector = 2   4   8   16   32   64   128   256

EFR 3.2: With the boxed function M-file below saved, MATLAB will give the following outputs:

```
function f = fact(n)
% FACT f = fact(n) returns the factorial n! of a nonnegative integer
n
f=1;
for i=1:n
        f=f*i;
end
```
>> fact(4), fact(10), fact(0)
→ans = 24, 3628800, 1

EFR 3.3: At any (non-endpoint) maximum or minimum value $y(x_0)$, a differentiable function has its derivative equaling zero. This means that the tangent line is horizontal, so that for small values of $\Delta x \equiv x - x_0$, $\Delta y / \Delta x$ approaches zero. Thus, the y-variations are much smaller than the x-variations as x gets close to the critical point in question. This issue will be revisited in detail in Chapter 6.

EFR 3.4: We have only considered the values of y at a discrete set of (equally spaced) x-values. It is possible for a function to oscillate wildly in intervals between sets of discrete points (think trig functions with large amplitudes). More analysis can be done to preclude such pathologies (e.g., checking to see that there are no other critical points).

EFR 3.5: The M-file for the function is straightforward:

```
function y = wiggly(x)
%Function M-file for the mathematical function of EFR 3.5
y=sin(exp(1./(x.^2+0.5).^2)).*sin(x);
```
(a) >> x=-2:.001:2;, plot(x,wiggly(x)) %plot is shown on left below
(b) >> quad(@wiggly,0,2,1e-5)   →ans = 1.03517910753379
(c) To better see what we are looking for, we create another plot of the function zoomed in near x = 0.
>> x=0:.001:.3;, plot(x,wiggly(x)) %plot is shown (w/ other additions) on right below.
We seek the x-coordinates of the red "x" and the green "x" in the figure below.

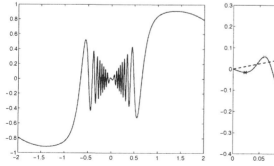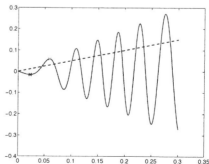

>> xmin=fminbnd(@wiggly,0,0.07,optimset('TolX',1e-5))
→xmin =0.02289435851906

```
>> xmax=fminbnd('-wiggly(x)',0,0.1,optimset('TolX',1e-5))
```
→xmax =0.05909071987402
The red and green x's can now be added to the graph as follows: >> hold on,
`plot(xmin,wiggly(xmin),'rx'), plot(xmin,wiggly(xmin),'gx')` (This also
gives us a visual check that we found what we were looking for.)
(d) To get a rough idea of the location of the x-value we are searching for, we now add the graph of the
line y = x/2 (as a black dotted line): >> `plot(x,x/2,'k--')`   From the graph, we see that the
intersection point we are looking for is the one closest to the midpoint of xmin and xmax.
`>> xcross=fzero('wiggly(x)-x/2',(xmin+xmax)/2)`
→ xcross =0.04479463640226
Let's do a quality check: >> `wiggly(xcross)-xcross/2`
→ans = 2.185751579730777e-016 (Very Good!)

CHAPTER 4: Programming in MATLAB

EFR 4.1: Simply run the code through MATLAB to see if you analyzed it correctly.

EFR 4.2: (a) The M-file is boxed below:

```
function [ ] = sum2sq(n)
%M-file for EFR 4.2
for a=1:sqrt(n)
    b=sqrt(n-a^2); %solve n=a^2+b^2 for b
    if b==floor(b); %checks to see if b is integer
        fprintf('the integer %d can be written as the sum of squares
of %d and %d', n,a,b)
        return
    end
end
fprintf('the integer %d cannot be written as the sum of squares', n)
```

(b) We now perform the indicated program runs:
>> sum2sq(5)   →the integer 5 can be written as the sum of squares of 1 and 2
>> sum2sq(25)   →the integer 25 can be written as the sum of squares of 3 and 4
>> sum2sq(12233) →the integer 12233 can be written as the sum of squares of 28 and 107
(c) The following modification of the above M-file will be more suitable to solving this problem:

```
function flag = sum2sqb(n)
%M-file for EFR 4.2b
flag=0; %will change to 1 if n can be written as a^2+b^2
for a=1:sqrt(n)
    b=sqrt(n-a^2); %solve n=a^2+b^2 for b
    if b==floor(b); %checks to see if b is integer
        flag=1;
        return
    end
end
```

The program has output 1 if and only if *n* is expressible as a sum of squares; otherwise the output is
zero. Now the following simple code will compute the desired integer *n*:
>> for n=99999:-1:1, flag=sum2sqb(n);
if flag==0
fprintf('%d is the largest integer less than 100,000 not expressible
as a sum of squares',n)
break
end
end
→99999 is the largest integer less than 100,000 not expressible as a sum of squares
(We did not have to go very far.)

(d) A minor modification to the above code will give us what we want; simply change the for loop to
`>> for n=1001:1:99999` (and the wording in the `fprintf` statement).
We then find the integer to be 1001.
(e) The following code will determine what we are looking for:
`>> for n=2:99999, flag=sum2sqb(n);, if flag==0, count=count+1;, end,`
`end`
`>> count`
→count =75972
Note: Part (e) took only a few seconds. If the programs were written less efficiently, for example, if we had run a nested loop by letting a and b run separately between all integers from 0 to $\sqrt{n}$ (or larger), some parts of this problem (notably, part (e)) could not be done in a reasonable amount of computer time.

EFR 4.3: (a) Before you run the indicated computations in a MATLAB session, try to figure out the output by hand. This will assure that you understand both the Collatz sequence generation process as well as the program. The reason for clearing the vector a at the end of the script is so that on subsequent runs, this vector will not start with old values from previous runs.
(b) The M-file is boxed below:

```
function n = collctr(an)
n=0;
while an ~= 1
    if ceil(an/2)==an/2    %tests if an is even
        an=an/2;
    else
        an=3*an+1;
    end
    n=n+1;
end
```

EFR 4.4: (a) The M-file is boxed below:

```
%raffledraw.m
%scriptfile for EFR 4.4

K = input('Enter number of players:    ');
N=zeros(K,26); %this allows up to 26 characters for each players
                %name.
n=input('Enter IN SINGLE QUOTES first player name:    ');
len(1)=length(n);
N(1,1:len(1))=n;
W(1)=input('Enter weight of first player:  ');
for i=2:K-1
    n=input('Enter IN SINGLE QUOTES next player name:   ');
    len(i)=length(n);
    N(i,1:len(i))=n;
    W(i)=input('Enter weight of this player:  ');
end
n=input('EnterIN SINGLE QUOTES last player name:   ');
len(K)=length(n);
N(K,1:len(K))=n;
W(K)=input('Enter weight of last player:  ');

totW = sum(W); %total weight of all players (=# of raffle tickets)

%the next four commands are optional, they only add suspense and
%drama to the raffle drawing which the computer can do in lightning
%time
fprintf('\r \r RANDOM SELECTION PROCESS INITIATED \r \r ...')
pause(1) %creates a 1 second pause
```

```
fprintf('\r \r ...SHUFFLING....\r \r')
pause(5) %creates a 5 second pause
%%%%%%%%%%%%%%%%%%%%%%%%%%%%

rand('state',sum(100*clock))
magic = floor(totW*rand); %this will be a random number between 0 and
                          %totW
count =W(1); %number of raffle tickets of player 1
if magic<=count
    fprintf('WINNER IS %s \r \r', char(N(1,1:len(1))))
    fprintf('CONGRATULATIONS %s!!!!!!!!!!!!!!', char(N(1,1:len(1))))
    return
else count = count + W(2);, k=2;
    while 1
        if magic <=count
          fprintf('WINNER IS %s \r \r', char(N(k,1:len(k))))
        fprintf('CONGRATULATIONS %s!!!!!!!!!!!!!!', char(N(k,1:len(k))))
        return
        end
        k=k+1;, count = count +W(k);
    end
end
```

(b) We now perform the indicated program runs:
>> raffledraw
Enter number of players: 4
Enter IN SINGLE QUOTES first player name: 'Alfredo'
Enter weight of first player: 4
Enter IN SINGLE QUOTES next player name: 'Denise'
Enter weight of this player: 2
Enter IN SINGLE QUOTES next player name: 'Sylvester'
Enter weight of this player: 2
Enter IN SINGLE QUOTES last player name: 'Laurie'
Enter weight of last player: 4

RANDOM SELECTION PROCESS INITIATED
... ...SHUFFLING....
→WINNER IS Laurie
→CONGRATULATIONS Laurie!!!!!!!!!!!!

On a second run the winner was Denise. If written correctly, and if this same raffledraw is run many times, it should turn out (from basic probability) that Alfredo and Laurie will each win roughly 4/12 or 33 1/3% of the time while Denise and Sylvester will win roughly 2/12 or 16 2/3% of the time.

CHAPTER 5: Floating Point Arithmetic and Error Analysis

EFR 5.1: For shorthand we write: FPA to mean "the floating point answer," EA to mean "the exact answer," E to mean the "error" = |FAP-EA|, and RE to mean the "relative error" = E/|EA|.
(a) FPA = 0.023, EA = 0.0225, E = 0.0005, RE = 0.02222 ⋯
(b) FPA = 370,000 × .45 = 170,000, EA = 164990.2536, E = 5009.7464, RE = 0.030363...
(c) FPA = 8000 ÷ 120 = 67 , EA = 65.04878... , E = 1.9512195121 ⋯, RE = 0.029996...

EFR 5.2: (a) As in the solution of Example 5.3, since the terms are decreasing, we continue to compute partial sums (in 2-digit rounded floating point arithmetic) until the terms get sufficiently small so as to no longer have any effect on the accumulated sum.

$S_1 = 1$, $S_2 = S_1 + 1/2 = 1 + .5 = 1.5$, $S_3 = S_2 + 1/3 = 1.5 + .33 = 1.8$, $S_4 = S_3 + 1/4 = 1.8 + .25 = 2.1$,
$S_5 = S_4 + 1/5 = 2.1 + .2 = 2.3$, $S_6 = S_5 + 1/6 = 2.3 + .17 = 2.5$, $S_7 = S_6 + 1/7 = 2.5 + .14 = 2.6$,
$S_8 = S_7 + 1/8 = 2.6 + .13 = 2.7$, $S_9 = S_8 + 1/9 = 2.7 + .11 = 2.8$, $S_{10} = S_9 + 1/10 = 2.8 + .1 = 2.9$.

This pattern continues until we reach S_{20}: In each such partial sum S_k, $1/k$ contributes 0.1 to the cumulative sum. As soon as we reach S_{21}, the terms (1/21 = 0.048) in floating point arithmetic become too small to have any effect on the cumulative sum so we have converged; thus the final answer is: $2.9 + 10 \times .1 = 3.9$.

(b) (i) $x^2 = 100$: Working in exact arithmetic, there are, of course, two solutions: $x = \pm 10$. These are also floating point solutions and any other floating point solutions will lie in some intervals about these two. Let's start with the floating point solution $x = 10$. In arithmetic of this problem, the next floating point number greater than 10 is 11 and (in floating point arithmetic) $11^2 = 120$, so there are no floating point solutions greater than 10. Similarly the floating point number immediately preceding 10 is 9.9 and (in floating point arithmetic) $9.9^2 = 98$, so there are no (positive) floating point solutions less than 10. Similarly, -10 is the only negative floating point solution. Thus there are exactly two floating point solutions (or more imprecisely: between 2 and 10 solutions).

(ii) $8x^2 = x^5$: In exact arithmetic, we would factor this $x^5 - 8x^2 = x^2(x^3 - 8) = 0$ to get the real solutions: $x = 0$ and $x = 2$. Because of underflow, near $x = 0$, we can get many (more than 10) floating point solutions. Indeed, since $e = 8$, if $|x| < 10^{-5}$, then both sides of the equation will underflow to zero so we will have a solution. Any number of form $\pm a.b \times 10^{-c}$, where a and b are any digits ($a \neq 0$) and $c = 6$, 7, or 8, will thus be a floating point solution, so certainly there are more than 10 solutions. (How many are there exactly?)

EFR 5.3: (a) As in the solution to Example 5.4, we may assume that $x \neq 0$ and write $x = .d_1 d_2 \cdots d_s d_{s+1} \cdots \times 10^e$. Now, since we are using s-digit rounded arithmetic, fl(x) is the closer of the two numbers $.d_1 d_2 \cdots d_s \times 10^e$ and $.d_1 d_2 \cdots d_s \times 10^e + 10^{-s} \times 10^e$ to x. Since the gap between these two numbers has length $10^{-s} \times 10^e$, we may conclude that $|x - \text{fl}(x)| \leq \frac{1}{2} \cdot 10^{-s} \times 10^e$. On the other hand, $|x| \geq .100 \cdots 0 \times 10^e = 10^{e-1}$. Putting these two estimates together, we obtain the following

estimate for the relative error: $\left| \dfrac{x - \text{fl}(x)}{x} \right| \leq \dfrac{\frac{1}{2} \cdot 10^{-s} \times 10^e}{10^{e-1}} = \dfrac{1}{2} \cdot 10^{1-s}$. Since equality is possible, we

conclude that $u = \dfrac{1}{2} \cdot 10^{1-s}$, as asserted. The floating point numbers are the same whether we are using chopped or rounded arithmetic, so the gap from 1 to the next floating point number is still 10^{1-s}, as explained in the solution of Example 5.4.

(b) If $x = 0$, we can put $\delta = 0$; otherwise put $\delta = [\text{fl}(x) - x]/x$.

EFR 5.4: (a) Since $N - i \leq N$ when i is nonnegative, we obtain from (6) that

$$| \text{fl}(S_N) - S_N | \leq u[(N-1)a_1 + (N-1)a_2 + (N-2)a_3 + \cdots + 2a_{N-1} + a_N]$$
$$\leq u[Na_1 + Na_2 + Na_3 + \cdots + Na_{N-1} + Na_N] = Nu \sum_{n=1}^{N} a_n.$$

(b) Simply divide both sides of the inequality in (a) by $\sum_{n=1}^{N} a_n$ obtain the inequality in (b).

EFR 5.5: From $1 - \dfrac{1}{3} + \dfrac{1}{5} - \dfrac{1}{7} + \cdots = \dfrac{\pi}{4}$, we can write $\pi = 4 - \dfrac{4}{3} + \dfrac{4}{5} - \dfrac{4}{7} + \cdots = \sum_{n=0}^{\infty} (-1)^n a_n$, where $a_n = 4/(2n+1)$. Letting, S_N denote the partial sum $\sum_{n=0}^{N}(-1)^n a_n$, Leibniz's theorem tells us

that Error $\equiv |\pi - S_N| \le a_{N+1} = 4/(2N+3)$. Since we want Error $< 10^{-7}$, we should take N large

enough to satisfy $4/(2N+3) < 10^{-7} \Rightarrow 2N+3 > 4 \cdot 10^7 \Rightarrow N > (4 \cdot 10^7 - 3)/2 = 19,999,998.5$.

Letting N = 19,999,999, we get MATLAB to perform the summation of the corresponding terms in
order of increasing magnitude:

```
>> format long
>> Sum=0;, N=19999999;
>> for n=N:-1:0
Sum=Sum+(-1)^n*4/(2*n+1);
end
>> Sum
```
→Sum = 3.14159260358979 (approximation to π)
```
>> abs(pi-Sum)
```
→ans = 4.999999969612645e-008 (exact error of approximation)

CHAPTER 6: Rootfinding

EFR 6.1: The accuracy of the approximation x7 is actually better than what was guaranteed from
(1). The actual accuracy is less than 0.001 (this can be shown by continuing with the bisection method
to produce an approximation xn with guaranteed accuracy less than 0.00001 (how large should n be?)
and then estimating $|x7 - \text{root}| \le |x7 - xn| + |xn - \text{root}| \le 9 \times 10^{-4} + 1 \times 10^{-5} < 0.001$. So actually, $|f(x7)|$ is
over 30 times as large as $|x7 - \text{root}|$. This can be explained by estimating $y'(\text{root}) \ge 30$ (do it
graphically, for example). Thus, for small values of $\Delta x \equiv x - \text{root}$, $\Delta y/\Delta x$ gets larger than 30. This is
why the y-variations turn out to be more than 30 times as large as the x-variations, when x gets close to
the root.

EFR 6.2: (a) Since $f(0) = 1 - 0 > 0$, $f(\pi/2) = 0 - \pi/2 < 0$, and f(x) is continuous, we know
from the intermediate value theorem that f(x) has a root in $[0, \pi/2]$. Since $f'(x) = \sin(x) - 1 < 0$ on
$(0, \pi/2)$, f(x) is strictly decreasing so it can have only one root on $[0, \pi/2]$.

(b) It is easy to check that the first value of n for which $\pi/(2 \cdot 2^n)$ $(= (b-a)/2^n)$ is less than 0.01 is n
= 8. Thus by (1), using $x0 = 0$, it will be sufficient to run through n = 8 iterations of the bisection
method to arrive at an approximation x8 of the root that has the desired accuracy. We do this with the
following MATLAB loop:

```
>> xn=0;, an=0;, bn=pi/2;, n=0;
>> while n<=8
xn=(an+bn)/2;, n=n+1;
if f(x)==0, root = xn;, return
elseif f(x)>0, an=xn;, bn=bn;
else, an=an;, bn=xn;
end
end
>> xn
```
→xn =0.73937873976088

(c) The following simple MATLAB loop will determine the smallest value of n for which $\pi/(2 \cdot 2^n)$
will be less than 10^{-12} (by (1) this would be the smallest number of iterations in the bisection method
for which we could be guaranteed the indicated accuracy). (This could certainly also be done using
logs.)

```
>> while pi/2/2^n>=1e-12, n=n+1;, end
>> n
```
→n = 41

```
>> pi/2/2^41, pi/2/2^40   %we perform a check
```
→ans = 7.143154683921678e-013 (OK)   1.428630936784336e-012 (too big, so it checks!)

EFR 6.3: (a) The condition `yn*ya > 0` mathematically translates to `yn` and `ya` having the same sign, so this is (mathematically) equivalent to our condition `sign(yn)==sign(ya)`.

(b) We are aiming for a root so at each iteration, `yn` and `ya` should be getting very small; thus their product `yn*ya` will be getting smaller much faster (e.g., if both are about 1e-175, then their product would be close to 1e-350 and this would underflow). Thus, with the modified loop we run the risk of a premature underflow destroying any further progress of the bisection method.

(c) Consider the function $f(x) = (x + .015)^{101}$, which certainly has a (unique) root $x = -0.015$ and satisfies the requirements for using the bisection method. As soon as the interval containing `xn` gets to within 1e-2 of the root, both y-values `yn` and `ya` would then be less than 1e-200; so their product would be less than 1e-400 and so would underflow to zero. This starts to occur already when $n = 2$ (`xn` = 0), and causes the modified if-branch to default to the else-if option—taking the left half subinterval as the new interval. From this point on, all approximations will be less than -0.5, making it impossible to reach the 0.001 accuracy goal.

EFR 6.4: The distance from x to e is less than MATLAB's unit roundoff and the minimum gap between floating point numbers (see Example 5.4 and Exercise for the Reader 5.3). Thus MATLAB cannot distinguish between the two numbers x and e, and (in the notation of Chapter 5) we have fl(x) = fl(e) = e (since important numbers like e are built in to MATLAB as floating point numbers). As a result, when MATLAB evaluates ln(x), it really computes ln(fl(x)) = ln(e) and so gets zero.

EFR 6.5: (a) If we try to work with quadratic polynomials (parabolas), cycling cannot occur unless the parabola did not touch the x-axis (this is easy to convince oneself of with a picture and not hard to show rigorously). If we allow polynomials that do not have roots, then an example with a quadratic polynomial is possible, as shown in the left-hand figure below. For a specific example, we could take $f(x) = x^2 + 1$. For cycling as in the picture, we would want $x_1 = x_0$. Putting this into Newton's formula and solving (the resulting quadratic) gives $x_0 = 1/\sqrt{3}$. One can easily run a MATLAB program to see that this indeed produces the asserted cycling. To get an example of polynomial cycling with a polynomial that actually has a root we need to use at least a third-degree polynomial. Working with $f(x) = x^3 - x = x(x-1)(x+1)$, which has the three (equally spaced) roots $x = 0, \pm 1$ the graph suggests that we can have a period-two cycling, so we put $x_1 = x_0$ into Newton's formula. The resulting cubic equation is easily solved exactly (it factors) or with the Symbolic Toolbox (see Appendix A) or approximately using Newton's method. The solution $x_0 = 1/\sqrt{5}$ produces the period-two cycling shown in the right-hand figure below, as can be checked by running Newton's method.

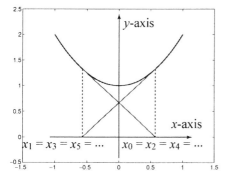
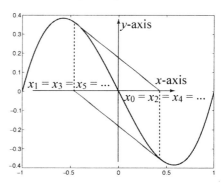

(b) On the right is an illustration of a period four cycle in Newton's method. If desired, one could make an explicit such example by taking, $f(x) = x^3 - x - 3$. The calculations would be, of course, more elaborate than those of part (a); it turns out that x_0 should be taken to be a bit less than zero. (More precisely, about -0.007446; you may wish to run a couple of hundred iterations of Newton's method using this value for x_0 to observe the cycling.) By contemplating the picture, it becomes clear that this function has cycles of any order.   Just move x_0 closer to the right toward the location where $f'(x)$ has a root.

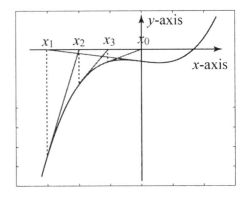

EFR 6.6: (a) The M-file is boxed below:

```
function [root, yval,niter] = secant(varfun,x0, x1, tol, nmax)
% input variables: varfun, x0, x1  tol, nmax
% output variables:  root, yval, niter
% varfun = the string representing a mathematical function (built-in,
% M-file, or inline) , x0  and x1 = the two (different) initial
% approx.
% The program will perform the Secant method to approximate a root of
% varfun near x=x0 until either successive approximations differ by
% less than tol or nmax iterations have been completed, whichever
% comes first.   If the tol and nmax variables are omitted default
% values of eps (approx. 10^(-16)) and 30 are used.
% We assign the default tolerance and maximum number of iterations if
% none are specified
if nargin < 4
    tol=eps;, nmax=50;
end


%we now initialize the iteration
xn=x0;,xnnext=x1;

%finally we set up a loop to perform the approximations
for n=1:nmax
    yn=feval(varfun, xn);, ynnext=feval(varfun, xnnext);
    if ynnext == 0
        fprintf('Exact root found\r')
        root = xnnext;, yval = 0;, niter=n;
        return
    end
    if yn == ynnext
        error('horizontal secant encountered, Secant method failed, try
changing x0, x1')
    end
    newx=xnnext-feval(varfun, xnnext)*(xnnext-
xn)/(feval(varfun,xnnext)-feval(varfun,xn));
    if abs(newx-xnnext)<tol
            fprintf('The secant method has converged\r')
            root = newx;, yval = feval(varfun, root);, niter=n;
```

```
              return
      elseif n==nmax
              fprintf('Maximum number of iterations reached\r')
              root = newx;, yval = feval(varfun, root);, niter=nmax
              return
      end
      xn=xnnext;, xnnext=newx;
end
```

(b) The syntax of this M-file is very close to that of newton:
```
>> f=inline('x^4-2');, [r y n] = secant(f, 2,1.5)
```
→The secant method has converged, r = 1.18920711500272,
y = -2.220446049250313e-016, n = 9
```
>> abs(r-2^(1/4))     →ans = 0
```
In conclusion, the secant method took nine iterations and the approximate root (r) had residual which was essentially zero (in floating point arithmetic) and coincided with the exact answer $\sqrt[4]{2}$ (in floating point arithmetic).

EFR 6.7:   (a) For shorthand we write: HOC to mean "the highest order of convergence," and AEC to mean "the asymptotic error constant." For each sequence, we determine these quantities if they exist:
(i) HOC = 1; AEC = 1 (linear convergence), (ii) HOC = 1, AEC = ½ (linear convergence), (iii) HOC = 3/2, AEC = 1, (iv) HOC = 2, AEC = 1 (quadratic convergence), (v) HOC does not exist. There is hyperconvergence for every order $\alpha < 2$, but the sequence does not have quadratic convergence.

(b) The sequence $e_n = e^{-3^n}$ has HOC = 3. In general, $e_n = e^{-k^n}$ has HOC = k whenever k is a positive number.

EFR 6.8: Write $f(x) = (x-r)^M h(x)$, where M is the order of the root (and so $h(r) \neq 0$). Differentiating, we see that the function $F(x) \equiv f(x)/f'(x)$ can be written as $F(x) = (x-r)H(x)$, where $H(x) = h(x)/[Mh(x)+(x-r)h'(x)]$. Since $H(r) = 1/M \neq 0$, we see that $x = r$ is a simple root of $F(x)$.   Since $F'(x) \equiv [(f'(x))^2 - f(x)f''(x)]/(f'(x))^2$, this method requires computing both $f'(x)$ and $f''(x)$. The roundoff errors can also get quite serious. For example, if we are converging to a simple root, then in the iterative computations of $F'(x_n) = [(f'(x_n))^2 - f(x_n)f''(x_n)]/(f'(x_n))^2$, $(f'(x_n))^2$ will be converging to a positive number, while $f(x_n)f''(x_n)$ will be converging to zero. Thus, when these two numbers are subtracted roundoff errors can be insidious. With higher-order roots each of $(f'(x_n))^2$ and $f(x_n)f''(x_n)$ will be getting small very fast and can underflow to zero causing Newton's method to stop. If the root is a multiple root and the order is known not to be too high then this method performs reasonably well. If the order is known, however, the newtonmr method is a better choice.

CHAPTER 7: Matrices And Linear Systems

EFR 7.1:   Abbreviate the matrices in (1) by $DE = P$, and write $P = [p_{ij}]$. Now, by definition, p_{ij} = (ith row of D) • (jth column of E) = $d_i \cdot e_{ij}$ (by diagonal form of D). But by the diagonal form of E, e_{ij} (and hence also p_{ij}) is zero unless $i = j$, in which case $e_{ij} = e_i$.   Thus $p_{ij} = \{d_i e_i,$ if $i = j$; $0,$ if $i \neq j$ and this is a restatement of (1).

EFR 7.2: (a) The M-file is boxed on the next page:

```
function A=randint(n,m,k)
%generates an n by m matrix whose entries are random integers whose
%absolute values do not exceed k
A=zeros(n,m);
for i=1:n
    for j=1:m
        x=(2*k+1)*rand-k; %produces a random real number in (-k,k+1)
        A(i,j)=floor(x);
    end
end
```

(b) In the random experiments below, we print out the matrices only in the first trial.
\>\> A=randint(6,6,9);, B=randint(6,6,9);, det(A*B), det(A*B)-
det(A)*det(B)

→A =
9	-5	2	0	7	5
-1	-9	6	-1	2	6
8	5	-6	-2	8	8
-2	7	-8	-3	6	-9
-7	-6	-6	2	-4	-6
-9	5	-1	8	-1	-2

→B =
7	0	-6	3	6	-9
3	-2	6	0	4	-1
-4	-6	-6	3	-4	1
-7	4	-2	7	7	2
0	8	6	3	6	3
-3	-4	-3	1	4	-4

→ det(A*B) =   -1.9436e+010   →ans =   0

\>\> A=randint(6,6,9);, B=randint(6,6,9);, det(A*B), det(A*B)-
det(A)*det(B)
→ans = 6.8755e+009, 0
\>\> A=randint(6,6,9);, B=randint(6,6,9);, det(A*B), det(A*B)-
det(A)*det(B)
→ ans = 8.6378e+010, 0
The last output 0 in each of the three experiments indicates that formula (4) checks.
(c) Here, because of their size, we do not print out any of the matrices.
\>\> A=randint(16,16,9);, B=randint(16,16,9);, det(A*B), det(A*B)-
det(A)*det(B)
→ans = -1.2268e+035, 1.8816e+021
\>\> A=randint(16,16,9);, B=randint(16,16,9);, det(A*B), det(A*B)-
det(A)*det(B)
→ans =1.4841e+035, -6.9913e+021
\>\> A=randint(16,16,9);, B=randint(16,16,9);, det(A*B), det(A*B)-
det(A)*det(B)
→ans = 3.3287e+035, ans = 7.0835e+021

The results in these three experiments are deceptive. In each, it appears that the left and right sides of (4) differ by something of magnitude 10^{21}. This discrepancy is entirely due to roundoff errors! Indeed, in each trial, the value of the determinant of AB was on the order of 10^{35}. Since MATLAB's (double precision IEEE) floating point arithmetic works with only about 15 significant digits, the much larger (35-digit) numbers appearing on the left and right sides of (4) have about the last 20 digits turned into unreliable "noise." This is why the discrepancies are so large (the extra digit lost came from roundoff errors in the internal computations of the determinants and the right side of (4)). Note that in part (b), the determinants of the smaller matrices in question had only about 10 significant digits, well within MATLAB's working precision.

EFR 7.3: Using the fill command as was done in the text to get the gray cat of Figure 7.3(b), you can get those other-colored cats by simply replacing the RGB vector for gray by the following: Orange → RGB = [1 .5 0], Brown → RGB = [.5 .25 0], Purple → RGB = [.5 0 .5]. Since each of these colors can have varying shades, your answers may vary. Also, the naked eye may not be able to distinguish between colors arising from small perturbations of these vectors (say by .001 or even .005). The RGB vector representing MATLAB's cyan is RGB = [0 1 1].

EFR 7.4: By property (10) (of linear transformations): $L(\alpha P_1) = \alpha L(P_1)$; if we put $\alpha = 0$, we get that $L(\vec{0}) = \vec{0}$ (where $\vec{0}$ is the zero vector). But a shift transformation $T_{V_0}(x, y) = (x, y) + V_0$ satisfies $T_{V_0}(\vec{0}) = \vec{0} + V_0 = V_0$. So the shift transformation T_{V_0} being linear would force $V_0 = \vec{0}$, which is not allowed in the definition of a shift transformation (since then T_{V_0} would then just be the identity transformation).

EFR 7.5: (a) As in the solution of Example 7.4, we individually multiply out the homogeneous coordinate transformation matrices (as per the instructions in the proof of Theorem 7.2) from right to left. The first transformation is the shift with vector (1,0) with matrix: $T_{(1,0)} \sim \begin{bmatrix} 1 & 0 & 1 \\ 0 & 1 & 0 \\ 0 & 0 & 1 \end{bmatrix} = H_1$. After

this we apply a scaling S whose matrix is given by $S \sim \begin{bmatrix} 2 & 0 & 0 \\ 0 & 1 & 0 \\ 0 & 0 & 1 \end{bmatrix} = H_2$. The homogeneous

cooordinate matrix for the composition of these two transformations is: $M = H_2 H_1 = \begin{bmatrix} 2 & 0 & 0 \\ 0 & 1 & 0 \\ 0 & 0 & 1 \end{bmatrix} \begin{bmatrix} 1 & 0 & 1 \\ 0 & 1 & 0 \\ 0 & 0 & 1 \end{bmatrix} = \begin{bmatrix} 2 & 0 & 2 \\ 0 & 1 & 0 \\ 0 & 0 & 1 \end{bmatrix}$. We assume (as in the text) that we have left in the

graphics window the first (white) cat of Figure 7.3(a) and that the CAT matrix A is still in our workspace. The following commands will now produce the new "fat CAT":
```
>> H1=[1 0 1;0 1 0; 0 0 1];, H2=[2 0 0;0 1 0;0 0 1];, M=H2*H1
>> AH=A;, AH(3,:)=ones(1,10);   %homogenize the CAT matrix
>> AH1=M*AH; % homogenized "fat CAT" matrix
>> hold on
>> plot(AH1(1,:), AH1(2,:), 'r')
>> axis([-2 10 -3 6]) % set wider axes to accommodate "fat CAT"
>> axis('equal')
```

The resulting plot is shown in the left-hand figure that follows.
(b) Each of the four cats needs to first get rotated by their specified angle about the same point (1.5, 1.5). As in the solution to Example 7.4, these rotations can be accomplished by first shifting this point to (0, 0) with the shift $T_{(-1.5,-1.5)}$, then performing the rotation, and finally shifting back with the inverse shift $T_{(1.5,1.5)}$. In homogeneous coordinates, the matrix representing this composition is (just like in the solution to Example 7.4):

$$M = \begin{bmatrix} 1 & 0 & 1.5 \\ 0 & 1 & 1.5 \\ 0 & 0 & 1 \end{bmatrix} \begin{bmatrix} \cos(\theta) & -\sin(\theta) & 0 \\ \sin(\theta) & \cos(\theta) & 0 \\ 0 & 0 & 1 \end{bmatrix} \begin{bmatrix} 1 & 0 & -1.5 \\ 0 & 1 & -1.5 \\ 0 & 0 & 1 \end{bmatrix}.$$

After this rotation, each cat gets shifted in the specified direction with $T_{(\pm 1, \pm 1)}$. For the colors of our cats let's use the following: black (rgb = [0 0 0]) light gray (rgb = [.7 .7 .7]) dark gray (rgb = [.3 .3 .3]), and brown (rgb = [.5 .25 0]). The following commands will then plot those cats:
```
>> clf, hold on   %prepare graphic window
>> %upper left cat,   theta = pi/6 (30 deg), shift vector = (-3, 3)
>> c = cos(pi/6);, s = sin(pi/6);
>> M=[1 0 1.5;0 1 1.5;0 0 1]*[c -s 0;s c 0;0 0 1]*[1 0 -1.5;0 1 -
1.5;0 0 1];
>> AUL=[1 0 -3;0 1 3;0 0 1]*M*AH;
>> fill(AUL(1,:), AUL(2,:), [0 0 0])
>> %upper right cat,   theta = -pi/6 (-30 deg), shift vector = (3, 1)
>> c = cos(-pi/6);, s = sin(-pi/6);
>> M=[1 0 1.5;0 1 1.5;0 0 1]*[c -s 0;s c 0;0 0 1]*[1 0 -1.5;0 1 -
1.5;0 0 1];
>> AUR=[1 0 1;0 1 1;0 0 1]*M*AH;
```

```
>> fill(AUR(1,:), AUR(2,:), [.7 .7 .7])
>> %lower left cat,  theta = pi/4 (45 deg), shift vector = (-3, -3)
>> c = cos(pi/4);, s = sin(pi/4);
>> M=[1 0 1.5;0 1 1.5;0 0 1]*[c -s 0;s c 0;0 0 1]*[1 0 -1.5;0 1 -
1.5;0 0 1];
>> ALL=[1 0 -3;0 1 -3;0 0 1]*M*AH;
>> fill(ALL(1,:), ALL(2,:), [.3 .3 .3])
>> %lower right cat,  theta = -pi/4 (-45 deg), shift vector = (3, -3)
>> c = cos(-pi/4);, s = sin(-pi/4);
>> M=[1 0 1.5;0 1 1.5;0 0 1]*[c -s 0;s c 0;0 0 1]*[1 0 -1.5;0 1 -
1.5;0 0 1];
>> ALR=[1 0 3;0 1 -3;0 0 1]*M*AH;
>> fill(ALR(1,:), ALR(2,:), [.5 .25 0])
>> axis('equal'), axis off %see graphic w/out distraction of axes.
```

EFR 7.6: (a) This first M-file is quite straightforward and is boxed below:

```
function B=mkhom(A)
B=A;
[n m]=size(A);
B(3,:)=ones(1,m);
```

(b) This M-file is boxed below:

```
function Rh=rot(Ah,x0,y0,theta)
%viz. EFR 7.6; theta should be in radians
%inputs a 3 by n matrix of homogeneous vertex coordinates, xy
coordinates
%of a point and an angle theta.  Output is corresponding matrix of
%vertices rotated by angle theta about (x0,y0).

%first construct homogeneous coordinate matrix for shifting (x0,y0)
to (0,0)
SZ=[1 0 -x0;0 1 -y0; 0 0 1];
%next the rotation matrix at (0,0)
R=[cos(theta) -sin(theta) 0; sin(theta) cos(theta) 0;0 0 1];
%finally the shift back to (x0,y0)
SB=[1 0 x0;0 1 y0;0 0 1];
%now we can obtain the desired rotated vertices:
Rh=SB*R*SZ*Ah;
```

EFR 7.7: (a) The main transformation that we need in this movie is vertical scaling. To help make the code for this exercise more modular, we first create, as in part (b) of the last EFR, a separate M-file for vertical scaling:

```
function Rh=vertscale(Ah,b,y0)
%inputs a 3 by n matrix of homogeneous vertex coordinates, a (pos.)
```

```
%numbers a for  y- scales, and an optional  arguments y0
%for center of scaling.  Output is homogeneous coor. matrix of scaled
%vertices.  default value of y0 is 0.

if nargin <3
    y0=0;
end
%first construct homogeneous coordinate matrix for shifting y=y0 to
%y=0
SZ=[1 0 0;0 1 -y0; 0 0 1];
%next the scaling matrix at (0,0)
S=[1 0 0; 0 b 0;0 0 1];
%finally the shift back to y=0
SB=[1 0 x0;0 1 y0;0 0 1];
%now we can obtain the desired scaled vertices:
Rh=SB*S*SZ*Ah;
```

Making use of the above M-file, the following script recreates the CAT movie of Example 7.4 using homogeneous coordinates:

```
%script for EFR 7.6(a):  catmovieNo1.m  cat movie creation
%sets up the basic CAT of Chapter 7 as in Figure 7.3(a), except in
color black and the
%vertices are stored as columns of a matrix A.  Screen is left with
"hold on".
clf, counter=1;

A=[0  0  .5  1  2  2.5  3  3  1.5  0; ...
     0  3  4  3  3  4  3  0  -1  0]; %Basic CAT matrix
%Ah = mkhom(A); %use the M-file from EFR 7.6

t=0:.02:2*pi; %creates time vector for parametric equations for eyes
xL=1+.4*cos(t);, y=2+.4*sin(t); %creates circle for left eye
LE=mkhom([xL; y]); %homogeneous coordinates for left eye
xR=2+.4*cos(t);, y=2+.4*sin(t); %creates circle for right eye
RE=mkhom([xR; y]); %homogeneous coordinates for right eye
xL=1+.15*cos(t);, y=2+.15*sin(t); %creates circle for left pupil
LP=mkhom([xL; y]); %homogeneous coordinates for left pupil
xR=2+.15*cos(t);, y=2+.15*sin(t); %creates circle for right pupil
RP=mkhom([xR; y]); %homogeneous coordinates for right pupil


for s=0:.2:2*pi
 factor = (cos(s)+1)/2;
 plot(A(1,:), A(2,:), 'k'), hold on
 axis([-2 5 -3 6]), axis('equal')
 LEtemp=vertscale(LE,factor,2);, LPtemp=vertscale(LP,factor,2);
 REtemp=vertscale(RE,factor,2);, RPtemp=vertscale(RP,factor,2);
 hold on
 fill(LEtemp(1,:), LEtemp(2,:),'y'), fill(REtemp(1,:),
REtemp(2,:),'y')
 fill(LPtemp(1,:), LPtemp(2,:),'k'), fill(RPtemp(1,:),
RPtemp(2,:),'k')
 M(:, counter) = getframe;
 hold off
 counter=counter+1;
end
```

(b) As in part (a), the following script M-file will make use of two supplementary M-files, AhR=reflx(Ah, x0) and, AhS=shift(Ah, x0, y0), that perform horizontal reflections and shifts in homogeneous coordinates, respectively. The syntaxes of these M-files are explained in Exercises 5 and 6 of this section. Their codes can be written in a fashion similar to the code vertscale but for completeness are can be downloaded from the ftp site for this text (see the beginning of this appendix). They can be avoided by simply performing the homogeneous coordinate transformations directly, but at a cost of increasing the size of the M-file below:

```
%coolcatmovie.m: script for making coolcat movie matrix M of EFR 7.7

%act one:  eyes shifting left/right
t=0:.02:2*pi;, counter=1
A=[0  0  .5  1  2  2.5  3  3  1.5  0; ...
   0  3  4  3  3  4  3  0  -1  0];
x=1+.4*cos(t);, y=2+.4*sin(t);,xp=1+.15*cos(t);, yp=2+.15*sin(t);
LE=[x;y]; LEh=mkhom(LE);, LP=[xp;yp]; LPh=mkhom(LP);
REh=reflx(LEh, 1.5);, RPh=reflx(LPh, 1.5);
LW=[.3 -1; .2 -.8];, LW2=[.25 -1.1;.25 -.6]; %left whiskers
LWh=mkhom(LW);, LW2h=mkhom(LW2);
RWh=reflx(LWh, 1.5);, RW2h=reflx(LW2h, 1.5); %reflect left whiskers
                                             %to get right ones
M=[1 1.5 2;.25 -.25 .25];, Mh=mkhom(M);   %matrix & homogenization of
                                          %cats mouth
Mhrefl=refly(Mh,-.25); %homogeneous coordinates for frown
for n=0:(2*pi)/20:2*pi
plot(A(1,:), A(2,:),'k')
axis([-2 5 -3 6]), axis('equal')
hold on
plot(LW(1,:), LW(2,:),'k'), plot(LW2(1,:), LW2(2,:),'k')
plot(RWh(1,:), RWh(2,:),'k')
plot(RW2h(1,:), RW2h(2,:),'k')
plot(Mhrefl(1,:), Mhrefl(2,:),'k')
fill(LE(1,:), LE(2,:),'y'), fill(REh(1,:), REh(2,:),'y')
LPshft=shift(LPh,-.25*sin(n),0);, RPshft=shift(RPh,-.25*sin(n),0);
fill(LPshft(1,:), LPshft(2,:),'k'), fill(RPshft(1,:),
RPshft(2,:),'k')
Mov(:, counter)=getframe;
hold off
counter = counter +1;
end


%act two:  eyes shifting up/down
for n=0:(2*pi)/20:2*pi
plot(A(1,:), A(2,:),'k')
axis([-2 5 -3 6]), axis('equal')
hold on
plot(LW(1,:), LW(2,:),'k'), plot(LW2(1,:), LW2(2,:),'k')
plot(RWh(1,:), RWh(2,:),'k')
plot(RW2h(1,:), RW2h(2,:),'k')
plot(Mhrefl(1,:), Mhrefl(2,:),'k')
fill(LE(1,:), LE(2,:),'y'), fill(REh(1,:), REh(2,:),'y')
LPshft=shift(LPh,0,.25*sin(n));, RPshft=shift(RPh,0,.25*sin(n));
fill(LPshft(1,:), LPshft(2,:),'k'), fill(RPshft(1,:),
RPshft(2,:),'k')
Mov(:, counter)=getframe;
hold off
counter = counter +1;
end


%act three:  whisker rotating up/down then smiling
for n=0:(2*pi)/10:2*pi
plot(A(1,:), A(2,:),'k')
axis([-2 5 -3 6]), axis('equal')
hold on
fill(LE(1,:), LE(2,:),'y'),fill(LP(1,:), LP(2,:),'k')
```

```
fill(REh(1,:), REh(2,:),'y'),fill(RPh(1,:), RPh(2,:),'k')
LWrot=rot(LWh,.3,.2,-pi/6*sin(n));, LW2rot=rot(LW2h, .25,.25,-
pi/6*sin(n));
RWrot=reflx(LWrot, 1.5);, RW2rot=reflx(LW2rot, 1.5);
plot(LWrot(1,:), LWrot(2,:),'k'), plot(LW2rot(1,:), LW2rot(2,:),'k')
plot(RWrot(1,:), RWrot(2,:),'k'),plot(RW2rot(1,:), RW2rot(2,:),'k')
if n == 2*pi
   plot(Mh(1,:), Mh(2,:),'k')
   for n=1:10, L(:,n)=getframe;, end
   Mov(:, counter:(counter+9))=L;
   break
else
   plot(Mhrefl(1,:), Mhrefl(2,:),'k')

end
Mov(:, counter)=getframe;

hold off
counter = counter +1;
end

%THE END
```

EFR 7.8: (a) Certainly the zeroth generation consists of $1 = 3^0$ triangles. Since the sidelength is one, and the triangle has each of its angles being $\pi/3$, its altitude must be $\sin(\pi/3) = \sqrt{3}/2$. Thus, the area of the single zeroth generation triangle is $\sqrt{3}/4$. Now, each time we pass to a new generation, each triangle splits into three (equilateral) triangles of half the length of the triangles of the current generation. Thus, by induction, the nth generation will have 3^n equilateral triangles of sidelength $1/2^n$ and hence each of these has area $(1/2) \cdot 1/2^n \cdot [\sqrt{3}/2]/2^n = \sqrt{3}/4^{n+1}$.

(b) From part (a), the nth generation of the Sierpinski carpet consists of 3^n equilateral triangles each having area $\sqrt{3}/4^{n+1}$. Hence the total area of this nth generation is $\sqrt{3}(3/4)^n/4$. Since this expression goes to zero as $n \to \infty$, and since the Sierpinski carpet is contained in each of the generation sets, it follows that the area of the Sierpinski carpet must be zero.

EFR 7.9: (a) The 2×2 matrices representing dilations: $\begin{bmatrix} s & 0 \\ 0 & s \end{bmatrix}$ ($s > 0$), and reflections with respect to the x-axis: $\begin{bmatrix} -1 & 0 \\ 0 & 1 \end{bmatrix}$ or y-axis: $\begin{bmatrix} 1 & 0 \\ 0 & -1 \end{bmatrix}$ are both diagonal matrices and thus commute with any other 2×2 matrices; i.e., if D is any diagonal matrix and A is any other 2×2 matrix, then $AD = DA$. In particular, these matrices commute with each other and with the matrix representing a rotation through the angle θ: $\begin{bmatrix} \cos\theta & -\sin\theta \\ \sin\theta & \cos\theta \end{bmatrix}$. By composing rotations and reflections, we can obtain transformations that will reflect about any line passing through $(0,0)$. Once we throw in translations, we can reflect about any line in the plane and (as we have already seen) rotate with any angle about any point in the plane. By the definition of similitudes, we now see that compositions of these general transformations can produce the most general similitudes. Translating into homogeneous coordinates (using the proof of Theorem 7.2) we see that the matrix for such a composition can be expressed as $\begin{bmatrix} s\cos\theta & -s\sin\theta & x_0 \\ \pm s\sin\theta & \pm s\cos\theta & y_0 \\ 0 & 0 & 1 \end{bmatrix}$ where s now is allowed to be any nonzero number. If the sign in the second row is negative, we have a reflection: If $s > 0$, it is a y-axis reflection; if $s < 0$, it is an x-axis reflection.

(b) Let T_1 and T_2 be two similar triangles in the plane. Apply a dilation, if necessary, to T_1 so that it has the same sidelengths as T_2. Next, apply a shift transformation to T_1 so that a vertex gets shifted to a corresponding vertex of T_2, and then apply a rotation to T_1 about this vertex so that a side of T_1 transforms into a corresponding side of T_2.

At this point, either T_1 and T_2 are now the same triangle, or they are reflections of one another across the common side. A final reflection about this line, if necessary, will thus complete the transformation of T_1 into T_2 by a similitude.

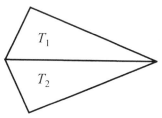

(c) It is clear that dilations, rotations, and shifts are essential. For an example to see why reflection is needed, simply take T_1 to be any triangle with three different angles and T_2 to be its reflection about one of the edges (see figure). It is clearly not possible to transform one of these two triangles into the other using any combination of dilations, rotations, and shifts.

EFR 7.10: (a) There will be only one generation; here are the outputs that were asked for (in format short):

```
A→      0    1.0000  2.0000        A1 →    0    0.5000  1.0000
        0    1.7321     0                  0    0.8660     0
     1.0000  1.0000  1.0000             1.0000  1.0000  1.0000

A2 →  1.0000  1.5000  2.0000        A3→   0.5000  1.0000  1.5000
         0    0.8660     0                0.8660  1.7321  0.8660
      1.0000  1.0000  1.0000             1.0000  1.0000  1.0000

A1([1 2],2) → 0.5000              A3([1 2],2) →  1.5000
              0.8660                            0.8660
```

(b) Since the program calls on itself and does so more than once (as long as `niter` is greater than zero), placing a `hold off` anywhere in the program will cause graphics created on previous runs to be lost, so such a feature could not be incorporated into the program.

(c) Since we want the program to call on itself iteratively with different vertex sets, we really need to allow vertex sets to be inputted. Different vertex inputs are possible, but in order for the program to function effectively, they should be vertices of a triangle to which the similitudes in the program correspond. (e.g., any of the triangles in any generation of the Sierpinski gasket).

EFR 7.11: (a) S2, S1, S3, S2, S3, S2

(b) We list the sequence of float points in nonhomogeneous coordinates and in `format short`:
[0.5000 0.8660],   [0.2500 0.4330],   [1.1250 0.2165],   [1.0625 0.9743],   [1.5313 0.4871],
[1.2656 1.1096]

(c) The program is designed to work for any triangle in the plane. The reader can check that the three similitudes are constructed in a way that uses midpoints of the triangle and the resulting diagram will look like that of Figure 7.15.

EFR 7.12: (a) As with `sgasket2`, the program `sgasket3` contructs future-generation triangles simply from the vertices and (computed) midpoints of the current-generation triangles. Thus, it can deal effectively with any triangle and produce Sierpinski-type fractal generations.

(b) For illustration purposes, the following trials were run on MATLAB's Version 5, so as to illustrate the flop count differences. The code is easily modified to work on the new version of MATLAB by simply deleting the "flops" commands.
```
V1=[0 0];,  V2=[1 sqrt(3)];,  V3=[2 0];   %vertices of an equilateral
triangle
```

```
test = [1 3 6 8 10];
```

`>> for i=1:5`
`flops(0), tic,`
`sgasket1(V1,V2,V3,test(i)), toc,`
`flops`
`end`
→ (ngen =1) elapsed_time = 0.0600,
ans =191
    (ngen =3) elapsed_time = 0.2500,
ans =2243
    (ngen =6) elapsed_time = 0.8510,
ans =62264
    (ngen =8) elapsed_time = 7.2310,
ans =560900
    (ngen =10) elapsed_time = 65.4640,
ans =5048624
	`>> for i=1:5`
	`flops(0), tic,`
	`sgasket3(V1,V2,V3,test(i)), toc,`
	`flops`
	`end`
	→ (ngen =1) elapsed_time = 0.1400,
	ans = 45
	    (ngen =3) elapsed_time = 0.1310,
	ans =369
	    (ngen =6) elapsed_time = 0.7210,
	ans =9846
	    (ngen =8) elapsed_time = 6.2990,
	ans =88578
	    (ngen =10) elapsed_time = 46.7260,
	ans =797166

We remind the reader that the times will vary, depending on the machine being used and other processes being run. The above tests were run on a rather slow machine, so the resulting times are longer than typical.

EFR 7.13: The M-file is boxed below:

```
function []=snow(n)
S=[0 1 2 0;0 sqrt(3) 0 0];
index=1;
while index <=n
    len=length(S(1,:));
    for i=1:(len-1)
delta=S(:,i+1)-S(:,i);
perp=[0 -1;1 0]*delta;
T(:,4*(i-1)+1)=S(:,i);
T(:,4*(i-1)+2)=S(:,i)+(1/3)*delta;
T(:,4*(i-1)+3)=S(:,i)+(1/2)*delta+(1/3)*perp;
T(:,4*(i-1)+4)=S(:,i)+(2/3)*delta;
T(:,4*(i-1)+5)=S(:,i+1);
end
index=index+1;
S=T;
end
plot(S(1,:),S(2,:), axis('equal')
```

The outputs of `snow(1)`, `snow(2)`, and `snow(6)` are illustrated in Figures 7.17 and 7.18.

EFR 7.14: For any pair of nonparallel lines represented by a two-dimensional linear system: $\begin{bmatrix} a & b \\ c & d \end{bmatrix}\begin{bmatrix} x \\ y \end{bmatrix} = \begin{bmatrix} e \\ f \end{bmatrix}$, the coefficient matrix will have nonzero determinant $\alpha = ad - bc$. The lines are also represented by the equivalent system $\begin{bmatrix} a/\alpha & b/\alpha \\ c & d \end{bmatrix}\begin{bmatrix} x \\ y \end{bmatrix} = \begin{bmatrix} e/\alpha \\ f \end{bmatrix}$, where now the coefficient matrix has determinant $(a/\alpha)d - (b/\alpha)c = 1$. This change simply amounts to dividing the first equation by α.

EFR 7.15: (a) As in the solution of Example 7.7, the interpolation equations $p(-2) = 4$, $p(1) = 3$, $p(2) = 5$, and $p(5) = -22$ (where $p(x) = ax^3 + bx^2 + cx + d$) translate into the linear system:

$$\begin{bmatrix} -8 & 4 & -2 & 1 \\ 1 & 1 & 1 & 1 \\ 8 & 4 & 2 & 1 \\ 125 & 25 & 5 & 1 \end{bmatrix}\begin{bmatrix} a \\ b \\ c \\ d \end{bmatrix} = \begin{bmatrix} 4 \\ 3 \\ 5 \\ -22 \end{bmatrix}$$ We solve this using left division, as in Method 1 of the solution of

Example 7.7:
```
>> format long
>> A=[-8 4 -2 1;1 1 1 1;8 4 2 1;125 25 5 1];, b=[4 3 5 -22]';
>> x=A\b
```
→x =   -0.47619047619048 (= a)
        1.05952380952381 (= b)
        2.15476190476190 (= c)
        0.26190476190476 (= d)

(b)   As in part (a) and the solution of Example 7.7, we create the matrix A and vector b of the corresponding linear system: $Ax = b$. A loop will facilitate the construction of A:
```
>> xvals = -3:5;, A = zeros(9) %initialize the 9 by 9 matrix A
>> for i =1:length(xvals)
A(i,:)=xvals(i).^(8:-1:0);
end
>> b = [-14.5  -12  15.5  2  -22.5  -112  -224.5  318  3729.5]'
```
We next go through each of the three methods of solving the linear system that were introduced in the solution of Example 7.7. We are working on an older and slower computer with MATLAB Version 5, so we will have flop counts, but the times will be slower than typical. The code is easily modified to work on the new version of MATLAB by simply deleting the "flops" commands. We show the output for x only for Method 1 (in format long) as the answers with the other two methods are essentially the same.
Method 1:

>> flops(0), tic, x=A\b, toc, flops	→x =	-0.00000000000000 0.00000000000000 0.50000000000000 -0.00000000000001 -6.00000000000000 -1.99999999999996 0.00000000000000 -17.00000000000003 2.00000000000000	→elapsed_time = 0.1300 →ans = 1125 (flops)

Method 2:

>> flops(0), tic, x=inv(A)*b, toc, flops	→elapsed_time = 0.3010 →ans = 1935 (flops)

Method 3:

>> Ab=A;, Ab(:,10)=b; >> flops(0), tic, rref(Ab), toc, flops	→elapsed_time = 3.3150 →ans = 2175 (flops)

The size of this problem is small enough so that all three methods produce essentially the same vector x. The computation times and flop counts begin to demonstrate the relative efficiency of the three methods. Reading off the coefficients of the polynomial in order (from x), we get (after taking into account machine precision and rounding): $a = b = d = g = 0$, $c = 1/2$, $e = -6$, $f = -2$, $h = -17$, and $k = 2$, so that the interpolating polynomial is given by $p(x) = \frac{1}{2}x^6 - 6x^4 - 2x^3 - 17x + 2$. It is readily checked that this function satisfies all of the interpolation requirements.

EFR 7.16:   As in Example 7.8, for a fixed n, if we let x denote the exact solution, we then have

$b_n = H_n x = c(n)\left(1 \ \frac{1}{2} \ \frac{1}{3} \ \cdots \ \frac{1}{n-1} \ \frac{1}{n}\right)'$.   In order for b_n to have all integer coordinates, we need to

have $c(n)$ be a multiple of each of the integers 1, 2, 3, ..., n. The smallest such $c(n)$ is thus the least common multiple of these numbers. We can use MATLAB's `lcm(a,b)` to find the lcm of any set of integers with a loop. Here is how it would work to find $c(n) = \text{lcm}(1,2,\dots,n)$:

```
>> cn=1   %initialize
>> for k=1:n,   c(n)=lcm(cn, k),   end
```

The remaining code for constructing the exact solution x, the numerical solution of Method 1 x_meth1, and the numerical solution of Method 2 x_meth2 are just as in Example 7.9. The flops commands in these codes should be omitted if you are using Version 6 or later. Also, since these computations were run on an older machine, the elapsed times will be larger than what is typical (but their ratios should be reasonably consistent). The loop below will give us the data we need for both parts (a) and (b):

```
>> for n=20:10:30
cn=1;   %initialize
for k=1:n,   c(n)=lcm(cn, k);,   end
x = zeros(n,1);,   x(1)=cn;
bn = hilb(n)*x;
flops(0), tic, x_meth1=hilb(n)\bn;, toc, flops
flops(0), tic, x_meth2=inv(hilb(n))*bn;, toc, flops
Pct_err_meth1=100*max(abs(x-x_meth1))/cn,
Pct_err_meth2=100*max(abs(x-x_meth2))/cn
end
```

Along with the expected output, we also got some warnings from MATLAB that the matrix is either singular or poorly conditioned (to be expected). The output is summarized in the following table:

	Computer Time: $n = 20/n = 30$	Flop Count: $n = 20/n = 30$	Percentage of Maximum Error: $n = 20/n = 30$
Method 1:	0/0 seconds	10,339/27,481	0%/0%
Method 2:	0/0.01 seconds	20,312/63,509	512.5%/5400%

Note: The errors may vary depending on which version of MATLAB you are using.
(c) The errors with Method 1 turn out to be undetectable as n runs well over 1000. The computation times become more of a problem than the errors. MATLAB's "left divide" is based on Gaussian elimination with partial pivoting. After we study this algorithm in the next section, the effectiveness of this algorithm on the problem at hand will become clear.

EFR 7.17: (a) & (b): The first two are in reduced row echelon form. The corresponding general solutions are as follows:

(for M_1): $x_1 = 3$, $x_2 = 2$; (for M_2): $x_1 = 2s - 3t - 2$, $x_2 = s$, $x_3 = 5t + 1$, $x_4 = t$, where s and t are any real numbers.

`>> rref([1 3 2 0 3;2 6 2 -8 4])`	→ans	1 3 0 -8 1 0 0 1 4 1

(c) From the outputted reduced row echelon form, we obtain the following general solution of the first system: $x_1 = 1 - 3s + 8t$, $x_2 = s$, $x_3 = 1 - 4t$, $x_4 = t$, where s and t are any real numbers. Because of the arithmetic nature of the algorithm being used (as we will learn in the next section), it is often advantageous to work in `format rat` in cases where the linear system being solved is not too large and has integer or fraction coefficients. We do this for the second system:

`>> format rat`
`>> rref([1 -2 1 1 2 2;...`
`         -2 4 2 2 -2 0;...`
`          3 -6 1 1 5 4;...`
`         -1 2 3 1 1 3])`
	→
	ans
		1	-2	0	0	3/2	1

		0	0	1	0	1	3/2
		0	0	0	1	-1/2	-1/2
		0	0	0	0	0	0

From the output, we obtain the following general solution to the second system:
$x_1 = 1 + 2s - 3t/2$, $x_2 = s$, $x_3 = 3/2 - t$, $x_4 = t/2 - 1/2$, $x_5 = t$, where s and t are any real numbers.

EFR 7.18: (a) The algorithm for forward substitution: $x_1 = b_1/a_{11}, \quad x_j = \left(b_j - \sum_{k=1}^{j-1} a_{jk}x_k\right)/a_{jj}$

(the first formula is redundant since the latter includes it as a special case) is easily translated into the following MATLAB code (cf. Program 7.4):

```
function x=fwdsubst(L,b)
%Solves the lower triangular system Lx=b by forward substitution
%Inputs:   L = lower triangular matrix,  b = column vector of same
%dimension
%Output:   x = column vector (solution)
[n m]=size(L);
x(1)=b(1)/L(1,1);
for j=2:n
    x(j)=(b(j)-L(j,1:j-1)*x(1:j-1)')/L(j,j);
end
x=x';
```

`>> L=[1 2 3 4;0 2 3 4;0 0 3 4;0 0 0 4]';,`
`b=[4 3 2 1]';`
`>> format rat`
`>> fwdsubst(L,b)`
	→ans =	4
		-5/2
		-5/6
		-5/12

EFR 7.19: The two M-files are boxed below:

```
function B=rowmult(A,i,c)
% Inputs:   A = any matrix, i = any row index, c = any nonzero number
% Output:   B = matrix resulting from A by replacing row i by this row
% multiplied by c.
[m,n]=size(A);
if i<1|i>m
    error('Invalid index')
end
B=A;
B(i,:)=c*A(i,:);
```

```
function B=rowcomb(A,i,j,c)
% Inputs:   A = any matrix,  i, j = row indices,  c = a number
% Output:   B = matrix resulting from A by adding to row j the number
% c times row i.
[m,n]=size(A);
if i<1|i>m|j<1|j>m
    error('Invalid index')
end
if i==j
    error('Invalid row operation')
end
B=A;
B(j,:)=c*A(i,:)+A(j,:);
```

EFR 7.20: If we use `gausselim` to solve the system of Example 7.13, we get the correct answer (with lightning speed) with a flop count of 104 (if you have access to Version 5). In the table below, we give the corresponding data for the linear systems of parts (a) and (b) of EFR 7.16 (compare with the table in the solution of that exercise):

	Computer Time: $n = 20/n = 30$	Flop Count: $n = 20/n = 30$	Percentage of Maximum Error: $n = 20/n = 30$
Program 7.6	0.03/0.06 seconds	9,906/31,201	0%/0%

We observe that the time is detectable, although it was not when we used MATLAB's "left divide". Similarly, if we solve the larger systems of part (c) of EFR 7.16, we still get 0% errors for large values of n, but the times needed for `gausselim` to do the job are much greater than they were for "left divide". MATLAB's "left divide" is perhaps its most important program. It is based on Gaussian elimination, but also relies on numerous other results and techniques from numerical linear algebra. A full description of "left divide" would be beyond the scope of this book; for the requisite mathematics, we refer to [GoVL-83].

EFR 7.21: Working just as in Example 7.14, but this time in rounded floating point arithmetic, the answers are as follows:
(a) $x_1 = 1$, $x_2 = .999$ and (b) $x_1 = .001$, $x_2 = .999$.

EFR 7.22: Looking at (28) we see that solving for x_j takes: 1 division + $(n - j)$ multiplications + $(n - j - 1)$ additions (if $j < n$) + 1 subtraction (if $j < n$).
Summing from $j = n$ to $j = 1$, we deduce that:

Total multiplications/divisions = $\sum_{j=1}^{n} n - j + 1 = n^2 + n - n(n+1)/2 = (n^2 + n)/2$,

Total additions/subtractions = $\sum_{j=1}^{n-1}[n - j - 1 + 1] = \sum_{j=1}^{n-1}[n - j] = \sum_{j=1}^{n-1} j = (n^2 - n)/2$.

Adding gives the grand total of n^2 flops, as asserted.

EFR 7.23: Here we let $x = (x_1, x_2, \cdots, x_n)$ denote any n-dimensional vector and $\|x\|$ denote its max norm $\|x\|_{\infty} = \max\{|x_1|, |x_2|, \cdots, |x_n|\}$. The first norm axiom (36A) is clear from the definition of the max norm.   The second axiom (36B) is also immediate: $\|cx\| = \max\{|cx_1|, |cx_2|, \cdots, |cx_n|\}$ $= |c| \max\{|x_1|, |x_2|, \cdots, |x_n|\} = |c| \|x\|$.   Finally, the triangle inequality (36C) for the max norm readily follows from the ordinary triangle inequality for real numbers:

$$\|x + y\| = \max\{|x_1 + y_1|, |x_2 + y_2|, \cdots, |x_n + y_n|\}$$
$$= \max\{|x_1| + |y_1|, |x_2| + |y_2|, \cdots, |x_n| + |y_n|\} \le \|x\| + \|y\| .$$

EFR 7.24: (a) We may assume that $B \ne 0$, since otherwise both sides of the inequality are zero. Using definition (38), we compute:

$$\|AB\| = \max\left\{\frac{\|ABx\|}{\|x\|}, x \ne 0(\text{vector})\right\} = \max\left\{\frac{\|A(Bx)\|}{\|x\|} \cdot \frac{\|Bx\|}{\|Bx\|}, Bx \ne 0(\text{vector})\right\}$$

$$= \max\left\{\frac{\|A(Bx)\|}{\|Bx\|} \cdot \frac{\|Bx\|}{\|x\|}, Bx \ne 0(\text{vector})\right\} \le \|A\|\|B\|$$

(b) First note that for any vector $x \ne 0$, the vector $y = Ax$ is also nonzero (since A is nonsingular), and $A^{-1}y = x$. Using this notation along with definition (38), we obtain:

$$\|A^{-1}\| = \max\left\{\frac{\|A^{-1}y\|}{\|y\|}, y \ne 0(\text{vector})\right\} = \left(\min\left\{\frac{\|y\|}{\|A^{-1}y\|}, y \ne 0(\text{vector})\right\}\right)^{-1}$$

$$\underset{y=Ax}{=} \left(\min\left\{\frac{\|Ax\|}{\|x\|}, x \ne 0(\text{vector})\right\}\right)^{-1} .$$

EFR 7.25: (a) We first store the matrix A with the following loop, and then ask MATLAB for its condition number:
```
>>A=zeros(12);,   for i=1:12,  A(i,:)=i.^(11:-1:0);,   end,
c1=cond(A,inf)
>> c1  →Warning: Matrix is close to singular or badly scaled.  Results may be inaccurate.
RCOND = 8.296438e-017.
→c1 = 1.1605e+016
```

(b) >> c=norm(double(inv(sym(A))),inf)*norm(A,inf) →c =1.1605e+016
>> c-c1 →ans = 3864432
The approximation $c1$ to the condition number c is quite different, but relatively it at least has the same order of magnitude.   If we choose a larger Vandermonde matrix here, we would begin to experience more serious problems as was the situation in Example 7.23.
(c) >> b = (-1).^(0:11).*(1:12);, b=b'; % first create the vector b
>> z=A\b;, r=b-A*z; (We get another warning as above.)
>> errest=c1*norm(r,inf)/norm(A,inf)   →errest = 0.0020
>> norm(z,inf)→ans =8.7156e+004
At first glance, the accuracy looks quite decent. The warnings, however, remove any guarantees that Theorem 7.7 would otherwise allow us to have.
(d) >> z2=inv(A)*b;, r2=b-A*z2;   -> Warning: Matrix is close to singular or badly scaled. Results may be inaccurate. RCOND = 8.296438e-017.
>> errest2=c1*norm(r2,inf)/norm(A,inf)   →errest2 = 2.3494
(e) As in Example 7.23, we solve the system symbolically and then get the norms that we asked for:
>> S=sym(A);, x=S\b;, x=double(x);
>> norm(x-z,inf)→ans =3.0347e-005
>> norm(x-z2,inf)→ans =3.0347e-005
Thus, despite the warning we received, the numerical results are much more accurate than the estimates of Theorem 7.7 had indicated.

EFR 7.26:   (a) Since $\lambda I - A$ is a triangular matrix, Proposition 7.3 tells us that the determinant $p_A(\lambda) = \det(\lambda I - A)$ is simply the product of the diagonal entries: $p_A(\lambda) = (\lambda - 2)^2(\lambda - 1)^2$. Thus A has two eigenvalues: $\lambda = 1, 2$, each having algebraic multiplicity 2.
(b) >> [V, D] = eig([2 1 0 0;0 2 0 0;0 0 1 0;0 0 0 1])

→V =	1.0000	-1.0000	0	0	→D =	2	0	0	0
	0	0.0000	0	0		0	2	0	0
	0	0	1.0000	0		0	0	1	0
	0	0	0	1.0000		0	0	0	1

From the output of eig, we see that the eigenvalue $\lambda = 1$ has two linearly independent eigenvectors: [0 0 1 0]' and [0 0 0 1]', and so has geometric multiplicity 2, while the eigenvalue $\lambda = 2$ has only one independent eigenvector [2 0 0 0]' , and so has geometric multiplicity 1.
(c) From the way in which part (a) was done, we see that the eigenvalues of any triangular matrix are simply the diagonal entries (with repetitions indicating algebraic multiplicities).

EFR 7.27: (a) The M-file is boxed below:

```
function [x, k, diff] = jacobi(A,b,x0,tol,kmax)
% performs the Jacobi iteration on the linear system Ax=b.
% Inputs:  the coefficient matrix 'A', the inhomogeneity (column)
% vector 'b', the seed (column) vector 'x0' for the iteration
% process, the tolerance 'tol' which will cause the iteration to stop
% if the 2-norms of differences of successive iterates becomes
% smaller than 'tol', and 'kmax' which is the maximum number of
% iterations to perform.
% Outputs:  the final iterate 'x', the number of iterations performed
% 'k', and a vector 'diff' which records the 2-norms of successive
% differences of iterates.
% If any of the last three input variables are not specified, default
% values of x0= zero column vector, tol=1e-10 and kmax=100 are used.

%assign default input variables, as necessary
if nargin<3, x0=zeros(size(b));, end
if nargin<4, tol=1e-10;, end
if nargin<5, kmax=100;, end
if min(abs(diag(A)))<eps
    error('Coefficient matrix has zero diagonal entries, iteration
```

```
cannot be performed.\r')
end

[n m]=size(A);
xold=x0;
k=1;, diff=[];

while k<=kmax
    xnew=b;
    for i=1:n
        for j=1:n
            if j~=i
                xnew(i)=xnew(i)-A(i,j)*xold(j);
            end
        end
        xnew(i)=xnew(i)/A(i,i);
    end
    diff(k)=norm(xnew-xold,2);
    if diff(k)<tol
      fprintf('Jacobi iteration has converged in %d iterations.\r', k)
        x=xnew;
        return
    end
    k=k+1;, xold=xnew;
end
fprintf('Jacobi iteration failed to converge.\r')
x=xnew;
```

(b) >> A=[3 1 -1;4 -10 1;2 1 5]; b=[-3 28 20]';
   >> [x, k, diff] = jacobi(A,b,[0 0 0]',1e-6);
→Jacobi iteration has converged in 26 iterations.
>> norm(x-[1 -2 4]',2)→ans = 3.9913e-007 (Error is in agreement with Example 7.26.)
>> diff(26)→ans = 8.9241e-007 (Last successive difference is in agreement with Example 7.26.)
>> [x, k, diff] = jacobi(A,b,[0 0 0]');
→Jacobi iteration has converged in 41 iterations. (With default error tolerance 1e-10)

EFR 7.28: (a) The M-file is boxed below:

```
function [x, k, diff] = sorit(A,b,omega, x0,tol,kmax)
% performs the SOR iteration on the linear system Ax=b.
% Inputs:  the coefficient matrix 'A', the inhomogeneity (column)
% vector 'b', the relaxation paramter 'omega', the seed (column)
% 'x0' for the iteration process, the tolerance 'tol' vector which
% will cause the iteration to stop if the 2-norms of successive
% iterates becomes smaller than 'tol', and 'kmax' which is the
% maximum number of iterations to perform.
% Outputs:  the final iterate 'x', the number of iterations performed
% 'k', and a vector 'diff' which records the 2-norms of successive
% differences of iterates.
% If any of the last three input variables are not specified, default
% values of x0= zero column vector, tol=1e-10 and kmax=100 are used.

%assign default input variables, as necessary
if nargin<4, x0=zeros(size(b));, end
if nargin<5, tol=1e-10;, end
if nargin<6, kmax=100;, end

if min(abs(diag(A)))<eps
    error('Coefficient matrix has zero diagonal entries, iteration
cannot be performed.\r')
```

```
end

[n m]=size(A);
xold=x0;
k=1;, diff=[];

while k<=kmax
    xnew=b;
    for i=1:n
        for j=1:n
            if j<i
                xnew(i)=xnew(i)-A(i,j)*xnew(j);
            elseif j>i
                xnew(i)=xnew(i)-A(i,j)*xold(j);
            end
        end
        xnew(i)=xnew(i)/A(i,i);
        xnew(i)=omega*xnew(i)+(1-omega)*xold(i);
    end
    diff(k)=norm(xnew-xold,2);
    if diff(k)<tol
        fprintf('SOR iteration has converged in %d iterations\r', k)
        x=xnew;
        return
    end
    k=k+1;, xold=xnew;
end
fprintf('SOR iteration failed to converge.\r')
x=xnew;
```

(b) We set the relaxation parameter equal to 1 for SOR to reduce to Gauss-Seidel:
```
>> A=[3 1 -1;4 -10 1;2 1 5]; b=[-3 28 20]';
>> [x, k, diff] = sorit(A,b,1,[0 0 0]',1e-6)
```
→ SOR iteration has converged in 17 iterations
```
>> norm(x-[1 -2 4]',2)
```
→ans =1.4177e-007 (This agrees exactly with the error estimate of Example 7.27.)
```
>> [x, k, diff] = sorit(A,b,.9,[0 0 0]',1e-6);
```
→SOR iteration has converged in 9 iterations

EFR 7.29: Below is the complete code needed to recreate Figure 7.41. After running this code,
follow the instructions of the exercise to create the key.
```
>> jerr=1;, n=1;
>> while jerr>=1e-6
x=jacobi(A,b,[0 0 0]',1e-7,n);
Jerr(n)=norm(x-[1 -2 4]',2);, jerr=Jerr(n);, n=n+1;
end
>> semilogy(1:n-1,Jerr,'bo-')
>> hold on

>> gserr=1;, n=1;
>> while gserr>=1e-6
x=gaussseidel(A,b,[0 0 0]',1e-7,n);
GSerr(n)=norm(x-[1 -2 4]',2);, gserr=GSerr(n);, n=n+1;
end
>> semilogy(1:n-1,GSerr,'gp-')

>> sorerr=1;, n=1;
>> while sorerr>=1e-6
x=sorit(A,b,0.9, [0 0 0]',1e-7,n);
SORerr(n)=norm(x-[1 -2 4]',2);, sorerr=SORerr(n);, n=n+1;
```

```
end
>> semilogy(1:n-1,SORerr,'rx-')
>> xlabel('Number of iterations'), ylabel('Error')
```

EFR 7.30: (a) By writing out the matrix multiplication and observing repeated patterns we arrive at the following formula for the vector $b \equiv Ax$ of size 2500×1. Introduce first the following two 1×50 vectors b', $\bar{b}$:

$$b' = [1 \ 4 \ -1 \ 4 \ -1 \ \cdots \ 4 \ -1 \ 5],$$
$$\bar{b} = [0 \ 2 \ -2 \ 2 \ -2 \ \cdots \ 2 \ -2 \ 3].$$

In terms of copies of these vectors, we can express b as the transpose of the following vector:

$$b = [b' \ \bar{b} \ \bar{b} \ \cdots \ \bar{b} \ \bar{b} \ b'].$$

(b) We need first to store the matrix A. Because of its special form, this can be expeditiously accomplished using some loops and the diag command as follows:

```
>> x=ones(2500,1);, x(2:2:2500,1)=2;
>> tic, A=4*eye(2500);, toc
→elapsed_time =0.6090
>> v1=-1*ones(49,1);, v1=[v1;0]; %seed vector for sub/super diagonals
tic, secdiag=v1;
for i=1:49
if i<49
secdiag=[secdiag;v1];
else
secdiag=[secdiag;v1(1:49)];
end
end, toc
→elapsed_time =0.1250
>> tic, A=A+diag(secdiag,1)+diag(secdiag,-1)-diag(ones(2450,1),50)-
diag(ones(2450,1),-50);, toc
→elapsed_time =12.7660
>> tic, bslow=A*x;, toc
→elapsed_time = 0.2340
```

(c): To see the general concepts behind the following code, read Lemma 7.16 (and the notes that precede it).

```
tic, bfast=4*x+[secdiag;0].*[x(2:2500);0]+...
[0;secdiag].*[0; x(1:2499)]-[x(51:2500); zeros(50,1)]-...
[zeros(50,1); x(1:2450)];, toc    →elapsed_time = 0.0310
```

(d) If we take $N = 100$, the size of A will be $10,000 \times 10,000$, and this is too large for MATLAB to store directly, so Part (b) cannot be done. Part (a) can be done in a similar fashion to how it was done when N was 50. The method of part (c), however, still works in about 1/100th of a second. Here is the corresponding code:

```
>>x=ones(10000,1);, x(2:2:10000,1)=2;
>>v1=-1*ones(99,1);, v1=[v1;0]; %seed vector for sub/super diagonals
>>tic, secdiag=v1;, for i=1:99, if i<99, secdiag=[secdiag;v1];
else, secdiag=[secdiag;v1(1:99)];, end
end, toc
>> tic, bfast=4*x+[secdiag;0].*[x(2:10000);0]+...
[0;secdiag].*[0; x(1:9999)]-[x(101:10000); zeros(100,1)]-...
[zeros(100,1); x(1:9900)];, toc    →elapsed_time = 0.0100
```

EFR 7.31: (a) The M-file is boxed below:

```
function [x, k, diff] = sorsparsediag(diags, inds,b,omega,
x0,tol,kmax)
% performs the SOR iteration on the linear system Ax=b in cases where
% the n by n coefficient matrix A has entries only on a sparse set of
% diagonals.
% Inputs:  The input variables are 'diags', an n by J matrix where
```

```
% eachcolumn consists of the entries of one of A's diagonals.  The
% first column of diags is the main diagonal of A (even if all zeros)
% and 'inds' , a 1 by n vector of the corresponding set of indices
% for the diagonals (index zero corresponds to the main diagonal).
% the relaxation paramter 'omega', the seed (column) vector 'x0' for
% the iteration process, the tolerance 'tol' which will cause the
% iteration to stop if the infinity-norms of successive iterates
% become smaller than 'tol', and 'kmax' which is the maximum number
% of iterations to perform.
% Outputs:  the final iterate 'x', the number of iterations performed
% 'k', and a vector 'diff' which records the 2-norms of successive
% differences of iterates.
% If any of the last three input variables are not specified, default
% values of x0= zero column vector, tol=1e-10 and kmax=1000 are used.

%assign default input variables, as necessary
if nargin<5, x0=zeros(size(b));, end
if nargin<6, tol=1e-10;, end
if nargin<7, kmax=1000;, end

if min(abs(diags(:,1)))<eps
    error('Coefficient matrix has zero diagonal entries, iteration
cannot be performed.\r')
end

[n D]=size(diags);
xold=x0;
k=1;, diff=[];

while k<=kmax
    xnew=b;
    for i=1:n
        for d=2:D %run thru non-main diagonals and scan for entries
that effect xnew(i)
            ind=inds(d);
            if ind<0&i>-ind %diagonal below main and j<i case
                aij=diags(i+ind,d);
                xnew(i)=xnew(i)-aij*xnew(i+ind);
            elseif ind>0&i<=n-ind %diagonal above main and j>i case
                aij=diags(i,d);
                xnew(i)=xnew(i)-aij*xold(i+ind);
            end
        end
        xnew(i)=xnew(i)/diags(i,1);
        xnew(i)=omega*xnew(i)+(1-omega)*xold(i);
    end
    diff(k)=norm(xnew-xold,inf);
    if diff(k)<tol
            fprintf('SOR iteration has converged in %d iterations\r',
k)
            x=xnew;
            return
    end
    k=k+1;, xold=xnew;
end
fprintf('SOR iteration failed to converge.\r')
x=xnew;
```

(b) In order to use this program, we must create the input matrix diags from the nontrivial diagonals of the matrix A. The needed vectors were constructed in the solution of EFR 7.30(b); we reproduce the relevant code:

```
>> v1=-1*ones(49,1);, v1=[v1;0]; %seed vector for sub/super diagonals
secdiag=v1;
for i=1:49
if i<49
secdiag=[secdiag;v1];
else
secdiag=[secdiag;v1(1:49)];
end
end
```

We now construct the columns of diags to be the nontrivial diagonals of A taken in the order of the vector:

```
>> inds = [0   1   -1   50   -50]
>> diags = zeros(2500,5);
>> diags(:,1)=4;, diags(1:2499,[2 3])=[secdiag  secdiag];
>> diags(1:2450,[4 5])= [-ones(2450,1)  -ones(2450,1)];
```

We will also need the vectors x and b; we assume they have been obtained (and entered in the workspace) in one of the ways shown in the solution of EFR 7.30. We now apply our new SOR program on this problem using the default tolerance:

```
>> tic
>> [xsor, k, diff]=sorsparsediag(diags, inds,b,2/(1+sin(pi/51)),
zeros(size(b)));, toc
```
→SOR iteration has converged in 222 iterations
→elapsed_time = 0.6510
```
>> max(abs(xsor-x))
```
→ans = 6.1213e-010

References

[Ame-77]   Ames, William F., *Numerical Methods for Partial Differential Equations*, Barnes and Noble, New York (1977).

[Ant-00] Anton, Howard, *Elementary Linear Algebra, Eighth Edition*, John Wiley & Sons, New York (2000).

[Atk-89] Atkinson, Kendall E., *An Introduction to Numerical Analysis, Second Edition*, John Wiley & Sons, New York (1989).

[Bar-93] Barnsley, Michael F., *Fractals Everywhere, Second Edition*, Academic Press, Boston, MA (1993).

[BaZiBy-02]   Barnett, Raymond A., Michael R. Ziegler, and Karl E. Byleen, *Finite Mathematics, For Business, Economics, Life Sciences and Social Sciences, Ninth Edition*, Prentice-Hall, Upper Saddle River, NJ (2002).

[Bec-71]   Beckmann, Petr, *A History of π, Second Edition*, Golem Press, Boulder, CO (1971).

[BPRR-53]   Bruce, G. H., D. W. Peaceman, H. H. Rachford, and J. D. Rice, "Calculations of Unsteady State Gas Flow through Porous Media," *Transactions of the American Institute of Mineral Engineers (Petrol Div.)*, vol. **198**, pp. 79–92 (1953).

[BuFa-01]   Burden, Richard, L., and J. Douglas Faires, *Numerical Analysis, Seventh Edition*, Brooks/Cole, Pacific Grove, CA (2001).

[Col-42]   Collatz, Lothar, "Fehlerabschätzung für das Iterationsverfahren zur Auflösung linearer Gleichungssysteme" (German), *Zeitschrift für Angewandte Mathematik und Mechanik. Ingenieurwissenschaftliche Forschungsarbeiten*, vol. **22**, pp. 357–361 (1942).

[Con-72] Conway, John H., "Unpredictable Iterations," *Proceedings of the 1972 Number Theory Conference*, University of Colorado, Boulder, pp. 49–52 (1972).

[Fal-85]   Falconner, Kenneth J., *The Geometry of Fractal Sets*, Cambridge University Press, Cambridge, UK (1985).

[GoVL-83]   Golub, Gene, H., and Charles F. Van Loan, *Matrix Computations*, Johns Hopkins University Press, Baltimore (1983).

[Gre-97], Greenbaum, Anne, *Iterative Methods for Solving Linear Systems*, SIAM, Philadelphia (1997).

[HaLi-00] Hanselman, Duane, and Bruce Littlefield, *Mastering MATLAB 6: A Comprehensive Tutorial and Reference*, Prentice Hall, Upper Saddle River, NJ (2001).

[HiHi-00] Higham, Desmond J., and Nicholas J. Higham, *MATLAB Guide*, SIAM, Philadelphia (2000).

[HoKu-71] Hoffman, Kenneth, and Ray Kunze, *Linear Algebra*, Prentice-Hall, Englewood Cliffs, NJ (1971).

[HuLiRo-01] Hunt, Brian R., Ronald L. Lipsman, and Jonathan M. Rosenberg, *A Guide to MATLAB: For Beginners and Experienced Users*, Cambridge University Press, Cambridge, UK (2001).

[Kah-66] Kahan, William M., "Numerical Linear Algebra," *Canadian Mathematical Bulletin*, vol. **9**, pp. 757–801 (1966).

[Kol-99] Kolman, Bernard, and David R. Hill, *Elementary Linear Algebra, Seventh Edition*, Prentice-Hall, Upper Saddle River, NJ (1999).

[Lag-85] Lagarias, Jeffrey C., "The $3x + 1$ Problem and Its Generalizations," *American Mathematical Monthly*, vol. **92**, pp. 3–23 (1985).

[Lam-91] Lambert, John D., *Numerical Methods for Ordinary Differential Systems, The Initial Value Problem*, John Wiley & Sons, New York (1991).

[Lau-91] Lautwerier, Hans A., *Fractals: Endlessly Repeated Geometrical Figures*, Princeton University Press, Princeton, NJ (1991).

[Mat-95] Matilla, Pertti, *Geometry of Sets and Measures in Euclidean Spaces. Fractals and Rectifiability*, Cambridge University Press, Cambridge, UK (1995).

[OeS-99] Oliveira e Silva, Tomás, "Maximum Excursion and Stopping Time Record-Holders for the 3x+1 Problem: Computational Results," *Mathematics of Computation*, vol. **68**, 371–384 (1999).

[Ort-90] Ortega, James M., *Numerical Analysis, A Second Course*, SIAM, Philadelphia (1990).

[PSMI-98] Pärt-Enander, Eva, Anders Sjöberg, Bo Melin, and Pernilla Isaksson, *The MATLAB Handbook*, Addison-Wesley, Harlow, UK (1998).

[PSJY-92] Peitgen, Heinz-Otto, Dietmar Saupe, H. Jurgens, and L. Yunker, *Chaos and Fractals: New Frontiers of Science*, Springer-Verlag, New York (1992).

[Ros-00] Rosen, Kenneth H., *Handbook of Discrete and Combinatorial Mathematics*, CRC Press, Boca Raton, FL (2000).

[Ros-96] Ross, Kenneth A., *Elementary Analysis: The Theory of Calculus, Eighth Edition*, Springer-Verlag, New York (1996).

[Ros-02] Ross, Sheldon, *Simulation, Third Edition*, Academic Press, San Diego (2002).

[Rud-64] Rudin, Walter, *Principles of Mathematical Analysis, Second Edition*, McGraw-Hill, New York (1964).

[Sta-05] Stanoyevitch, Alexander, *Introduction to Numerical Ordinary and Partial Differential Equations Using MATLAB*, John Wiley & Sons, New York (2005).

[StBu-93] Stoer, Josef, and Roland Bulirsch, *Introduction to Numerical Analysis*, Springer-Verlag: TAM Series #12, New York (1993).

[Str-88] Strang, Gilbert, *Linear Algebra and Its Applications, Third Edition*, Prentice-Hall, Englewood Cliffs, NJ (1988).

[TrBa-97] Trefethen, Lloyd N., and David Bau, III, *Numerical Linear Algebra*, SIAM, Philadelphia (1997).

[Wil-88] Wilkinson, James H., *The Algebraic Eigenvalue Problem*, Clarendon Press, Oxford, UK (1988).

[You-71], Young, David M., *Iterative Solution of Large Linear Systems*, Academic Press, New York (1997).

MATLAB Command Index

FAMILIAR MATHEMATICAL FUNCTIONS OF ONE VARIABLE

Algebraic	`sqrt(x)` $(=\sqrt{x}\,)$, `abs(x)` $(=|x|)$
Exponential/ Logarithmic	`exp(x)` $(=e^{x})$, `log(x)` $(=\ln(x))$, `log10`
Trigonometric	`sin, cos, tan, sec,` etc, `asin,` etc, `asin,` etc, `sinh, cosh,` etc., `asinh,` etc.,

MATLAB COMMANDS AND M-FILES: NOTE: Textbook-constructed M-files are indicated with an asterisk after page reference. Optional input variables are underlined.

SYMBOLIC TOOLBOX COMMANDS:

General Index